Solutions Manual

The Systematic Identification of Organic Compounds

Eighth Edition

Ralph L. Shriner
Southern Methodist University

Christine K. F. Hermann
Radford University

Terence C. Morrill
Rochester Institute of Technology

David Y. Curtin
University of Illinois, Urbana

Reynold C. Fuson
University of Nevada

Prepared by

Christine K. F. Hermann
Radford University

WILEY

John Wiley & Sons, Inc.

To order books or for customer service call 1-800-CALL-WILEY (225-5945).

ISBN 0-471-46690-5

10 9 8 7 6 5 4 3 2 1

Preface

For the first six editions of this textbook, no solutions manual was available. As some of the problems presented throughout the textbook are quite difficult, I wrote the solutions manual as an aid to users of the textbook. The textbook contains no problems in Chapters 1, 2, and 12.

I would like to thank Terence Morrill for the answers to Problem Sets 6 - 20 in Chapter 11. These answers had been in handwritten form for many years. I would like to thank Danielle Davis for her assistance in writing the solutions manual for the seventh edition. I would like to thank Samuel Overstreet for his patience and support during the preparation of this edition.

Christine K. F. Hermann
Radford University

Table of Contents

CHAPTER 3
Preliminary Examination, Physical Properties, and Elemental Analysis

1. Use the equation for a nonassociated liquid.

$$\text{corr bp} = 167 + \frac{760 - 650}{10 \text{ mm}} \left[\left\{ \frac{0.56 - 0.50}{200 - 150} \times \left(167 - 150 \right) \right\} + 0.50 \right]$$

$$= 172.72$$

1,3-dichlorobenzene

2. Use the equation for an associated liquid.

$$\text{corr bp} = 180 + \frac{760 - 725}{10 \text{ mm}} \left[\left\{ \frac{0.46 - 0.42}{200 - 150} \times \left(180 - 150 \right) \right\} + 0.42 \right]$$

$$= 181.55$$

2,3-dichoro-1-propanol

$$\begin{array}{c} CH_2-CH-CH_2 \\ | \quad\;\; | \quad\;\; | \\ Cl \quad Cl \quad OH \end{array}$$

3. 280°C

4. 105°C

5. $\text{sp gr} \dfrac{20}{20} = \dfrac{0.989}{0.834} = 1.186$

$\text{sp gr} \dfrac{20}{4} = \dfrac{0.989}{0.834} \times 0.99823 = 1.184$

methyl salicylate

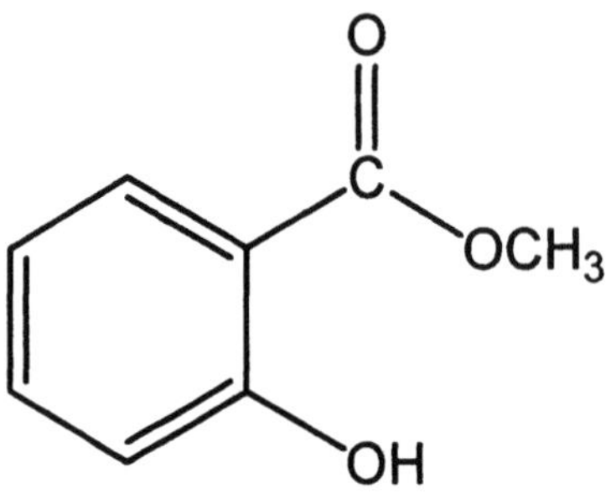

6. $n\,\dfrac{20}{D} = 1.430 + [(35 - 20)(0.00045)]$

$= 1.437$

7. $[\alpha] = \dfrac{6.65^{o}}{10 \text{ cm } \times \dfrac{1 \text{ dm}}{10 \text{ cm}} \times \dfrac{5 \text{ g}}{50 \text{ mL}}}$

$= 66.5^{o}$

sucrose

8.   Methylene chloride with ether, toluene, or hydrocarbons as co-solvents.

9.   Methanol or ethanol.

10. $\text{mg of C} = 10.71 \text{ mg of } CO_2 \times \dfrac{12.011 \ C}{44.010 \ CO_2}$

$= 2.92 \text{ mg of C in original sample}$

$\% \ C = \dfrac{2.92 \text{ mg of C}}{13.66 \text{ mg of sample}} \times 100 = 21.38 \% \text{ of C}$

$\text{mg of H} = 3.28 \text{ g of } H_2O \times \dfrac{2.016 \ (2 \ H)}{18.015 \ (H_2O)}$

$= 0.367 \text{ mg of H in original sample}$

$\% \ H = \dfrac{0.367 \text{ mg of H}}{13.66 \text{ mg of sample}} \times 100 = 2.69 \% \text{ of H}$

$\% \ Br = \dfrac{3.46 \text{ mg of Br}}{4.86 \text{ mg of sample}} \times 100 = 71.19 \% \text{ of Br}$

$\% \ O = 100 - (21.38 + 2.69 + 71.19) = 4.74\% \text{ of O}$

$C = \dfrac{21.38}{12.011} = 1.78 \qquad \dfrac{1.79}{0.296} = 6.04$

$H = \dfrac{2.69}{1.008} = 2.69 \qquad \dfrac{2.69}{0.296} = 9.09$

$Br = \dfrac{71.19}{79.906} = 0.891 \qquad \dfrac{0.891}{0.296} = 3.01$

$O = \dfrac{4.74}{16.000} = 0.296 \qquad \dfrac{0.296}{0.296} = 1.00$

Empirical formula = $C_6H_9Br_3O$
Empirical weight = 336.86
Molecular formula = $C_{12}H_{18}Br_6O_2$

11. **a.** $C_6H_{12}O$
$U = 6 + 1 - \frac{1}{2}(12) + \frac{1}{2}(0) = 1$
one double bond; *or* one ring

b. $C_5H_{10}Cl_2$
$U = 5 + 1 - \frac{1}{2}(12) + \frac{1}{2}(0) = 0$
no double bonds, triple bonds, *or* rings

c. $C_7H_{13}N$
$U = 7 + 1 - \frac{1}{2}(13) + \frac{1}{2}(1) = 2$
two rings; *or* two double bonds; *or* one ring and one double bond; *or* one triple bond

d. $C_{12}H_{10}$
$U = 12 + 1 - \frac{1}{2}(10) + \frac{1}{2}(0) = 8$
benzene plus three rings; *or* benzene plus three double bonds; *or* benzene plus two rings and one double bond; *or* benzene plus two double bonds and one ring; *or* benzene plus one triple bond and one double bond; *or* benzene plus one triple bond and one ring

CHAPTER 4
Separation of Mixtures

1. chlorobenzene, *N,N*-dimethylaniline, 2-naphthol,
    2-methoxybenzaldehyde

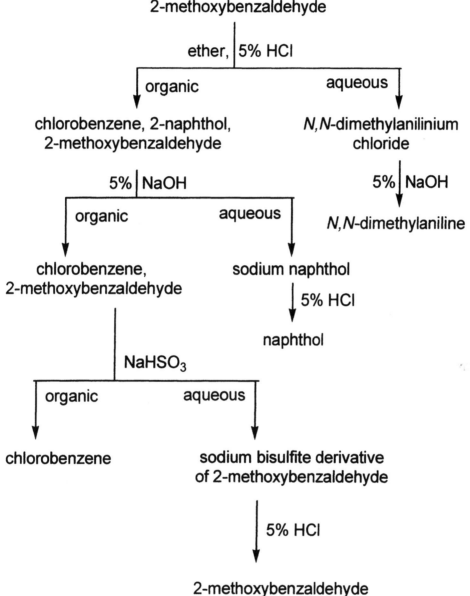

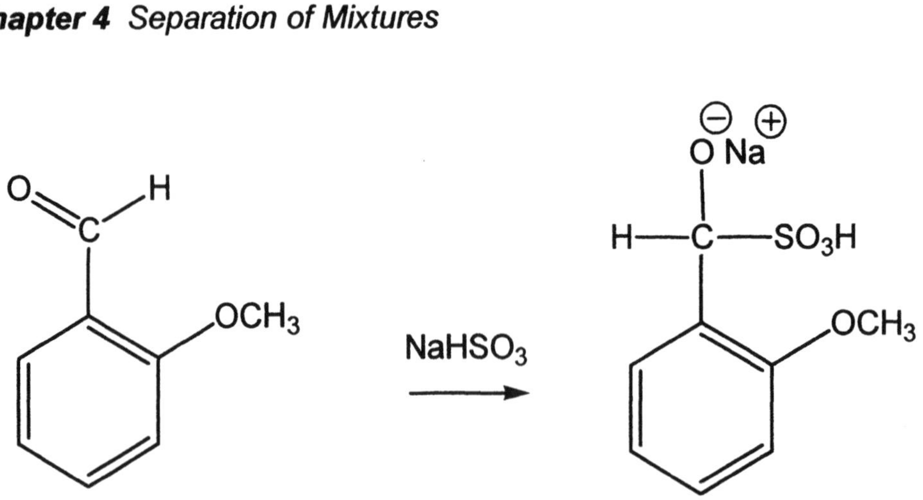

2-methoxybenzaldehyde

NaHSO₃

sodium bisulfite
derivative of
2-methoxybenzaldehyde

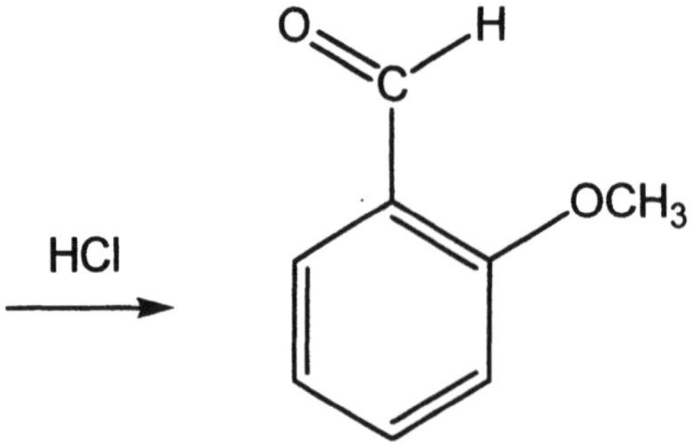

HCl

2-methoxybenzaldehyde

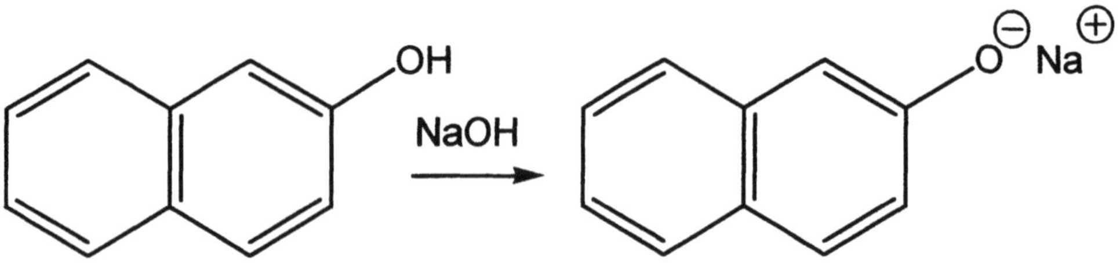

2-naphthol

NaOH

sodium naphtholate

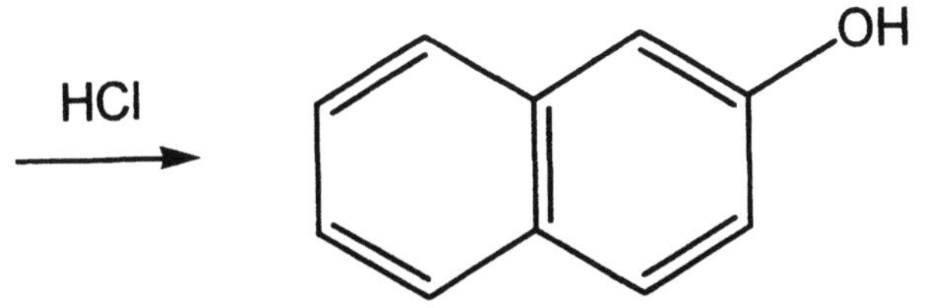

$$\xrightarrow{\text{HCl}}$$

2-naphthol

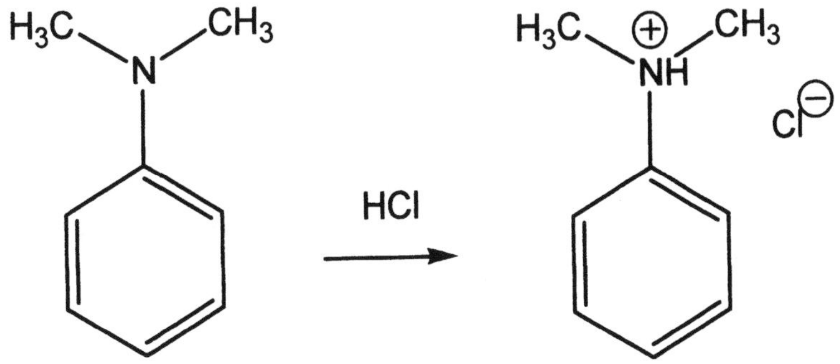

N,N-dimethylaniline $\xrightarrow{\text{HCl}}$ N,N-dimethylanilinium chloride

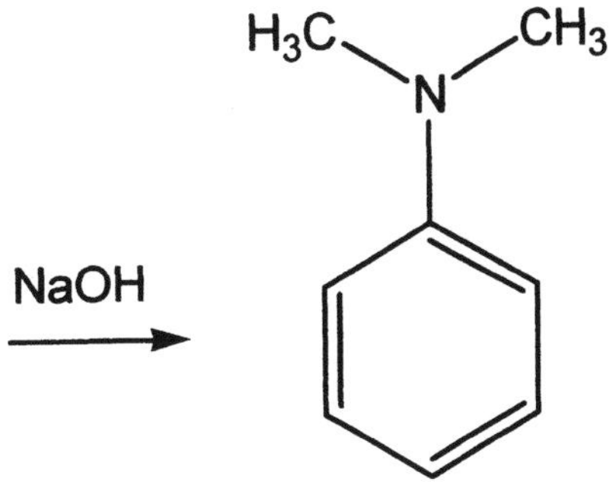

$\xrightarrow{\text{NaOH}}$

N,N-dimethylaniline

2.

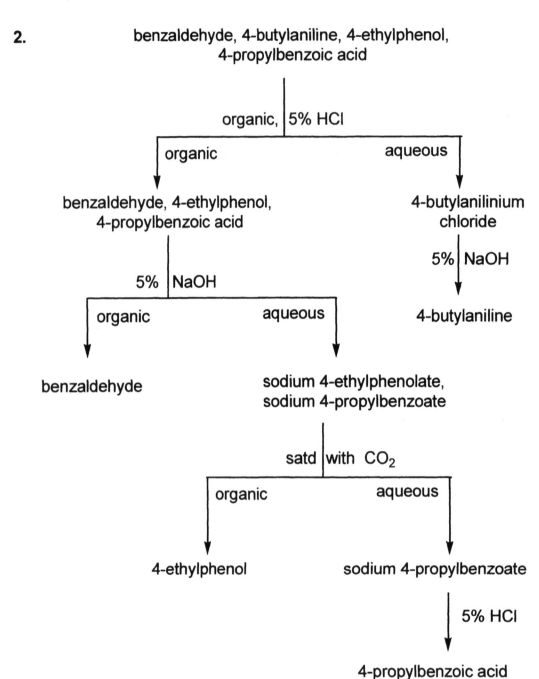

4-ethylphenol → NaOH → sodium 4-ethylphenolate

satd CO$_2$ → 4-ethylphenol

4-propylbenzoic acid → NaOH → sodium 4-propylbenzoate

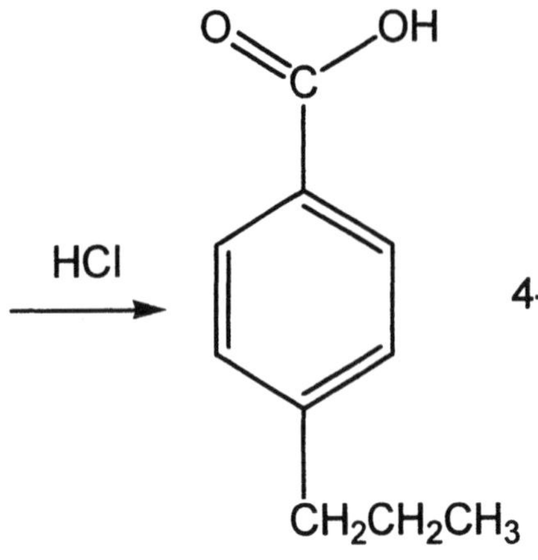

$$\xrightarrow{\text{HCl}}$$

4-propylbenzoic acid

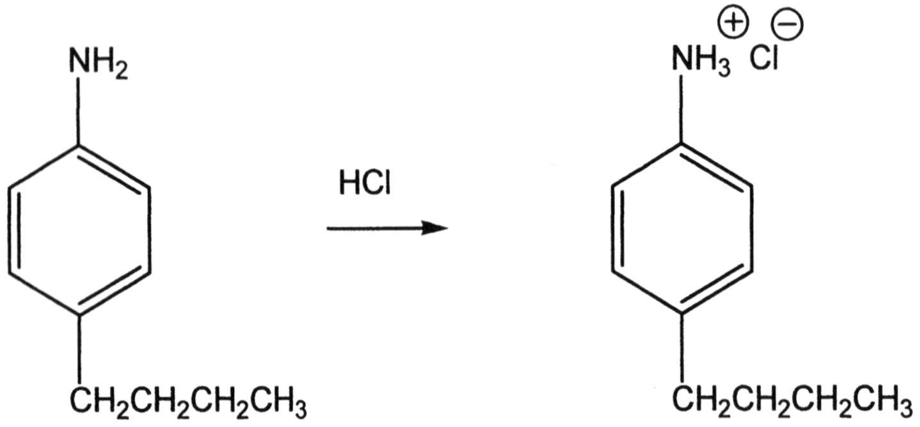

(structure of 4-propylbenzoic acid with COOH group and CH₂CH₂CH₃ substituent)

$$\xrightarrow{\text{HCl}}$$

4-butylaniline

4-butylanilium chloride

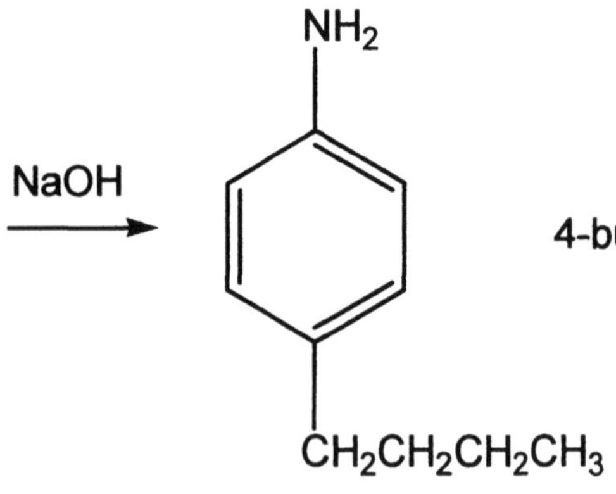

$$\xrightarrow{\text{NaOH}}$$

4-butylaniline

3. Since glucose is also insoluble in ether, it would be filtered out as Residue 1.

4. lactic acid, piperidine, acetic acid, 2-propanol

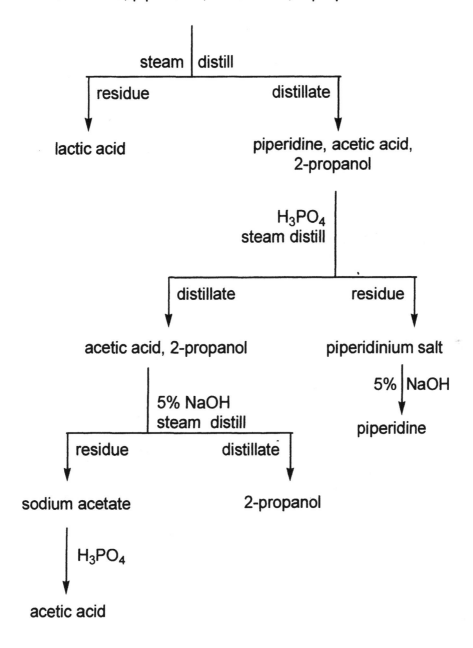

acetic acid   NaOH   sodium acetate

HCl   acetic acid

piperidine   H$_3$PO$_4$ piperidium dihydrogenphosphate   NaOH   piperidine

5. a.

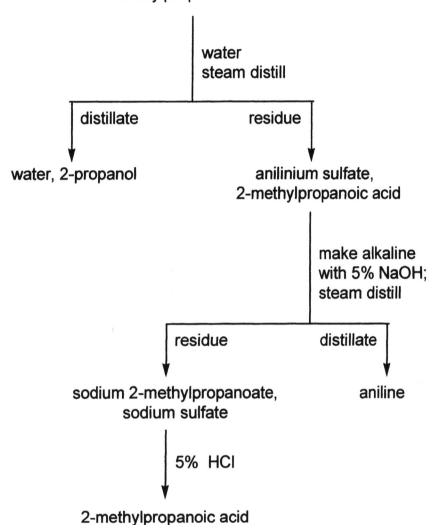

b.

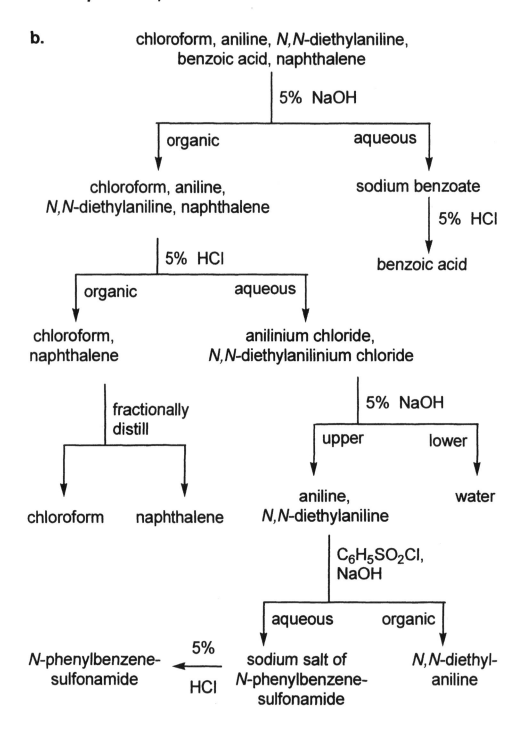

c.

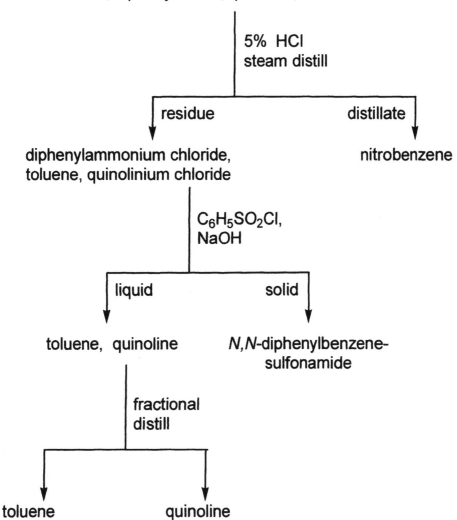

d.

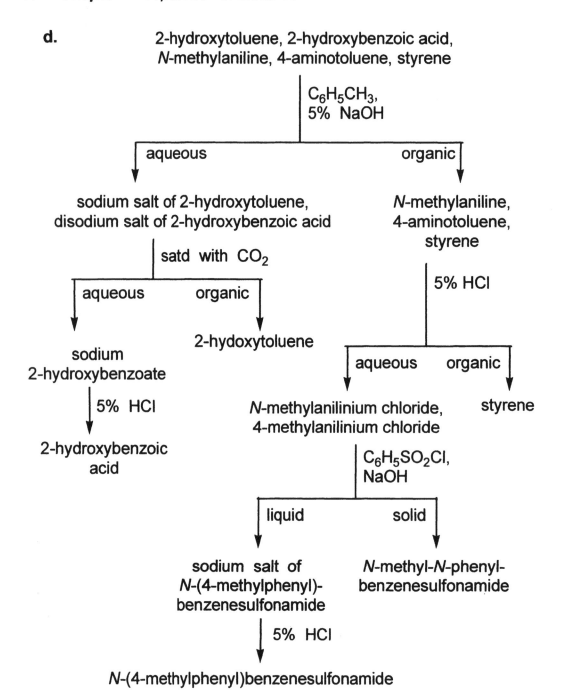

2-hydroxytoluene, 2-hydroxybenzoic acid,
N-methylaniline, 4-aminotoluene, styrene

$C_6H_5CH_3$, 5% NaOH

aqueous — sodium salt of 2-hydroxytoluene, disodium salt of 2-hydroxybenzoic acid

organic — *N*-methylaniline, 4-aminotoluene, styrene

satd with CO_2

aqueous — sodium 2-hydroxybenzoate

organic — 2-hydoxytoluene

5% HCl

2-hydroxybenzoic acid

5% HCl

aqueous — *N*-methylanilinium chloride, 4-methylanilinium chloride

organic — styrene

$C_6H_5SO_2Cl$, NaOH

liquid — sodium salt of *N*-(4-methylphenyl)-benzenesulfonamide

solid — *N*-methyl-*N*-phenyl-benzenesulfonamide

5% HCl

N-(4-methylphenyl)benzenesulfonamide

e.

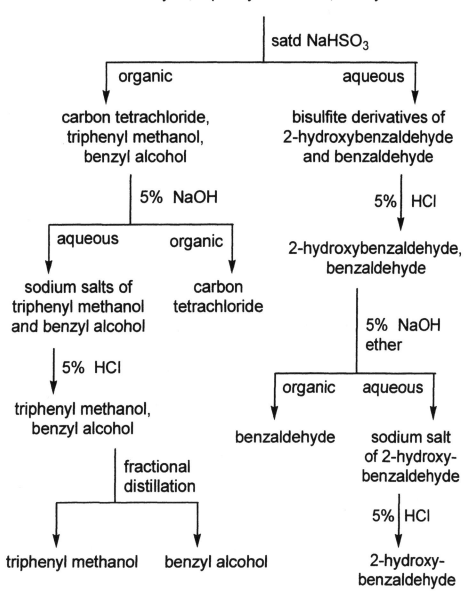

carbon tetrachloride, 2-hydroxybenzaldehyde,
benzaldehyde, triphenyl methanol, benzyl alcohol

satd NaHSO₃

organic

carbon tetrachloride,
triphenyl methanol,
benzyl alcohol

5% NaOH

aqueous

sodium salts of
triphenyl methanol
and benzyl alcohol

5% HCl

triphenyl methanol,
benzyl alcohol

fractional
distillation

triphenyl methanol   benzyl alcohol

organic

carbon
tetrachloride

aqueous

bisulfite derivatives of
2-hydroxybenzaldehyde
and benzaldehyde

5% HCl

2-hydroxybenzaldehyde,
benzaldehyde

5% NaOH
ether

organic   aqueous

benzaldehyde   sodium salt
of 2-hydroxy-
benzaldehyde

5% HCl

2-hydroxy-
benzaldehyde

f. diethyl ether, 3-pentanone, diethylamine, acetic acid

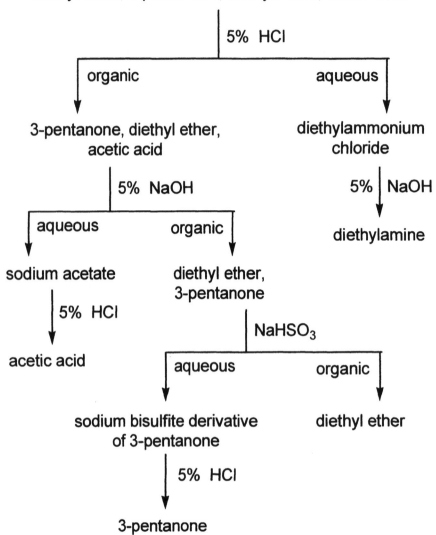

g. cyclohexane, 4-methylaniline, toluene,
3-hydroxytoluene, 2-chloro-4-aminobenzoic acid

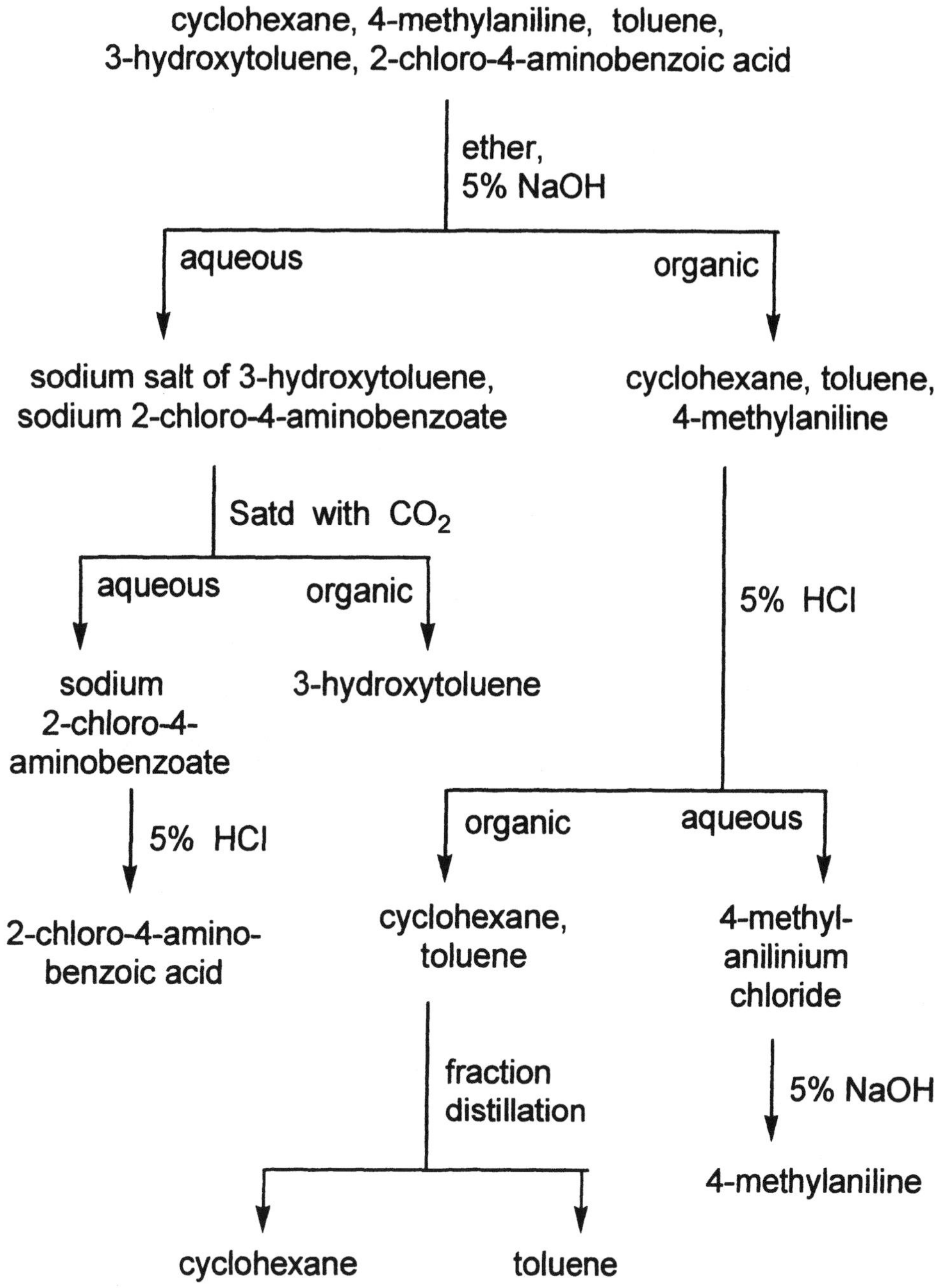

| ether,
5% NaOH

aqueous ←→ organic

aqueous:
sodium salt of 3-hydroxytoluene,
sodium 2-chloro-4-aminobenzoate

organic:
cyclohexane, toluene,
4-methylaniline

Satd with CO_2

aqueous → sodium 2-chloro-4-aminobenzoate

organic → 3-hydroxytoluene

5% HCl

5% HCl (on sodium 2-chloro-4-aminobenzoate) →
2-chloro-4-amino-benzoic acid

organic → cyclohexane, toluene

aqueous → 4-methyl-anilinium chloride

fraction distillation →
cyclohexane   toluene

5% NaOH (on 4-methyl-anilinium chloride) →
4-methylaniline

6. To solve this problem, the number of moles of each substance must be determined.

Compound	grams	MW	moles
2-methyl-2-propanol	2.47	74.63	0.033
benzyl alcohol	3.60	108.81	0.033
benzaldehyde	3.53	106.79	0.033
acetyl chloride	5.20	78.69	0.066
acetophenone	4.00	120.92	0.033
N,N-dimethylaniline	28.2	121.95	0.231

The acetyl chloride will react with the 2-methyl-2-propanol and the benzyl alcohol to form the 1,1-dimethylethyl acetate and the benzyl acetate, respectively.

2-methyl-2-propanol   acetyl chloride   1,1-dimethylethyl acetate (*t*-butyl acetate) benzyl alcohol   acetyl chloride   benzyl acetate

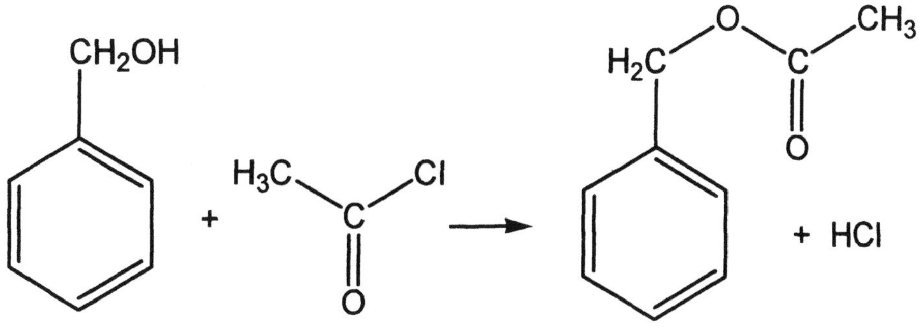

The hydrogen chloride, produced from the above reactions, reacts with *N,N*-dimethylaniline to form *N,N*-dimethylanilinium chloride.

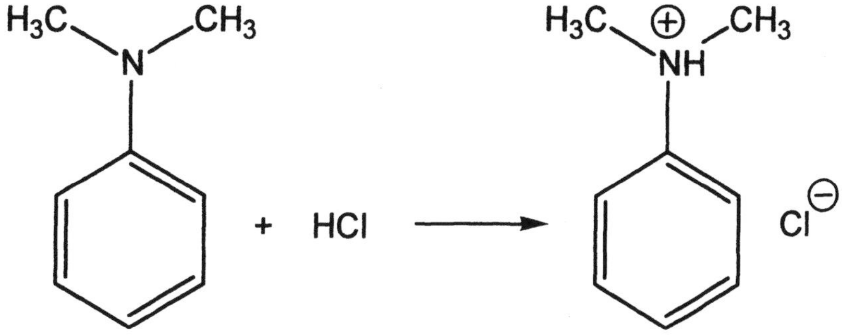

N,N-dimethylaniline   *N,N*-dimethylanilinium chloride

If complete reactions are assumed, then the following compounds would be present at the end of a week.

Compound	moles
1,1-dimethylethyl acetate	0.033
benzyl acetate	0.033
benzaldehyde	0.033

acetophenone	0.033
N,N-dimethylanilinium chloride	0.066
N,N-dimethylaniline	0.165

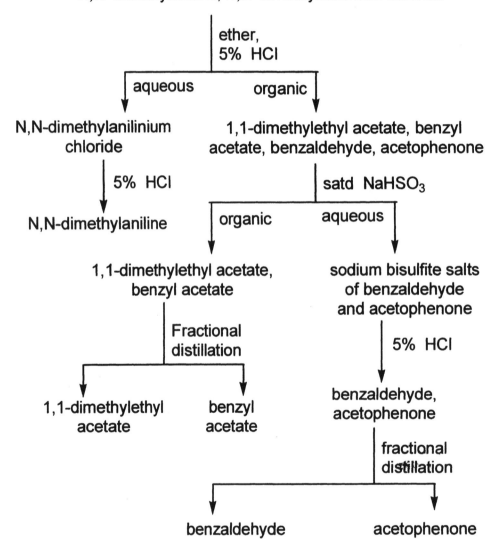

7. **a.** Since these compounds are alcohols, gas chromatography could be utilized. These compounds would be separated by bp.

    b. Extraction methods would be the preferred method of separating these compounds.

    c. Gas chromatography could be utilized to separate these compounds. These compounds would be separated by bp.

8. The compounds must be soluble in a solvent and/or be able to be volatized in order to be separated by GC or LC. Esters are more volatile and are more easily dissolved than carboxylic acids.

CHAPTER 5
Classification of Organic Compounds by Solubility

1. a. N -- alcohols, aldehydes, ketones, esters with one functional group and more than five but fewer than nine carbons, ethers, epoxides, alkenes, alkynes, some aromatic compounds (especially those with activating groups).

   b. S_1 -- monofunctional alcohols, aldehydes, ketones, esters, nitriles, and amides with five carbons or fewer.

   c. A_2 -- weak organic acids; phenols, enols, oximes, imides, sulfonamides, thiophenols, all with more than five carbons; β-diketones; nitro compounds with α-hydrogens.

2. a. 1-chlorobutane $\qquad$ $CH_3CH_2CH_2CH_2Cl$

   1-Chlorobutane is insoluble in water since it is an haloalkane. It is insoluble in 5% sodium hydroxide solution, 5% hydrochloric acid solution, and 96% sulfuric acid solution. Thus 1-chlorobutane is in solubility class I.

   b. 4-methylaniline

   4-Methylaniline is insoluble in water, since it is an aniline. It is basic, due to the presence of the amino group, and therefore is insoluble in 5% sodium hydroxide solution, but soluble in 5% hydrochloric acid solution, and thus 4-methylaniline is in solubility class B.

   c. 1-nitroethane $\qquad$ $CH_3CH_2NO_2$

   1-Nitroethane is insoluble in water, since it is a weak organic acid. It is soluble in 5% sodium hydroxide solution, but insoluble in 5% sodium bicarbonate solution, and thus 1-nitroethane is in solubility class A_2.

d. alanine

Alanine is an amino acid, thus is amphoteric, and soluble in water. Alanine is insoluble in ether and thus in solubility class S_2.

e. benzophenone

Benzophenone is insoluble in water, because it has more than five carbon atoms. Since it is a neutral compound, containing only carbon, hydrogen, and oxygen, it is insoluble in 5% sodium hydroxide solution, insoluble in 5% hydrochloric acid solution, and soluble in 96% sulfuric acid solution. Therefore, benzophenone is in solubility class **N**.

f. benzoic acid

As a strong organic acid, benzoic acid is insoluble in water, soluble in 5% sodium hydroxide solution, and soluble in 5%

sodium bicarbonate solution. Therefore, benzoic acid is in solubility class A_1.

g.   hexane $CH_3CH_2CH_2CH_2CH_2CH_3$

As a saturated hydrocarbon, hexane is insoluble in water, 5% sodium hydroxide solution, 5% hydrochloric acid solution, and 95% sulfuric acid. Hexane is in solubility class I.

h.   4-methylbenzyl alcohol H_3C—⟨benzene ring⟩—CH_2OH

4-Methylbenzyl alcohol is insoluble in water, since it has more than five carbon atoms. It is a neutral compound with only carbon, hydrogen, and oxygen. It is insoluble in 5% sodium hydroxide solution and 5% hydrochloric acid solution. However, 4-methylbenzyl alcohol is soluble in 96% sulfuric acid and thus in solubility class N.

i.   ethyl methyl amine $CH_3NHCH_2CH_3$

As an amine with less than six carbons, ethyl methyl amine is soluble in water. It is soluble in ether and its aqueous solutions turn litmus paper blue, therefore it is basic. Ethyl methyl amine is in solubility class S_B.

j.   propoxybenzene   ⟨benzene ring⟩—$OCH_2CH_2CH_3$

Propoxybenzene is insoluble in water because it has more than five carbon atoms. It is a neutral compound with only carbon, hydrogen, and oxygen and is insoluble in 5% sodium hydroxide solution, and 5% hydrochloric acid solution. However, propoxybenzene is soluble in 96% sulfuric acid solution and in solubility class N.

k. propanal

$$CH_3CH_2-\overset{\overset{\displaystyle O}{\|}}{C}-H$$

Propanal is soluble in water, because it has less than five carbons. As a neutral compound, it is soluble in ether and its aqueous solutions do not change the color of litmus paper. Propanal is in solubility class S_1.

l. 1,3-dibromobenzene

As an aryl halide, 1,3-dibromobenzene is insoluble in water, 5% sodium hydroxide solution, 5% hydrochloric acid solution, and 96% sulfuric acid. 1,3-Dibromobenzene is in solubility class I.

m. propanoic acid

$$CH_3CH_2-\overset{\overset{\displaystyle O}{\|}}{C}-OH$$

Propanoic acid is soluble in water, because it contains less than five carbon atoms. It is soluble in ether, but its aqueous solutions turn litmus paper red. Propanoic acid is in solubility class S_A.

n. benzenesulfonamide

As a weak organic acid, benzenesulfonamide is insoluble in water. It is soluble in 5% sodium hydroxide solution, but insoluble in 5% sodium bicarbonate solution. Benzene-sulfonamide is in solubility class A_2.

o.   1-butanol $CH_3CH_2CH_2CH_2OH$

1-Butanol is soluble in water because it contains less than five carbon atoms. It is soluble in ether and its aqueous solutions do not change the color of litmus. 1-Butanol is in solubility class S_1.

p.   methyl propanoate

Methyl propanoate is soluble in water, since it contains less than five carbon atoms. It is soluble in ether and its aqueous solutions do not change the color of litmus. Methyl propanoate is in solubility class S_1.

q.   4-methylcyclohexanone

4-Methylcyclohexanone is insoluble in water since it contains more than five carbon atoms. It is also insoluble in 5% sodium hydroxide solution and 5% hydrochloric acid solution, but soluble in 96% sulfuric acid. Therefore, 4-methylcyclo-hexanone is in solubility class N.

r.   4-aminobiphenyl

Since 4-aminobiphenyl is an aniline, it is insoluble in water. It is insoluble in 5% sodium hydroxide solution, but soluble in 5% hydrochloric acid solution. 4-Aminobiphenyl is in solubility class B.

s.   4-methylacetophenone

Since 4-methylacetophenone contains more than five carbon atoms, it is insoluble in water. It is also insoluble in 5% sodium hydroxide solution and 5% hydrochloric acid solution, but soluble in 96% sulfuric acid. Therefore, 4-methylacetophenone is in solubility class N.

t.   naphthalene

Naphthalene is insoluble in water since it contains more than five carbon atoms. It is insoluble in 5% sodium hydroxide solution and 5% hydrochloric acid solution. However, naphthalene is soluble in 96% sulfuric acid, and therefore in solubility class N.

u.   phenylalanine

Phenylalanine is an amino acid, thus is amphoteric, and

soluble in water. Phenylalanine is insoluble in ether and thus in solubility class S_2.

v. benzoin

Benzoin is insoluble in water, since it has more than five carbon atoms. It is insoluble in 5% sodium hydroxide solution and 5% hydrochloric acid solution. However, benzoin is soluble in 96% sulfuric acid, and therefore in solubility class N.

w. 4-hydroxybenzenesulfonic acid

As a strong organic acid, 4-hydroxybenzenesulfonic acid is insoluble in water. However, it is soluble in 5% sodium hydroxide solution and 5% sodium bicarbonate solution. 4-Hydroxybenzenesulfonic acid is in solubility class A_1.

3. Listed in order from least basic to most basic.

and

c. diphenylamine

a. benzanilide

and

f. butanamide

<

e. 4-methylaniline

<

and $CH_3CH_2CH_2CH_2CH_2NH_2$

d. benzylamine

b. pentylamine

To solve this problem, the solubility class of each compound must be determined (see question 4).

$$N, S_1 < B < S_B$$

4.				
	a.	Benzanilide	Solubility class	N
	b.	Pentylamine	Solubility class	S_B
	c.	Diphenylamine	Solubility class	N
	d.	Benzylamine	Solubility class	S_B
	e.	4-Methylaniline	Solubility class	B
	f.	Butanamide	Solubility class	S_1

5. Listed in order from least soluble to most soluble in water.

    a. Smaller alcohols are more soluble.

$$CH_3CH_2CH_2CH_2OH \quad < \quad (CH_3)_2CHOH$$

1-butanol   2-propanol

$$< \quad CH_3CH_2OH \quad < \quad CH_3OH$$

ethanol   methanol

b. More hydroxyl groups increase solubility.

$$CH_3CH_2CH_2CH_3 \quad < \quad CH_3CH_2CH_2CH_2OH$$

butane   1-butanol

$$< \quad HOCH_2CH_2CH_2CH_2OH$$

1,2-butanediol

c. Branching increases solubility.

$$CH_3CH_2CH_2CH_2OH \quad < \quad CH_3CH_2\underset{\underset{\displaystyle OH}{|}}{C}HCH_3 \quad < \quad CH_3\underset{\underset{\displaystyle CH_3}{|}}{\overset{\overset{\displaystyle CH_3}{|}}{C}}OH$$

1-butanol   2-butanol   2-methyl-2-propanol

d. $N < S_1 - N < S_B < S_2$.

benzaldehyde   3-pentanone

triethylamine   ammonium butanoate

6.	**a.**	Methanol	Solubility class	S_1
		Isopropyl alcohol	Solubility class	S_1
		Ethanol	Solubility class	S_1
		1-Butanol	Solubility class	S_1
	b.	Butane	Solubility class	I
		1,4-Butanediol	Solubility class	S_2
		1-Butanol	Solubility class	S_1
	c.	1-Butanol	Solubility class	S_1
		2-Methyl-2-propanol	Solubility class	S_1
		2-Butanol	Solubility class	S_1
	d.	Ammonium butanoate	Solubility class	S_2
		3-Pentanone	Solubility class	S_1–N
		Benzaldehyde	Solubility class	N
		Trimethylamine	Solubility class	S_B

7.   The following compounds are listed in order of decreasing activity.

a. *meso*-tartaric acid (S$_2$)   **d.** 4-toluenesulfonic acid (S$_A$)

f. 2-bromo-6-nitrophenol (A$_1$)   **j.** 1-naphthoic acid (A$_1$)

e. 4-toluenesulfonamide (A$_2$)   **c.** benzohydroxamic acid (MN)

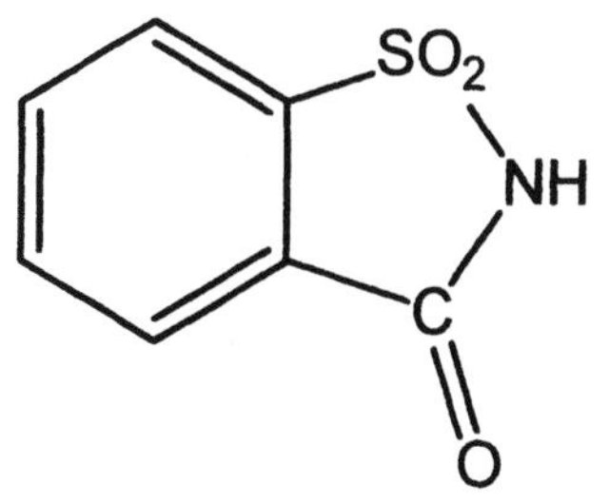

h. saccharin (MN)

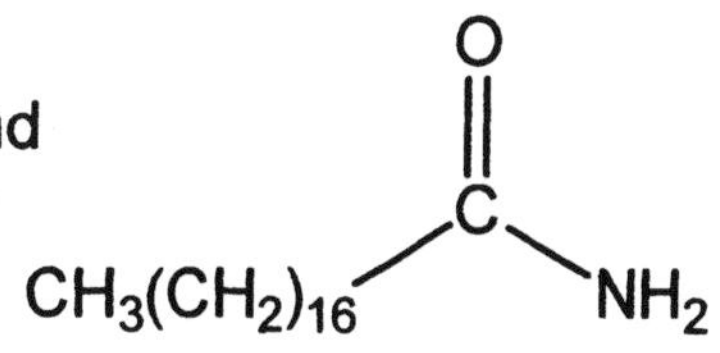

and

g. octadecanamide (MN)

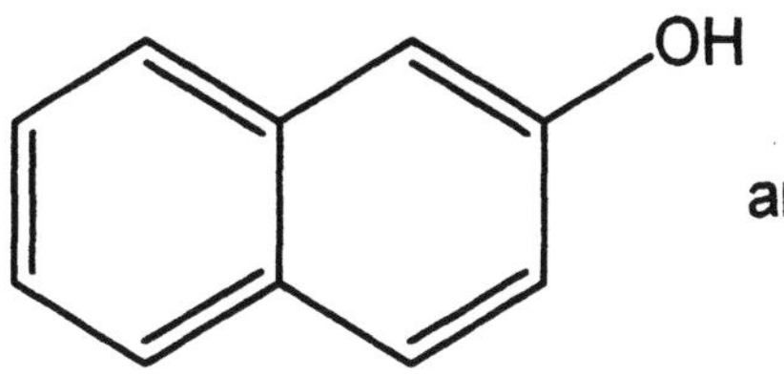

>

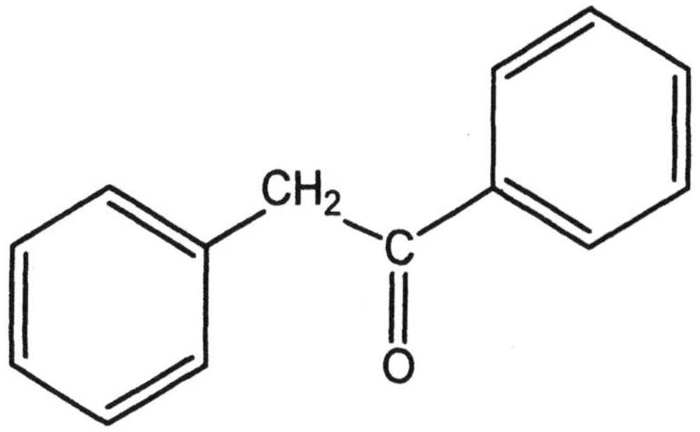

b. 2-naphthol (N)

and

i. benzyl phenyl ketone (N)

CHAPTER 6
Nuclear Magnetic Resonance Spectrometry

For some of the hydrogen and carbon calculations, the closest structure in the tables were used.

1. Compound with a formula of C_3H_6BrCl.

^1H NMR Spectrum:

	Chemical Shift (Actual)	(Calc)	Splitting	Integration	Interpretation
a.	2.28	2.1	quintet	2 H	CH_2 adjacent to 4 H
b.	3.57	3.2	t	2 H	CH_2 adjacent to CH_2
c.	3.71	3.3	t	2 H	CH_2 adjacent to CH_2

Calculations:

$$Br-CH_2-CH_2-CH_2-Cl$$

0.9	0.9	0.9
0.2	0.5	2.2
2.1	0.7	0.2
–––	–––	–––
3.2	2.1	3.3

Labeled Structure:

$$Br-CH_2-CH_2-CH_2-Cl$$
$$b\quad\ a\quad\ c$$

1-bromo-3-chloropropane

^{13}C NMR Spectrum:

	Chemical Shift (Actual)	(Calc)	Interpretation
a.	29.97	31.6	alkyl bromide
b.	34.89	38.1	alkyl
c.	42.61	43.6	alkyl chloride

Calculations:

$$Br-CH_2-CH_2-CH_2-Cl$$

15.6	16.1	15.6
-4	11	31
20	11	-3

| 31.6 | 38.1 | 43.6 |

Labeled Structure:

$$Br-CH_2-CH_2-CH_2-Cl$$
$$\quad\ a \qquad b \qquad c$$

2. Compound with a formula of $C_5H_{12}O$.

Unsaturation Number:

$U = 5 + 1 - \frac{1}{2}(12) + \frac{1}{2}(0) = 0$
no double bonds, triple bonds, or rings

^{13}C NMR and DEPT Spectra:

	Chemical Shift (Actual)	(Calc)	DEPT	Interpretation
a.	9.79	8.7	2 CH_3	alkyl
b.	29.49	32.6	2 CH_2	alkyl
c.	73.95	82.5	CH	alcohol

The methyl and methylenes must be doubled to obtain the correct number of carbons and hydrogens.

Calculations:

$$CH_3-CH_2-CH-CH_2-CH_3$$
$$|$$
$$OH$$

13.7	22.6	34.5	22.6	13.7
-5	10	48	10	-5

| 8.7 | 32.6 | 82.5 | 32.6 | 8.7 |

Labeled Structure:

$$\overset{a}{CH_3}-\overset{b}{CH_2}-\overset{c}{CH}-\overset{b}{CH_2}-\overset{a}{CH_3} \qquad \text{3-pentanol}$$
$$|$$
$$OH$$

3. Compound with a formula of $C_4H_{10}O_2$.

Unsaturation Number:

$U = 4 + 1 - \frac{1}{2}(10) + \frac{1}{2}(0) = 0$
no double bonds, triple bonds, or rings

^1H NMR Spectrum:

	Chemical Shift (Actual)	(Calc)	Splitting	Integration	Interpretation
a.	3.27	3.0	s	3 H	CH_3 isolated
b.	3.43	3.6	s	2 H	CH_2 isolated

The methyl and methylene cannot be next to each other since there is no splitting.

Calculations:

$$CH_3-O-CH_2-CH_2-O-CH_3$$

0.9	1.2	1.2	0.9
2.1	2.1	2.1	2.1
	0.3	0.3	
3.0	3.6	3.6	3.0

Labeled Structure:

a   b   b   a

$$CH_3-O-CH_2-CH_2-O-CH_3$$ 1,2-dimethoxyethane

^{13}C NMR and DEPT Spectra:

	Chemical Shift		DEPT	Interpretation
	(Actual)	(Calc)		
a.	58.95	56.1	CH_3	ether
b.	72.76	71.9	CH_2	ether

Calculations:

$$CH_3-O-CH_2-CH_2-O-CH_3$$

-2.1	5.9	5.9	-2.1
58	58	58	58
	8	8	
56.1	71.9	71.9	56.1

Labeled Structure:

a   b   b   a

$$CH_3-O-CH_2-CH_2-O-CH_3$$

4. Compound with a formula of $C_5H_{10}O$.

Unsaturation number:

$U = 5 + 1 - \frac{1}{2}(10) + \frac{1}{2}(0) = 1$
one double bond or one ring

^1H NMR Spectrum:

	Chemical Shift (Actual)	(Calc)	Splitting	Integration	Interpretation
a.	0.87	0.9	t	3 H	CH_3 adjacent to CH_2
b.	1.54	1.5	sextet	2 H	CH_2 adjacent to 5 H
c.	2.05	2.1	s	3 H	CH_3 isolated
d.	2.40	2.4	t	2 H	CH_2 adjacent to CH_3

The ^1H NMR spectrum indicates the presence of a propyl group and a methyl group. One carbon, one oxygen, and one double bond are missing.

Calculations:

$$CH_3-CH_2-CH_2\diagdown \underset{\underset{O}{\overset{\|}{C}}}{}\diagup CH_3$$

0.9	1.2	1.2	0.9
0.0	0.3	1.2	1.2
0.9	1.5	2.4	2.1

Labeled Structure:

a   b   d   c

CH₃–CH₂–CH₂⟍C⟋CH₃   2-pentanone
                |
                ‖
                O

^{13}C NMR and DEPT Spectra:

	Chemical Shift (Actual)	(Calc)	DEPT	Interpretation
a.	12.75	13.6	CH_3	alkyl
b.	16.55	17.1	CH_2	alkyl
c.	28.43	28.1	CH_3	alkyl
d.	44.41	45.6	CH_2	alkyl
e.	205.78		C	ketone

Calculations:

CH₃–CH₂–CH₂⟍C⟋CH₃
                |
                ‖
                O

15.6	16.1	15.6	-2.1
-2	1	30	30

| 13.6 | 17.1 | 45.6 | 28.1 |

Labeled Structure:

a   b   d   c

CH₃–CH₂–CH₂⟍C⟋CH₃
                |
                ‖
                O

5. Compound with a formula of C_8H_9NO.

Unsaturation number:

$U = 8 + 1 - \frac{1}{2}(9) + \frac{1}{2}(1) = 5$
benzene plus one double bond or one ring

^1H NMR Spectrum:

	Chemical Shift (Actual)	(Calc)	Splitting	Integration	Interpretation
a.	2.07	1.9	s	3 H	CH_3 isolated
b.	6.97-7.36	7.17, 7.30	m	3 H	aromatic
c.	7.55-7.67	7.60	m	2 H	aromatic
d.	9.81	-------	s	1 H	amide

The unknown is a monosubstituted aromatic compound, since there are five hydrogens in the aromatic range. The ^{13}C NMR spectrum confirms the presence of an amide.

Calculations:

1	2	3	4
0.9	7.27	7.27	7.27
1.0	0.33	0.03	-0.10
1.9	7.60	7.30	7.17

Labeled Structure:

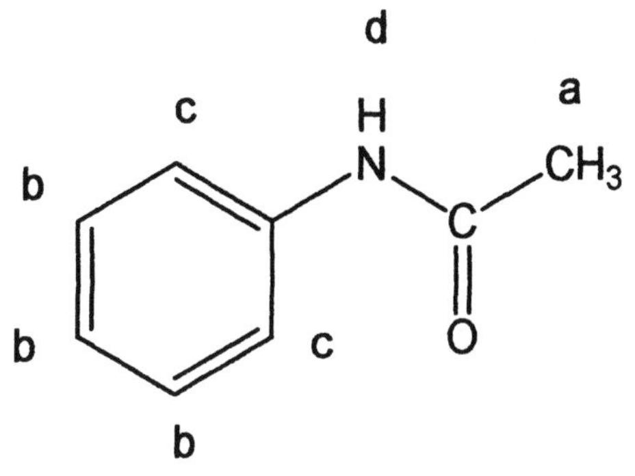

acetanilide

^{13}C NMR and DEPT Spectra:

	Chemical Shift (Actual)	(Calc)	DEPT	Interpretation
a.	24.30	28.1	CH_3	alkyl
b.	119.77	118.8	2 CH	aromatic
c.	123.47	123.1	CH	aromatic
d.	128.83	128.9	2 CH	aromatic
e.	139.60	139.8	C	aromatic
f.	168.98	-------	C	amide

Calculations:

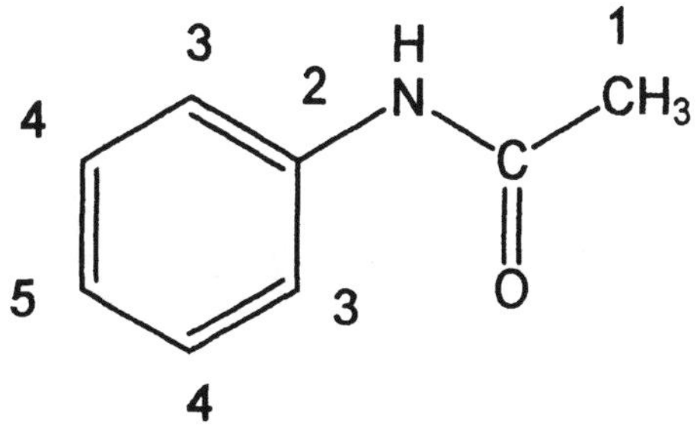

1	2	3	4	5
-2.1	128.7	128.7	128.7	128.7
30	11.1	-9.9	0.2	-5.6
28.1	139.8	118.8	128.9	123.1

Labeled Structure:

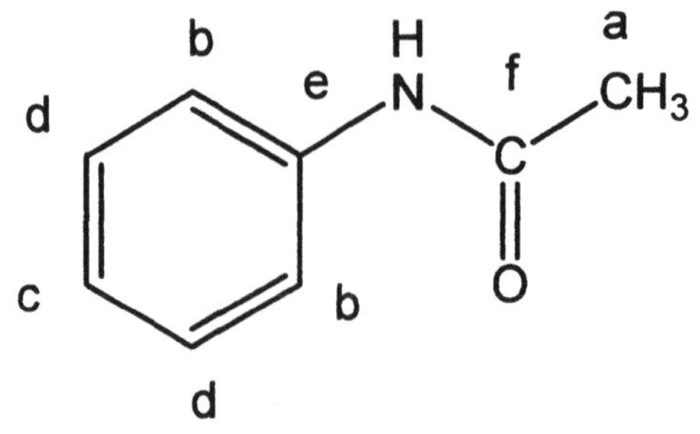

HETCOR:

^{13}C	f (168.98)	e (139.60)	d (128.83)	c (123.47)	b (119.77)	a (24.30)
^{1}H						
a (2.073)						x
b (6.97-7.36)				x	x	
c (7.55-7.67)					x	
d (9.81)						

Hydrogens **a** are attached to carbon **a**; hydrogens **c** are attached to carbons **b**; and hydrogens **b** are attached carbons **c** and **d**.

Labeled Structure:

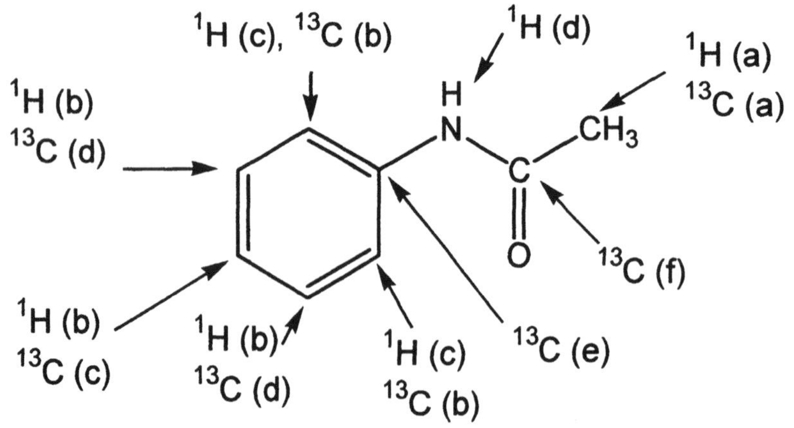

6. Compound with a formula of $C_{10}H_{12}O_2$.

Unsaturation Number:
$U = 10 + 1 - \frac{1}{2}(12) + \frac{1}{2}(0) = 5$
benzene plus one ring or double bond

^{1}H NMR Spectrum:

	Chemical Shift (Actual)	(Calc)	Splitting	Integration	Interpretation
a.	0.93	1.1	t	3 H	CH_3 adjacent to CH_2
b.	1.70	1.7	sextet	2 H	CH_2 adjacent to 5 H
c.	4.23	4.3	t	3 H	CH_3 adjacent to CH_2
d.	7.11-7.44	7.34, 7.47	m	3 H	aromatic
e.	8.02-8.16	8.01	m	2 H	aromatic

The compound is monosubstituted, since there are five aromatic hydrogens. A propyl group is present. A carbon, two oxygens, and a double bond are present.

Calculations:

1	2	3	4	5	6
0.9	1.2	1.2	7.27	7.27	7.27
0.2	0.5	3.1	0.74	0.07	0.20
1.1	1.7	4.3	8.01	7.34	7.47

Labeled Structure:

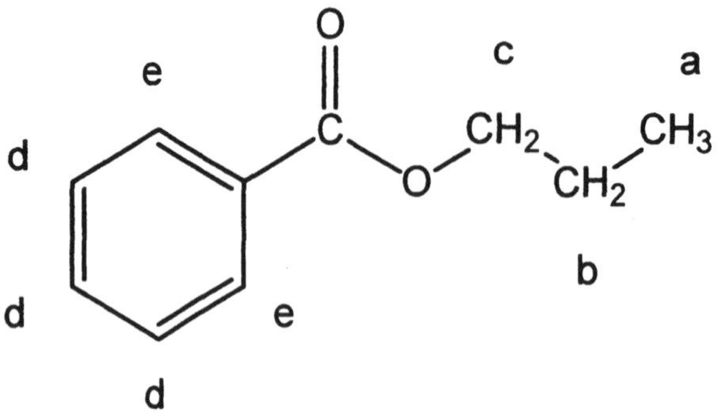

propyl benzoate

^{13}C NMR and DEPT Spectra:

	Chemical Shift (Actual)	(Calc)	DEPT	Interpretation
a.	9.61	12.6	CH_3	alkyl
b.	21.49	22.1	CH_2	alkyl
c.	65.50	66.6	CH_2	alkyl
d.	127.68	128.6	2 CH	aromatic
e.	128.59	129.9	2 CH	aromatic
f.	130.06	130.7	C	aromatic
g.	132.00	133.5	CH	aromatic
h.	165.18	--------	C	ester

Calculations:

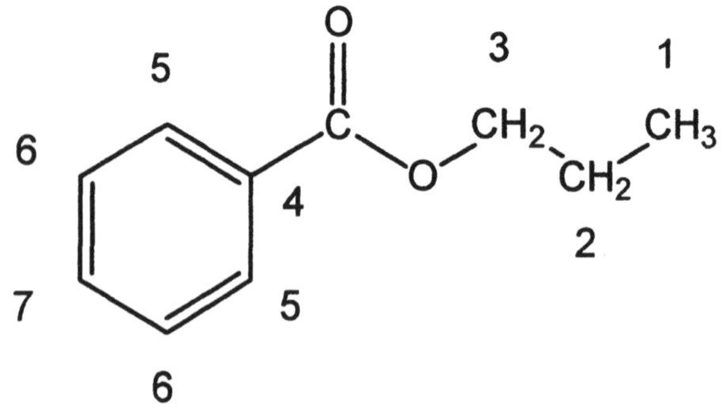

1	2	3	4	5	6	7
15.6	16.1	15.6	128.7	128.7	128.7	128.7
-3	6	51	2.0	1.2	-0.1	4.8
12.6	22.1	66.6	130.7	129.9	128.6	133.5

Labeled Structure:

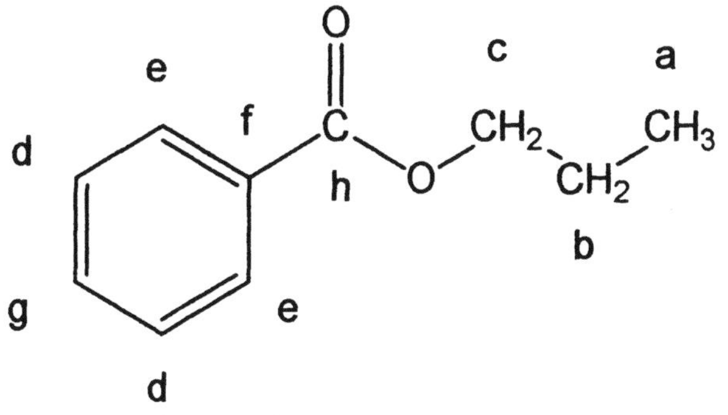

COSY:

¹H	e (8.02-8.16)	d (7.11-7.44)	c (4.23)	b (1.70)	a (0.93)
¹H					
a (0.93)				X	
b (1.70)			X		X
c (4.23)				X	
d (7.11-7.44)	X				
e (8.02-8.16)		X			

Hydrogens **b** are adjacent to hydrogens **a** and **c**; and hydrogens **d** are adjacent to hydrogens **e**. Refer to the top of page 46 for the labeled structure.

HETCOR:

^{13}C	g (132.00)	f (130.06)	e (128.59)	d (127.68)	c (65.50)	b (21.49)	a (9.61)
^{1}H							
a (0.93)							x
b (1.70)						x	
c (4.23)					x		
d (7.11-7.44)	x			x			
e (8.02-8.16)			x				

Hydrogens **a** are attached to carbon **a**; hydrogens **b** are attached to carbon **b**; hydrogens **c** are attached to carbon **c**; hydrogens **d** are attached to carbons **d** and **g**; and hydrogens **e** are attached to carbons **e**.

Labeled Structure:

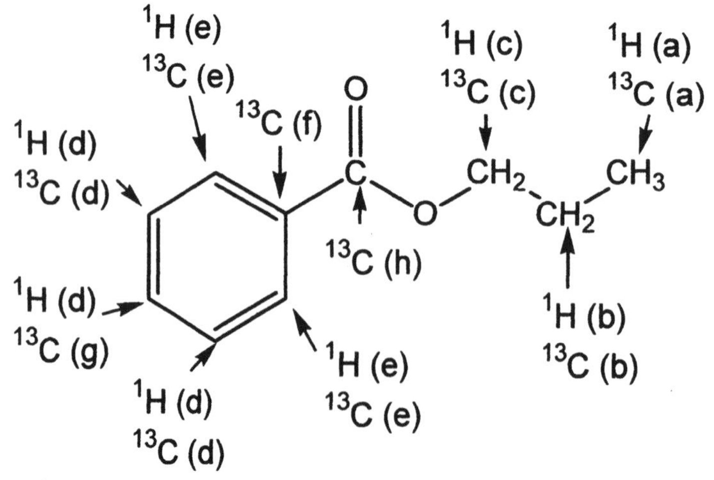

7. Compound with a formula of $C_{10}H_{13}NO_2$.

Unsaturation Number:

$U = 10 + 1 - \frac{1}{2}(13) + \frac{1}{2}(1) = 5$
benzene plus one ring or double bond

^1H NMR Spectrum:

	Chemical Shift (Actual)	(Calc)	Splitting	Integration	Interpretation
a.	1.30	1.4	t	3 H	CH_3 adjacent to CH_2
b.	2.03	1.9	s	3 H	CH_3 isolated
c.	3.96	4.0	q	2 H	CH_2 adjacent to CH_3
d.	6.83	7.03	d	2 H	aromatic
e.	7.51	7.52	d	2 H	aromatic
f.	9.78	-----	s	1 H	amide

A *para*-disubstituted benzene ring is indicated, along with the presence of an ethyl group.

Calculations:

1	2	3	4	5
0.9	7.27	7.27	1.2	0.9
1.0	0.33	0.03	2.8	0.5
	-0.08	-0.27		
1.9	7.52	7.03	4.0	1.4

Labeled Structure:

phenacetin

^{13}C NMR and DEPT Spectra:

	Chemical Shift		DEPT	Interpretation
	(Actual)	(Calc)		
a.	14.71	13.9	CH_3	alkyl
b.	23.82	28.1	CH_3	alkyl
c.	63.10	63.9	CH_2	ether
d.	114.34	114.5	CH	aromatic
e.	120.62	119.8	CH	aromatic
f.	132.59	131.1	C	aromatic
g.	154.43	154.5	C	aromatic
h.	167.74	------	C	amide

Calculations:

1	2	3	4	5	6	7
-2.1	128.7	128.7	128.7	128.7	5.9	5.9
30	11.1	-9.9	0.2	-5.6	58	8
	-7.7	1.0	-14.4	31.4		
28.1	131.1	119.8	114.5	154.5	63.9	13.9

Labeled Structure:

COSY:

¹H	f (9.78)	e (7.51)	d (6.83)	c (3.96)	b (2.03)	a (1.30)
a (1.30)				x		
b (2.03)						
c (3.96)						x
d (6.83)		x				
e (7.51)			x			
f (9.78)						

Hydrogens **a** are adjacent to hydrogens **c**; hydrogens **d** are adjacent to hydrogens **e**. Refer to structure on the top of page 50.

HETCOR:

^{13}C	g	f	e	d	c	b	a
	(154.43)	(132.59)	(120.62)	(114.34)	(63.10)	(23.82)	(14.71)

^{1}H	g	f	e	d	c	b	a
a (1.30)							x
b (2.03)						x	
c (3.96)					x		
d (6.83)				x			
e (7.51)			x				
f (9.78)							

Hydrogens **a** are attached to carbon **a**; hydrogens **b** are attached to carbon **b**; hydrogens **c** are attached to carbon **c**; hydrogens **d** are attached to carbons **d**; and hydrogens **e** are attached to carbons **e**.

Labeled Structure:

CHAPTER 7
Infrared Spectrometry

1. Compound of formula C_4H_9NO.

    Unsaturation Number:

    $U = 4 + 1 - \frac{1}{2}(9) + \frac{1}{2}(1) = 1$
    one double bond or one ring

    IR Spectrum:

Frequency	Bond	Compound Type
1390	C–N stretch	1° amide
1649	N–H bend	1° amide
1660	C=O stretch	1° amide
2985	C–H stretch	CH_3
3201	N–H stretch	1° amide
3354	N–H stretch	1° amide

    Structure:

    butanamide

2. Compound of formula $C_4H_{10}O$

    Unsaturation Number:

    $U = 4 + 1 - \frac{1}{2}(10) + \frac{1}{2}(0) = 0$
    no double bonds, triple bonds, or rings

IR Spectrum:

Frequency	Bond	Compound Type
1102	C–O stretch	2° alcohol, saturated
1326	O–H bend	2° alcohol
1373	C–H bend	CH_3
1455	C–H bend	CH_3
2872	C–H stretch	alkyl
2919	C–H stretch	alkyl
2955	C–H stretch	alkyl
3342	O–H stretch	alcohol

Structure:

H₃C — CH — CH₂CH₃ with OH on CH   2-butanol

3. Compound of formula C_7H_7Br.

Unsaturation Number:

$$U = 7 + 1 - \tfrac{1}{2}(8) + \tfrac{1}{2}(0) = 4$$
benzene

IR Spectrum:

Frequency	Bond	Compound Type
591	C–Br stretch	bromide
797	C–H out-of-plane bend	*p*-disubstituted benzene
1484	C=C stretch	aromatic
1596	C=C stretch	aromatic
2919	C–H stretch	alkyl
3013	C–H stretch	aromatic

Structure:

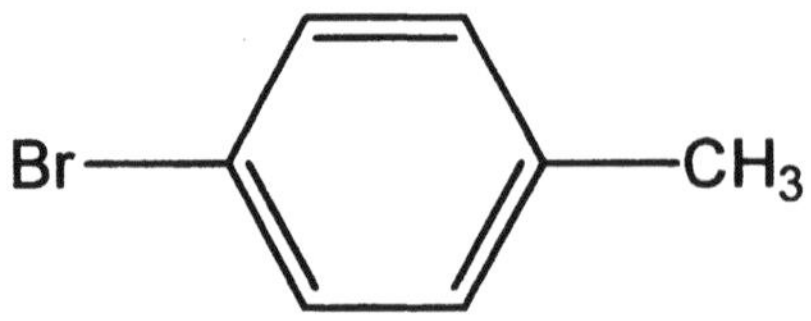

Br—⟨ring⟩—CH₃   4-bromotoluene

4. Compound of formula C_4H_8O.

Unsaturation Number:

$U = 4 + 1 - ½(8) + ½(0) = 1$
one double bond or one ring

IR Spectrum:

Frequency	Bond	Compound Type
1173	C=O stretch and bend	aliphatic ketone
1375	C–H bend	CH_3
1455	C–H bend	CH_3
1713	C=O stretch	aliphatic ketone
2978	C–H stretch	alkane

Structure:

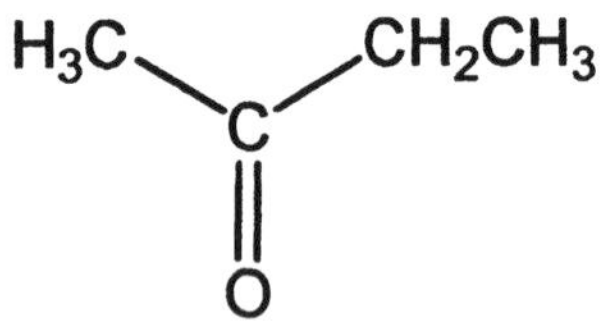

H₃C—C—CH₂CH₃   butanone
       ‖
       O

CHAPTER 8
Mass Spectrometry

1. **$C_2H_4N_2$**

    % [M + 1] = [1.08 x 2] + [0.0115 x 4] + [0.369 x 2]

            = 2.944

    % [M + 2] = $\left[\dfrac{(1.08 \times 2)^2}{200} \right]$

            = 0.0233

    C_3H_4O

    % [M + 1] = [1.08 x 3] + [0.0115 x 4] + [0.0381 x 1]

            = 3.324

    % [M + 2] = $\left[\dfrac{(1.08 \times 3)^2}{200} \right]$ + [0.205 x 1]

            = 0.257

    C_4H_8

    % [M + 1] = [1.08 x 4] + [0.0115 x 8]

            = 4.412

    % [M + 2] = $\left[\dfrac{(1.08 \times 4)^2}{200} \right]$

            = 0.0933

2. **$C_6H_{12}Br_2$**

% [M + 1] = [1.08 x 6] + [0.0115 x 12]

= 6.618

$$\% \ [M + 2] \ = \ \left[\frac{(1.08 \ x \ 6)^2}{200} \right] \ + \ [97.3 \ x \ 2]$$

= 194.810

$C_6H_{12}BrCl$

% [M + 1] = [1.08 x 6] + [0.0115 x 12]

= 6.618

$$\% \ [M + 2] \ = \ \left[\frac{(1.08 \ x \ 6)^2}{200} \right] \ + \ [32.0 \ x \ 1] \ + \ [97.3 \ x \ 1]$$

= 129.510

$C_6H_{12}Cl_2$

% [M + 1] = [1.08 x 6] + [0.0115 x 12]

= 6.618

$$\% \ [M + 2] \ = \ \left[\frac{(1.08 \ x \ 6)^2}{200} \right] \ + \ [32.0 \ x \ 2]$$

= 64.210

3. *m/z*   fragment

27       $H_2C{=}CH^+$

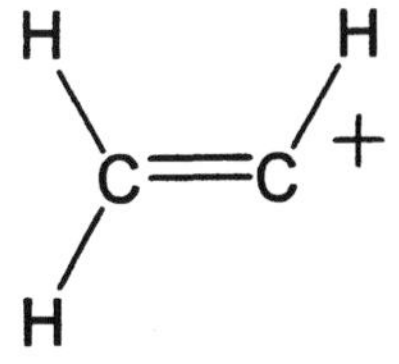

29       $CH_3CH_2{}^+$

31       $H_2C{=}\overset{+}{O}H$

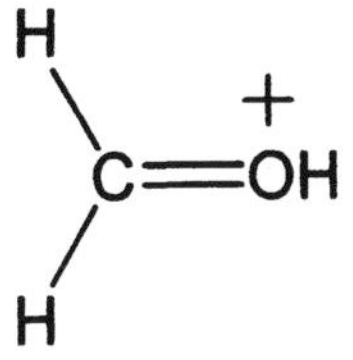

41       $^+H_2C{-}CH{=}CH_2$

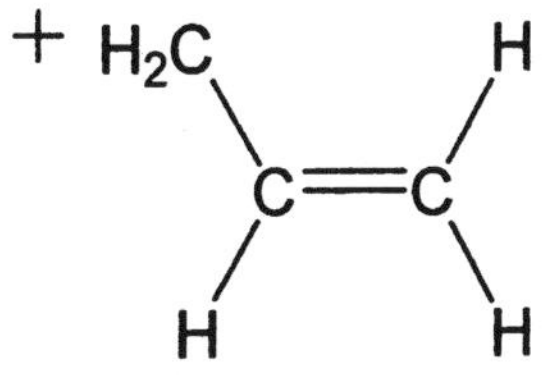

43       $CH_3CH_2CH_2{}^+$

45       $^+CH_2{-}CH_2{-}OH$

56       $CH_3CH_2CH{=}CH_2{}^{+\bullet}$ base peak

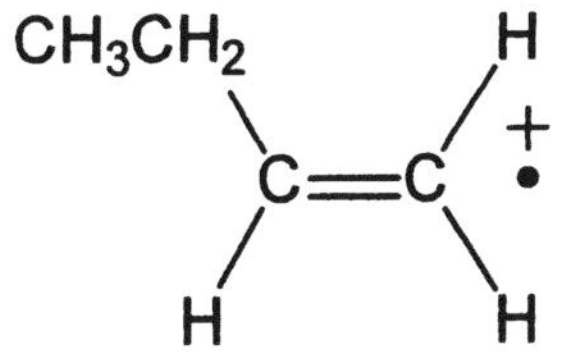

74       $CH_3{-}CH_2{-}CH_2{-}CH_2{-}OH^{+\bullet}$ molecular ion

4.   *m/z*   fragment

28
$$\overset{+}{\underset{}{CH_2}} \\ \| \\ CH_2$$

43     $H_3C-\overset{+}{C}\equiv O$ $CH_3CH_2\overset{+}{CH_2}$ base peak

58

71     $H_3C-\underset{\|}{\underset{O}{C}}-CH_2\overset{+}{CH_2}$ $\overset{+}{\underset{\|}{\underset{O}{C}}}-CH_2CH_2CH_3$

86     $H_3C-\underset{\|}{\underset{O}{C}}-CH_2CH_2CH_3$ molecular ion

5.   *m/z*   fragment

31        $H_2C \!\!=\!\! \overset{+}{O}H$

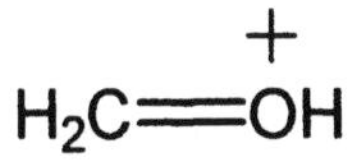

57

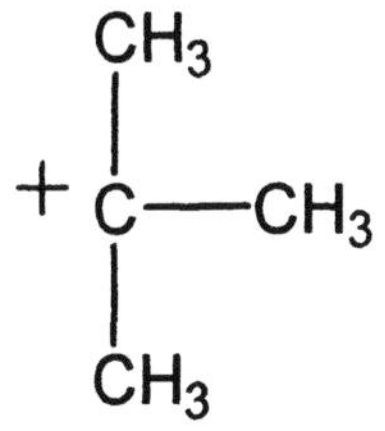

73 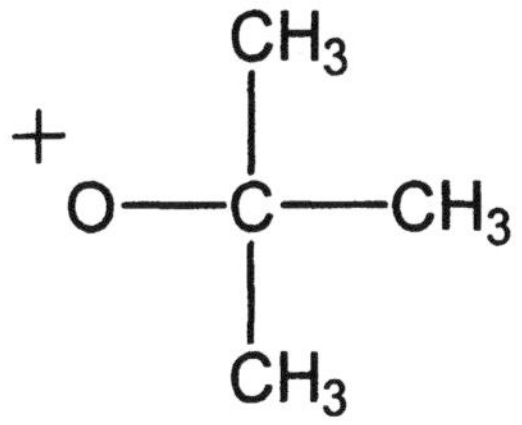 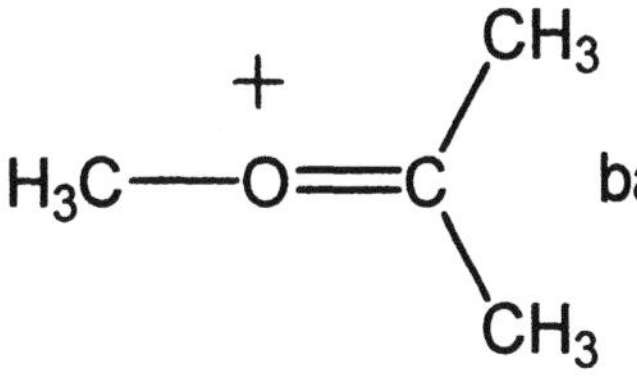   base peak

88 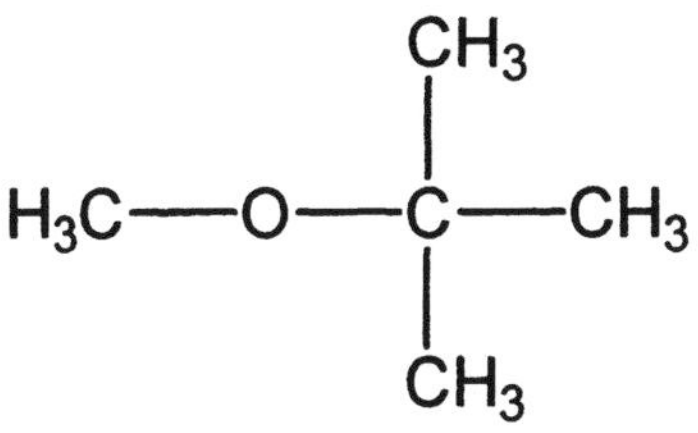   molecular ion

6.   *m/z*   fragment

29   CH_3CH_2

43          $CH_3CH_2CH_2$

71          $CH_3CH_2CH_2CH_2CH_2$ base peak

93          $^{79}BrCH_2$

95          $^{81}BrCH_2$

107          $^{79}BrCH_2CH_2$

109          $^{81}BrCH_2CH_2$

150          $^{79}BrCH_2CH_2CH_2CH_2CH_3$ ⎫
                                               ⎬ molecular ion
152          $^{81}BrCH_2CH_2CH_2CH_2CH_3$ ⎭

CHAPTER 9
Chemical Tests for Functional Groups

1. The sodium hydroxide liberates the aniline from the anilinium salt.

   From the anhydride:

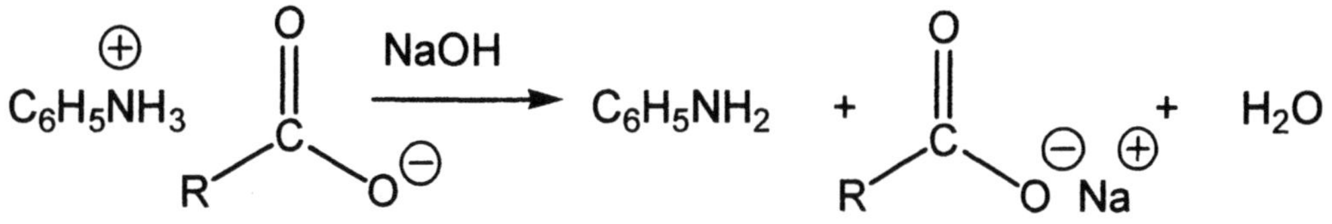

   anilinium salt   aniline

   From the acyl halide:

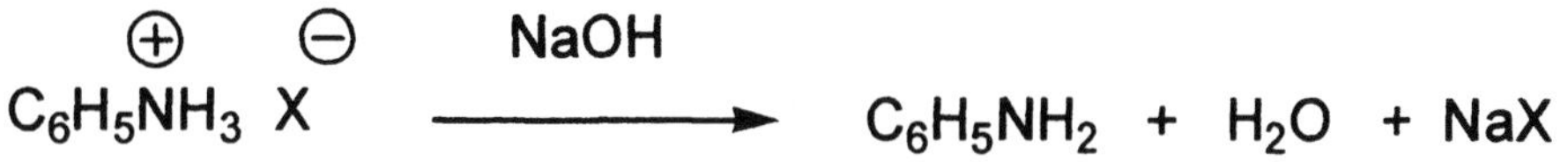

   $$C_6H_5\overset{\oplus}{N}H_3 \ \overset{\ominus}{X} \xrightarrow{\text{NaOH}} C_6H_5NH_2 \ + \ H_2O \ + \ NaX$$

   anilinium salt   aniline

2. *Reaction with propanoyl chloride:*

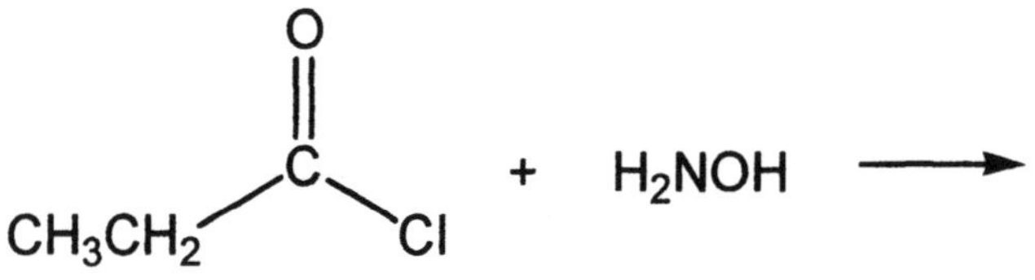

   propanoyl chloride   hydroxylamine

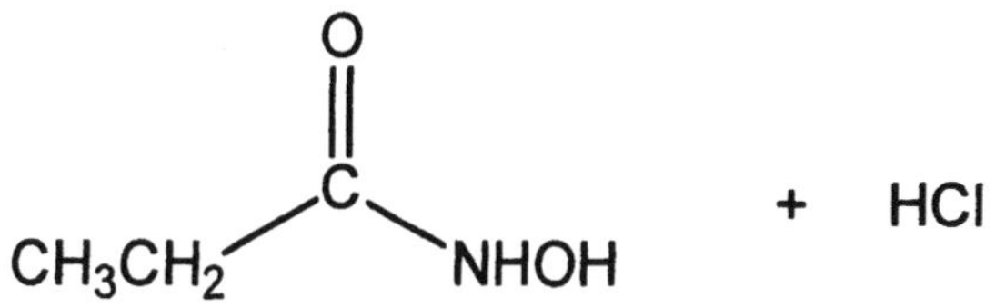

   hydroxamic acid

Reaction with propanoic anhydride:

propanoic anhydride + hydroxylamine ⟶

propanoic acid + hydroxamic acid

Reaction of the hydroxamic acid:

hydroxamic acid + FeCl₃ ⟶ ferric hydroxamate complex (burgundy or magenta color) + 3 HCl

To distinguish between propanoyl chloride and propanoic anhydride, use a test that detects the presence of chlorine. The silver nitrate test (Experiment 33) or the sodium iodide test (Experiment 34) detect the presence of chlorine.

3. *Mechanism of acetic anhydride with hydroxylamine:*

acetic anhydride   hydroxylamine

Mechanism of acetyl chloride with hydroxylamine:

acetyl chloride   hydroxylamine

The acyl halides are more reactive than the anhydrides because the lone pairs on the central oxygen on anhydrides are shared by two carbonyl groups, thereby forming other resonance structures. The contribution of the other resonance structures reduces the positive charge on the carbonyl carbon, making it less reactive than the acyl halides.

4. Sodium is used in testing neutral compounds. Therefore, it will react very little or not at all with phenol, benzoic acid, oximes, nitromethane, and benzenesulfonamide.

     The test is never used with these compounds for the following reasons. Benzoic acid is more acidic than phenol. Phenol and benzenesulfonamide have approximately the same acidity, and are both more acidic than alcohols. Oximes, like most bases, are basic. Nitromethane exists in tautomeric equilibrium with the corresponding nitronic acids through the carbanion.

5. If sodium comes into contact with a large amount of water, a dangerous explosion may occur, due to the formation of hydrogen, a flammable gas. Additionally, the sodium must be cleaned prior to use by carefully scraping off the outer oxide layer. Any utensils should be placed in alcohol to decompose any traces of sodium.

6. *Mechanism of acetyl chloride with an alcohol:*

acetyl chloride   alcohol

As seen in the mechanism, a smaller alcohol would react faster than a larger alcohol. Therefore, the trend is primary > secondary > tertiary.

A competing reaction for tertiary alcohols is the formation of the alkyl chloride.

7. Primary alcohols do not react perceptibly.

H₃C CH₂ CH₂ OH structure — 1-pentanol

CH₃ / H₃C CH CH₂ OH structure — 2-methyl-1-butanol

H₃C CH—CH₂ / H₃C CH₂–OH structure — 3-methyl-1-butanol

CH₃ / H₃C—C—CH₂ / CH₃ OH structure — 2,2-dimethyl-1-propanol

Secondary alcohols are between the primary and tertiary alcohols in reactivity.

H₃C CH₂ CH₃ / CH₂ CH / OH structure — 2-pentanol

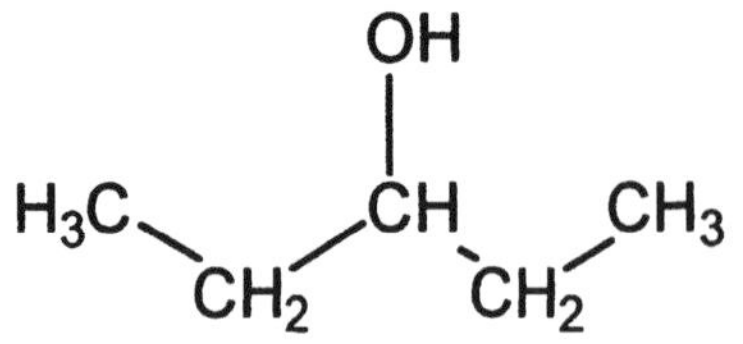

3-pentanol

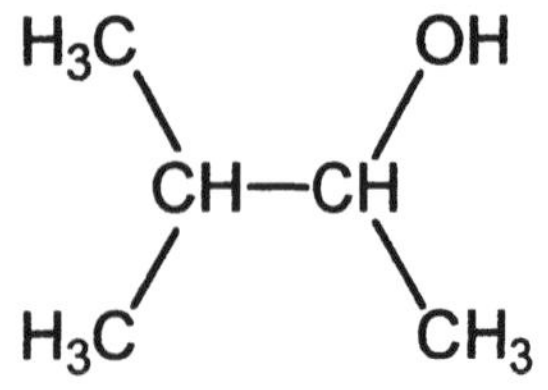

3-methyl-2-butanol

The tertiary alcohol reacts the fastest.

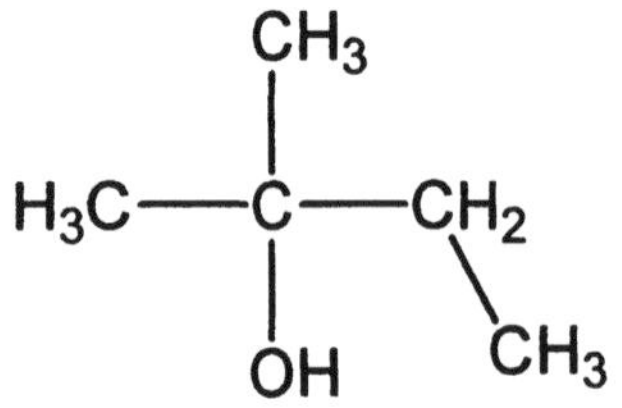

2-methyl-2-butanol

8. Listed below are the relative stabilities of the carbocations.

3° benzylic ≈ 3° allylic > 2° benzylic ≈ 2° allylic ≈ 3° alkyl > 1° benzylic ≈ 1° allylic ≈ 2° alkyl > 1° alkyl > vinylic ≈ aryl

Allyl alcohol will produce a 1° allylic carbocation, which is roughly equivalent to a 2° alkyl carbocation. The 1° allylic carbocation will react faster than a 1° propyl carbocation, produced from 1-propanol.

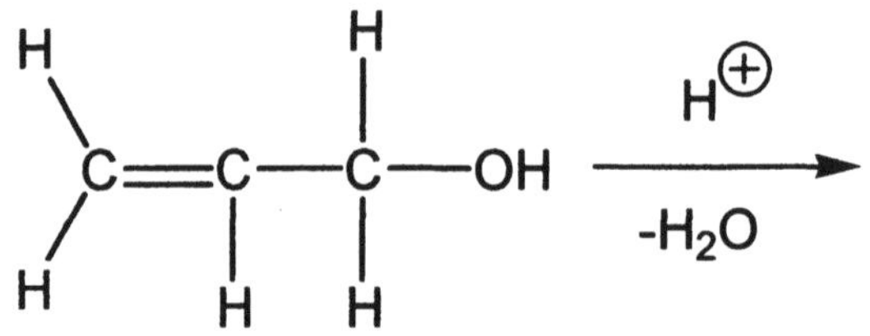

allyl alcohol

1⁰ allylic carbocation

1-propanol   1⁰ propyl carbocation

Benzyl alcohol will produce a 1° benzylic carbocation, which is roughly equivalent to a 2° alkyl carbocation. The 1° benzylic carbocation will react faster than a 1° pentyl carbocation, produced from 1-pentanol.

benzyl alcohol   1° benzylic carbocation

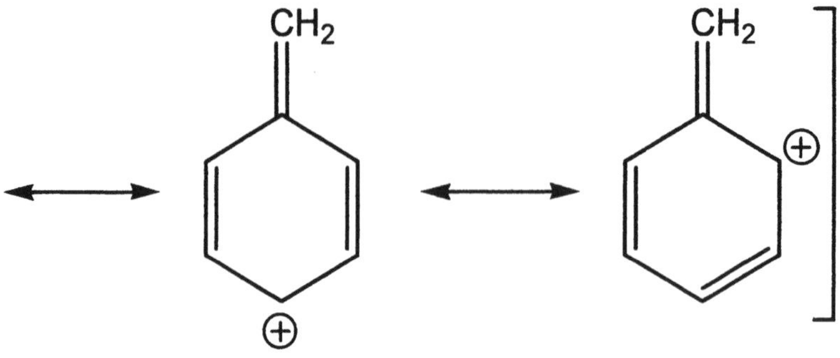

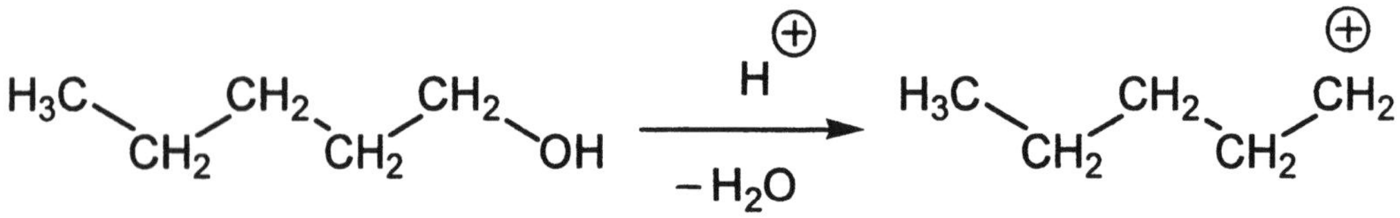

H₃C—CH₂—CH₂—CH₂—OH → H₃C—CH₂—CH₂—CH₂⁺

1-pentanol   1⁰ pentyl carbocation

9.   Butanal and 1-butanol will give a positive test with Jones reagent.

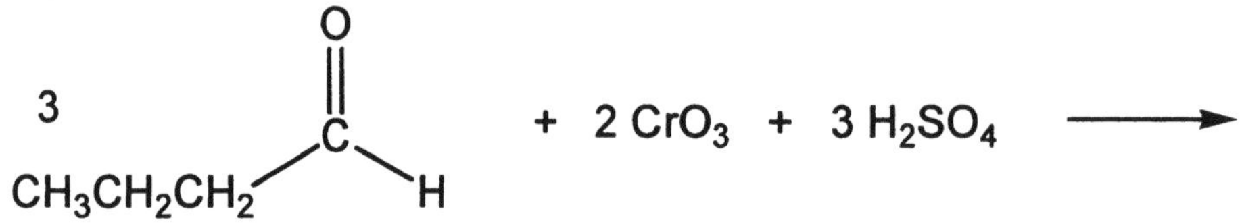

3 CH₃CH₂CH₂—CHO + 2 CrO₃ + 3 H₂SO₄ ⟶

butanal

3 CH₃CH₂CH₂—COOH + 3 H₂O + Cr₂(SO₄)₃

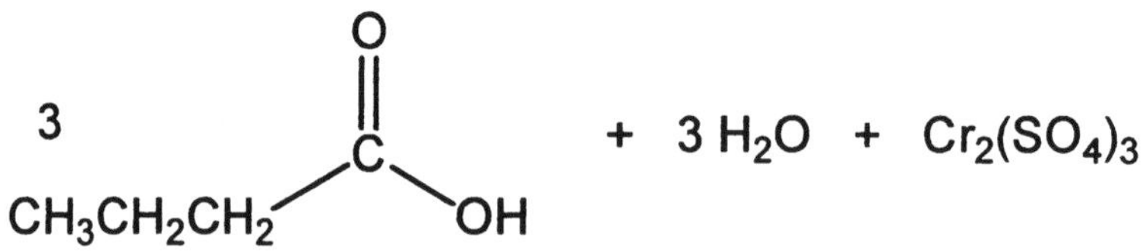

butanoic acid

$$3\ CH_3CH_2CH_2CH_2OH\ +\ 4\ CrO_3\ +\ 6\ H_2SO_4\ \longrightarrow$$

1-butanol

$$3 \quad \underset{CH_3CH_2CH_2}{\overset{\overset{\displaystyle O}{\|}}{C}}\overset{}{\underset{}{}}OH \quad +\ 9\ H_2O\ +\ 2\ Cr_2(SO_4)_3$$

butanoic acid

10. Acetyl chloride reacts with 1-butanol, but not butanal.

$$CH_3CH_2CH_2CH_2OH \quad + \quad \underset{H_3C}{\overset{\overset{\displaystyle O}{\|}}{C}}\overset{}{\underset{}{}}Cl \quad \longrightarrow$$

1-butanol   acetyl chloride

$$\underset{H_3C}{\overset{\overset{\displaystyle O}{\|}}{C}}OCH_2CH_2CH_2CH_3 \quad +\ HCl(g)$$

butyl acetate

$$\underset{CH_3CH_2CH_2}{\overset{\overset{\displaystyle O}{\|}}{C}}H \quad + \quad \underset{H_3C}{\overset{\overset{\displaystyle O}{\|}}{C}}Cl \quad \longrightarrow \quad \text{no reaction}$$

butanal   acetyl chloride

2,4-Dinitrophenylhydrazine reacts with butanal, but not 1-butanol.

butanal   2,4-dinitrophenylhydrazine

2,4-dinitrophenylhydrazone of butanal

CH₃CH₂CH₂CH₂OH   +   (2,4-dinitrophenylhydrazine)

1-butanol   2,4-dinitrophenylhydrazine

$\xrightarrow{H_2SO_4}$ no reaction

11. The sodium bisulfite test reacts with methyl ketones and cyclic ketones up to cyclooctanone. Because this reaction is very sensitive to steric hindrance, the cyclohexanone would react with sodium bisulfite, whereas the 3-pentanone would not.

12. Pinacolone and acetophenone have the following structures.

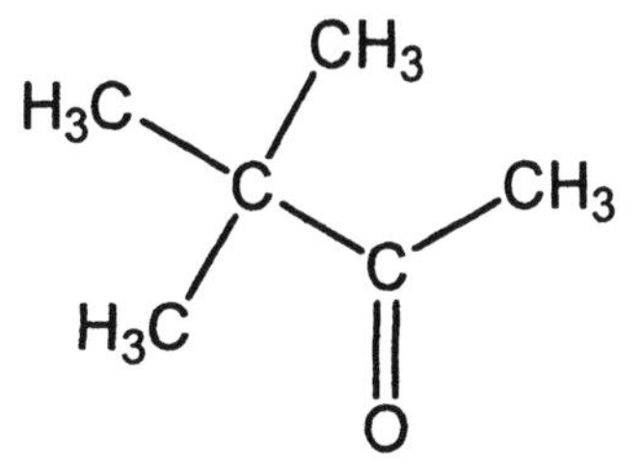

pinacolone
(3,3-dimethyl-2-butanone)

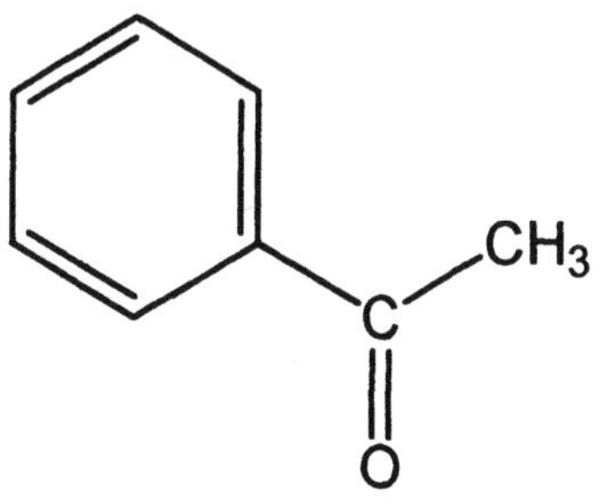

acetophenone

Both pinacolone and acetophenone are sterically hindered on one side. This reaction does not work with sterically hindered structures.

13. Sodium bisulfite will add to an α,β-unsaturated carbonyl compound via a Michael addition. Cinnamaldehyde is an α,β-unsaturated carbonyl compound. A second equivalent of sodium bisulfite would add to the carbonyl.

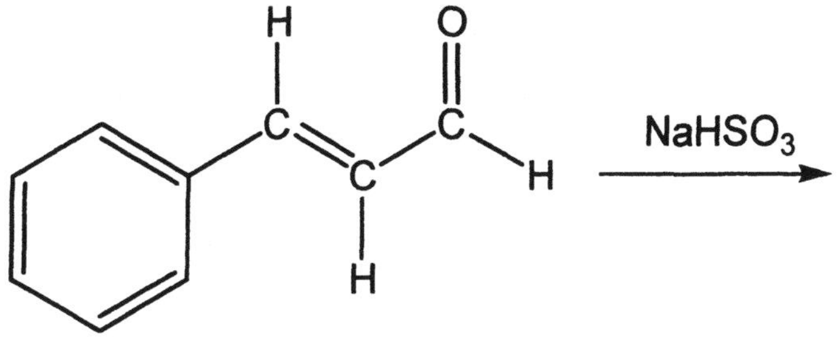

$NaHSO_3$ →

trans-cinnamaldehyde
(*trans*-3-phenyl-2-propenal)

14.   The bisulfite addition products that are derived from low molecular weight aldehydes and ketones are soluble in water, but not in ethanol. Therefore, ethanol is used as a solvent. Acetone would be considered a low molecular weight ketone and its bisulfite addition product would be soluble in water.

15.   The silver ion could react with a reactive halogen to form a precipitate.

$$Ag^{\oplus} + X^{\ominus} \longrightarrow AgX\ (s)$$

16.   The Purpald Test reacts with hexanal, but not 3-hexanone.

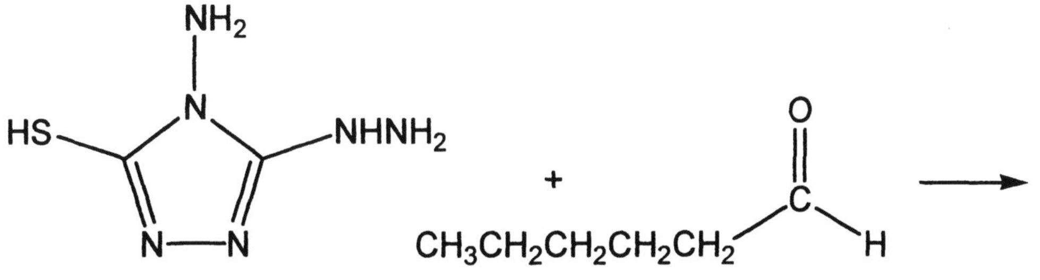

Purpald + hexanal →

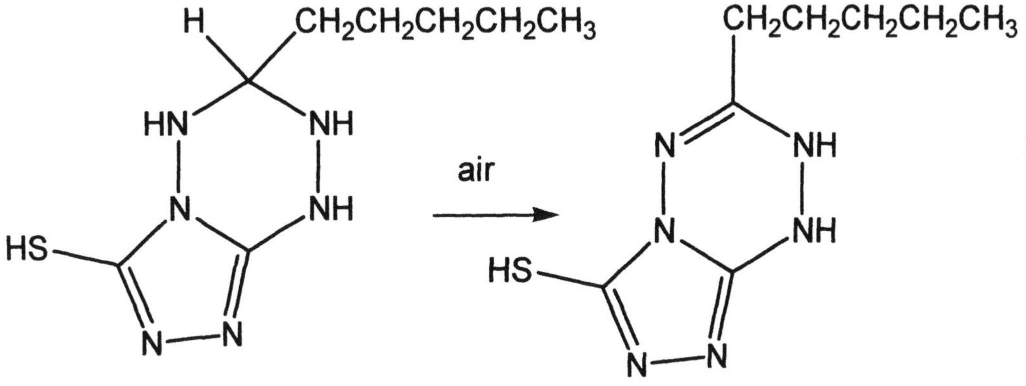

air →

6-mercapto-*s*-triazolo- [4,3,b]-*s*-tetrazine

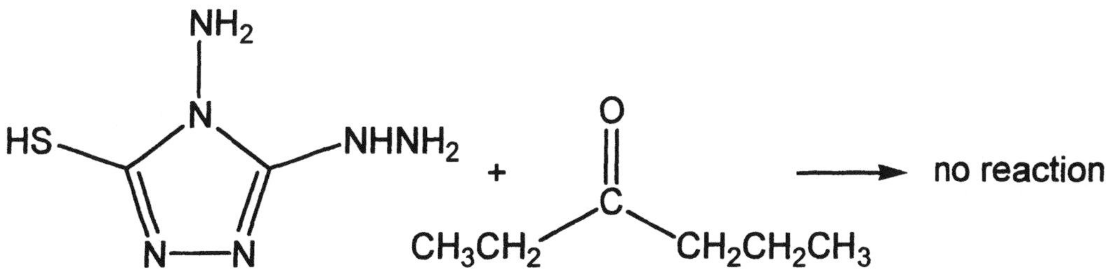

Purpald + 3-hexanone → no reaction

Tollens Test reacts with hexanal, but not 3-hexanone.

$$CH_3CH_2CH_2CH_2CH_2 \overset{\overset{\displaystyle O}{\|}}{C} {-} H \quad + \quad 2 \ Ag(NH_3)_2OH \longrightarrow$$

hexanal   Tollens reagent

$$CH_3CH_2CH_2CH_2CH_2 \overset{\overset{\displaystyle O}{\|}}{C} {-} O^{\ominus} \ NH_4^{\oplus}$$

ammonium butanoate

$$+ \ H_2O \ + \ 3 \ NH_3 \ + \ 2 \ Ag(s)$$

$$CH_3CH_2 \overset{\overset{\displaystyle O}{\|}}{C} CH_2CH_2CH_3 \quad + \quad 2 \ Ag(NH_3)_2OH \longrightarrow \begin{array}{c} no \\ reaction \end{array}$$

3-hexanone   Tollens reagent

17. Propylamine reacts with benzenesulfonyl chloride to yield a soluble salt; acidification results in an insoluble *N*-propylbenzenesulfonamide.

$$CH_3CH_2CH_2NH_2 \quad + \quad C_6H_5SO_2Cl \quad + \quad 2 \ NaOH \longrightarrow$$

propyl amine   benzenesulfonyl
chloride

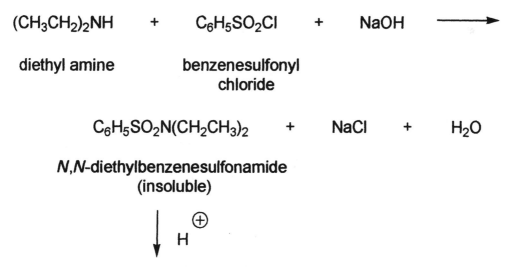

$\overset{\ominus}{C_6H_5SO_2NCH_2CH_2CH_3} \quad \overset{\oplus}{Na} \quad + \quad NaCl \quad + \quad 2\ H_2O$

sodium salt of *N*-propyl
benzenesulfonamide
(soluble)

$\downarrow \overset{\oplus}{H}$

$C_6H_5SO_2NHCH_2CH_2CH_3$

N-propyl benzenesulfonamide
(insoluble)

Diethylamine reacts with benzenesulfonyl chloride to give an insoluble *N,N*-diethylbenzenesulfonamide; this product does not dissolve in acid.

$(CH_3CH_2)_2NH \quad + \quad C_6H_5SO_2Cl \quad + \quad NaOH \quad \longrightarrow$

diethyl amine   benzenesulfonyl
chloride

$C_6H_5SO_2N(CH_2CH_3)_2 \quad + \quad NaCl \quad + \quad H_2O$

N,N-diethylbenzenesulfonamide
(insoluble)

$\downarrow \overset{\oplus}{H}$

No Reaction

Triethylamine forms an intermediate salt with benzenesulfonyl chloride, which breaks apart to the soluble sodium salt of benzenesulfonic acid and the insoluble triethylamine. Acidification results in the soluble triethylammonium salt.

$$(CH_3)_3N \quad + \quad C_6H_5SO_2Cl \quad \longrightarrow \quad C_6H_5SO_2N(CH_3)_3^{\oplus}\ Cl^{\ominus}$$

triethylamine   benzenesulfonyl chloride

$$\xrightarrow{\text{2 NaOH}} \quad C_6H_5SO_3^{\ominus}\ Na^{\oplus} \quad + \quad N(CH_3)_3 \quad + \quad NaCl \quad + \quad H_2O$$

sodium salt of benzenesulfonic acid (soluble)   triethylamine (insoluble)

$$\Big\downarrow H^{\oplus}$$

$$C_6H_5SO_3H \quad + \quad (CH_3)_3NH^{\oplus}\ Cl^{\ominus} \quad + \quad 2\ NaCl$$

benzenesulfonic acid   triethylammonium chloride (soluble)

18. Propylamine reacts with nitrous acid to form a diazonium salt, which decomposes spontaneously.

$$CH_3CH_2CH_2NH_2 \ + \ HONO \ + \ 2 \ HCl \longrightarrow [CH_3CH_2CH_2 \overset{\oplus}{N_2} \ \overset{\ominus}{Cl} \ ]$$

propylamine   nitrous   diazonium salt
                                  acid   (unstable at 0^0)

$$\xrightarrow[\text{spontaneous}]{H_2O} \quad N_2 \ (g) \ + \ CH_3CH_2CH_2OH \ + \ CH_3CH_2CH_2Cl$$

$$+ \ CH_3CH_2CH_2OCH_2CH_2CH_3 \ + \quad \text{alkene}$$

Diethylamine reacts with nitrous acid to form a *N,N*-diethyl-*N*-nitrosoamine.

$$(CH_3CH_2)_2NH \ + \ HONO \longrightarrow \begin{matrix} CH_3CH_2 \\ \diagdown \\ N-N=O \\ \diagup \\ CH_3CH_2 \end{matrix} \ + \ H_2O$$

diethylamine   nitrous acid   *N,N*-diethyl-*N*-nitrosoamine
                                     (yellow oil or solid)

Triethylamine reacts with the acid to form a soluble triethyl-ammonium salt.

$$(CH_3)_3N \quad + \quad \overset{\oplus}{H} \longrightarrow (CH_3)_3\overset{\oplus}{NH}$$

triethylamine   triethylammonium
                                           salt
                                          (soluble)

Aniline reacts with nitrous acid to form an aromatic diazonium salt, which loses nitrogen on warming to room temperature.

$$C_6H_5NH_2 \quad + \quad HONO \quad + \quad HCl \longrightarrow \quad C_6H_5\overset{+}{N_2} \ \overset{-}{Cl}$$

aniline   nitrous   diazonium salt
                            acid   (stable at 0°C)

$$\xrightarrow{\quad H_2O \quad} \qquad N_2 \text{ (g)} \quad + \quad C_6H_5OH \quad + \quad HCl$$

N-Methylaniline reacts with nitrous acid to form a *N*-methyl-*N*-nitroso-*N*-phenylamine.

N-methylaniline   nitrous acid

N-methyl-*N*-nitroso-
N-phenylamine
(yellow oil or solid)

N,N-Dimethylaniline reacts with nitrous acid to give *N,N*-dimethyl-*p*-nitrosoaniline.

N,N-dimethylaniline   nitrous acid

hydrochloride salt of *N,N*-dimethyl-*p*-nitrosoaniline
(orange color)

NaOH

N,N-dimethyl-*p*-nitrosoaniline
(green)

19. Ninhydrin is the monohydrate of 1,2,3-indanetrione (1,2,3-triketohydrindene).

ninhydrin
(monohydrate of 1,2,3-indanetrione
or 1,2,3-triketohydrindene)

1,2,3-indanetrione
(1,2,3-triketohydrindene)

20. Carboxylic acids, due to hydrogen bonding, have higher melting points and boiling points than esters. For example, benzoic acid has a melting point of 122°C, whereas ethyl benzoate has a boiling point of 213°C. If a compound is a liquid, it is more volatile.

21. For ethers in which the groups are either methyl or primary, the cleavage of the ether follows a S_N2 mechanism, as illustrated below. The alkyl halide is formed from the smaller alkyl group.

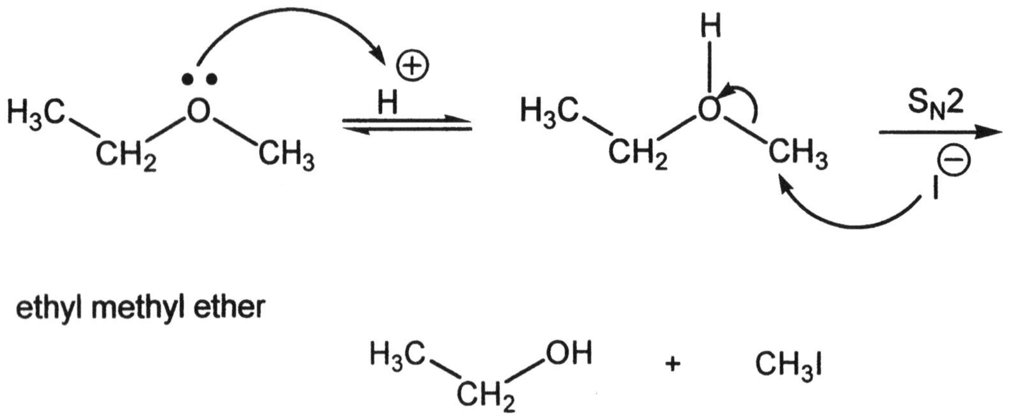

ethyl methyl ether

ethanol   iodomethane

For ethers in which one of the groups is a tertiary alkyl group, the cleavage of the ether follows a S_N1 mechanism. The alkyl halide is formed from the larger alkyl group.

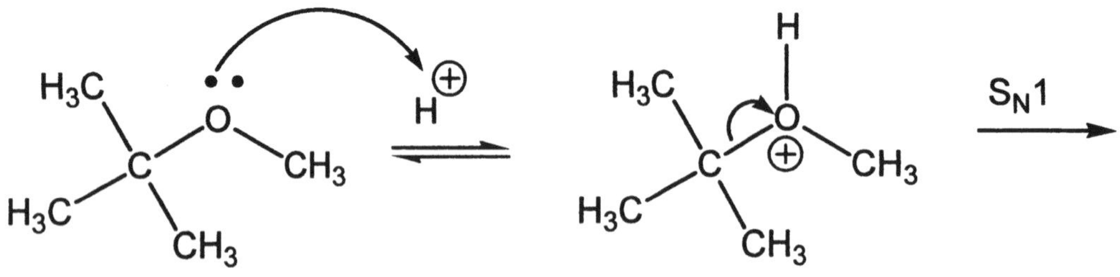

methyl *t*-butyl ether

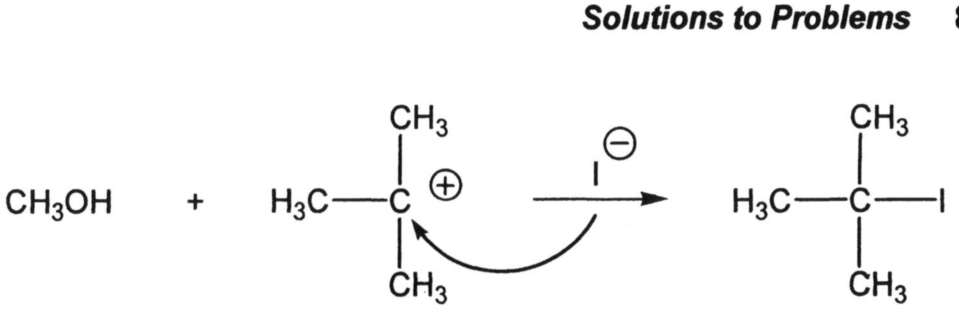

methanol   2-iodo-2-methylpropane

22.   Halides that react with silver nitrate follow an S_N1 mechanism. The formation of the carbocation is shown below. The carbocation is stabilized through resonance.

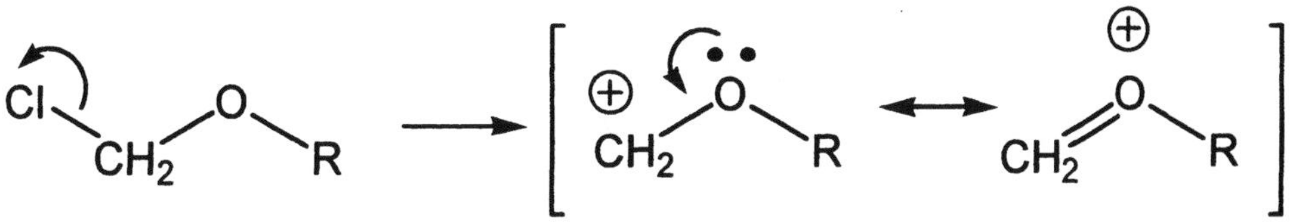

23.   Listed below are the relative stabilities of the carbocations.

3° benzylic ≈ 3° allylic > 2° benzylic ≈ 2° allylic ≈ 3° alkyl > 1° benzylic ≈ 1° allylic ≈ 2° alkyl > 1° alkyl > vinylic ≈ aryl

Benzyl chloride forms a 1° benzylic carbocation, which is stabilized through resonance. The 1° benzylic carbocation is more stable than the 1° cyclohexyl carbocation that is formed from cyclohexylmethyl chloride.

benzyl chloride   1⁰ benzylic carbocation

cyclohexylmethyl chloride   1⁰ cyclohexyl carbocation

24.   The low reactivity of aryl halides and vinyl halides toward ethanolic silver nitrate is attributed to the electronegativity of the sp^2-hybridized carbon and to the delocalization of the electrons through resonance. The double bonded carbon is sp^2-hybridized and is more

electronegative than an sp³-hybridized carbon. It is less likely to donate electrons to a halogen to form a carbocation and a halide ion. In some of the resonance structures listed below, there is a positive charge on the halogen and a negative charge on the carbon. These structures do not favor the placement of a negative charge on the halogen.

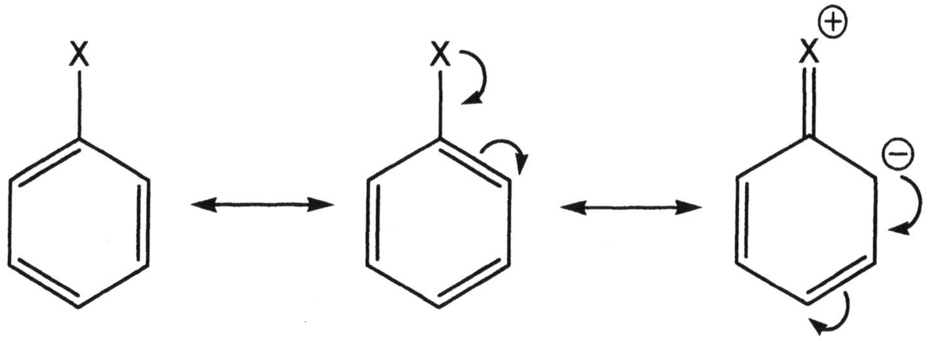

aryl halide

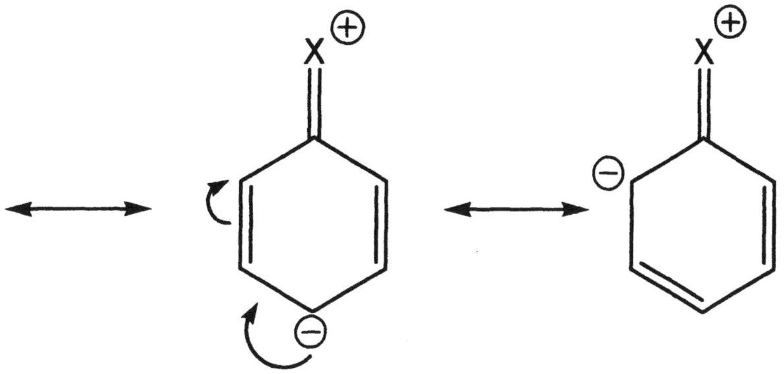

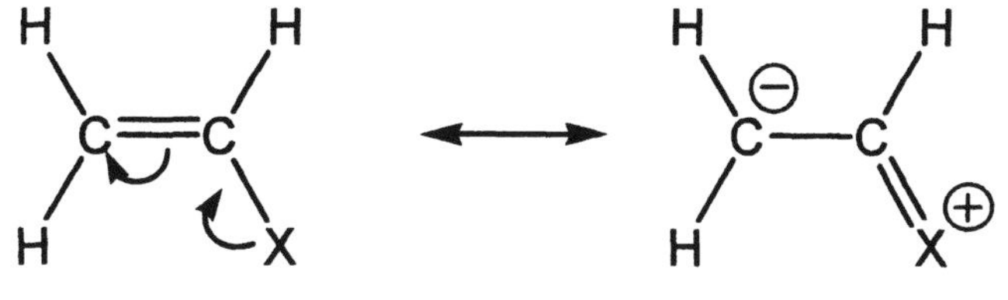

vinyl halide

25. Formation of the tertiary carbocation is sterically inhibited due to the cage-shaped skeleton.

1-chloronorbornane
(1-chlorobicyclo[2.2.1]heptane)

26. Bromine and potassium permanganate solutions are decolorized in the presence of carbon-carbon double bonds and carbon-carbon triple bonds.

27. The presence of electron-withdrawing groups attached to a doubly bonded carbon may prevent the addition of bromine across the double bond. Potassium permanganate is more versatile in this respect, because it is not affected by electron-withdrawing groups.

28. Carbonyl compounds which decolorize bromine solutions usually give a negative potassium permanganate test. Alcohols and compounds such as benzaldehyde and formaldehyde decolorize potassium permanganate solutions, but not bromine solutions. Therefore, to confirm the identity of a carbon-carbon double bond or carbon-carbon triple bond, both tests should be used.

29. 1-Butyne is a terminal alkyne and will react with sodium to produce hydrogen gas and sodium butynide.

$$2 \ CH_3CH_2C \equiv CH \ + \ 2 \ Na \longrightarrow$$

1-butyne

$$2 \ CH_3CH_2C \equiv C^{\ominus} \ Na^{\oplus} \ + \ H_2 \ (g)$$

sodium butynide

2-Butyne does not react with sodium.

$$2 \ CH_3C≡CCH_3 \ + \ 2 \ Na \longrightarrow \text{no reaction}$$
2-butyne

30. Since this reaction is to test for the presence of an aromatic ring, perhaps with a deactivating group, only compounds with the solubility class of I could be used. If the compound contains a strongly activating group such as –OH or –NH$_2$, a violent reaction may occur.

31. The following reactions occur in concentrated sulfuric acid:

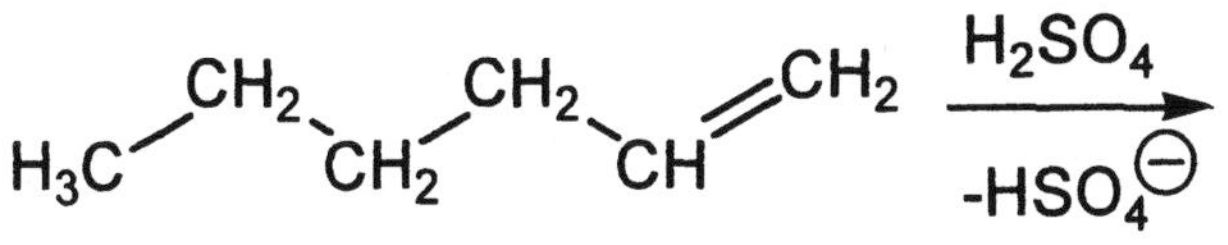

1-hexene

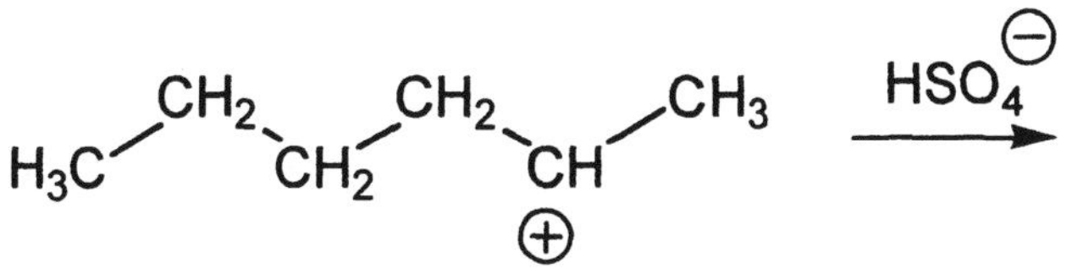

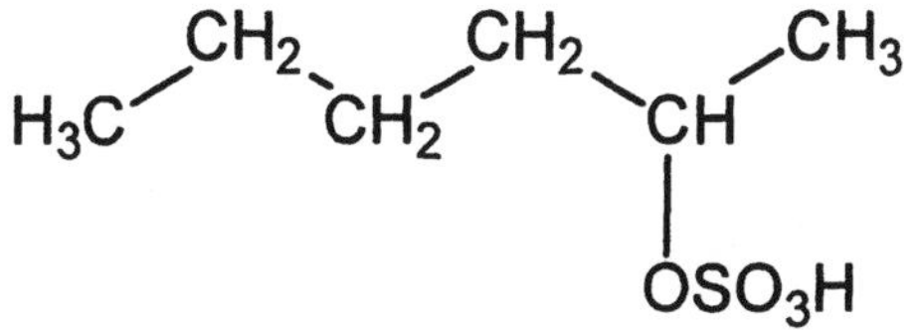

The following reactions occur in fuming sulfuric acid:

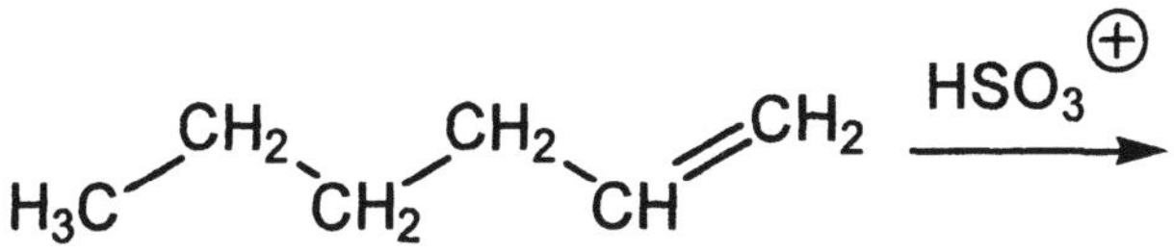

1-hexene

32.   To form a tetraarylmethane, a proton transfer would have to occur. Carbon does not have five bonds.

33.   2,4,6-Tribromoaniline will not react with bromine water. The amino group is a strongly activating *ortho*, *para* director and there are no *ortho* or *para* positions open for substititution.

34.   Yes, the bromine water is decolorized as the bromine reacts with phenol; one bromine atom substitutes onto the ring and the other bromine atom pairs with the hydrogen that has been substituted. The hydrogen bromide dissolves in the water.

35.   The bromine is not hydrolyzed in water. The brominating agent is molecular bromine.

36. The benzene ring attacks the bromine through electrophilic substitution. Electron-donating groups make the ring more reactive. An $-O^-$ would donate more electrons to the ring than an $-OH$. The $-OH$ is a strongly activating group.

CHAPTER 10
The Preparation of Derivatives

1. The neutralization equivalent is calculated by dividing the molecular weight by the number of –COOH. The calculations are shown below for benzoic acid and phthalic acid.

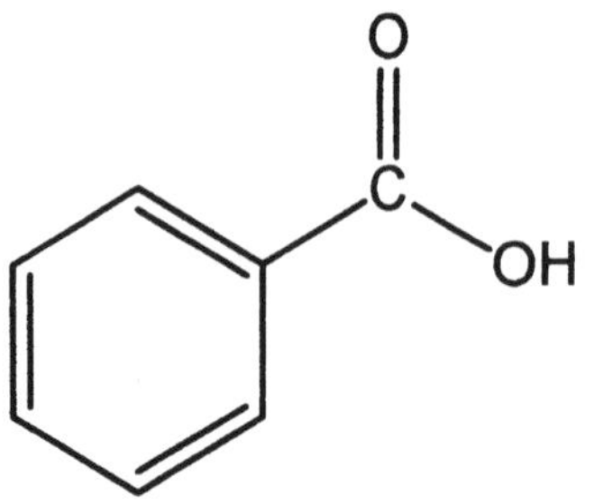

benzoic acid
MW = 122

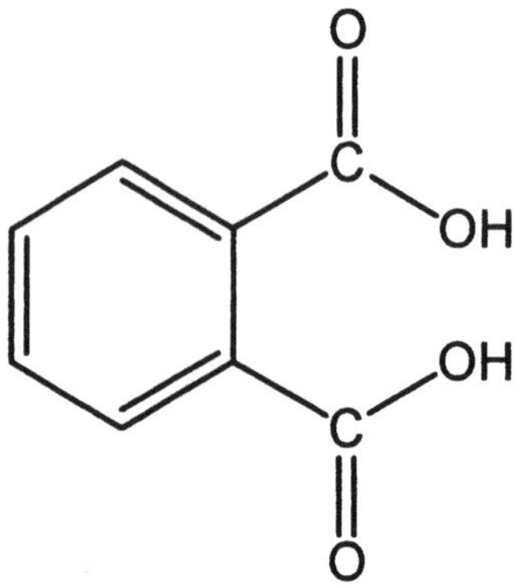

phthalic acid
MW = 166

$$NE = \frac{122}{1} = 122$$

$$NE = \frac{166}{2} = 83$$

2. If the sample is wet, the neutralization equivalent would be too high. The extra water weight would be counted as the weight of the sample and thus would result in the neutralization equivalent being too high.

3. Base does not have any effect on the amine, and thus would not change the neutralization equivalent if the solution was titrated with a base. However, amines can be titrated with hydrochloric acid and the neutralization equivalent determined.

4. A neutralization equivalent of a phenol is only useful if the K_a value is greater than 1×10^{-6}. Examples are phenols with two electron-withdrawing groups such as dinitrophenols.

5. The neutralization equivalent can only be used on acids which have

92

K_a values greater than 1×10^{-6}.

6. a.

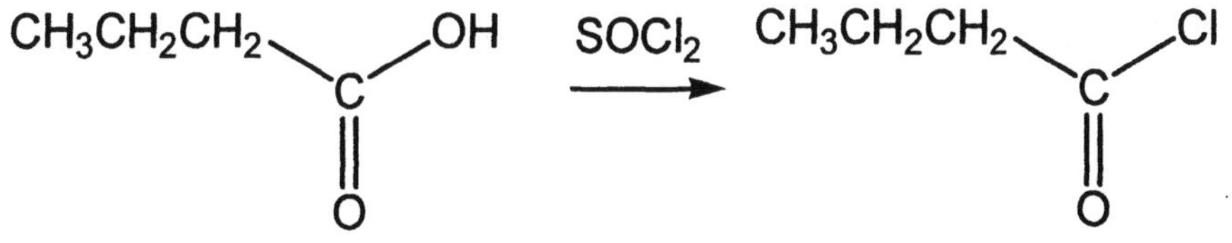

butanoic acid   butanoyl chloride

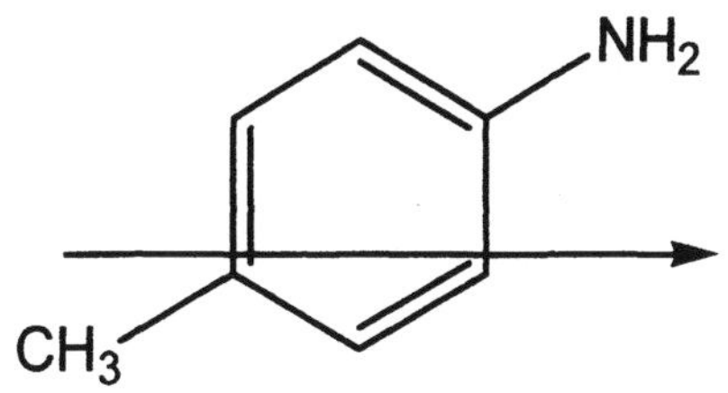

4-toluidine
(*p*-tolylamine)

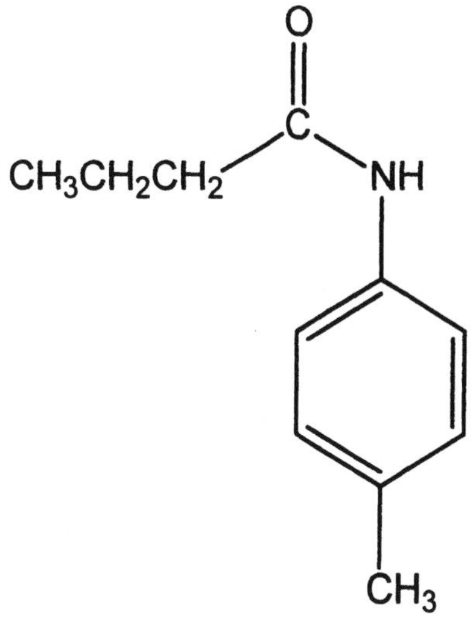

N-*p*-tolylbutyramide

b.

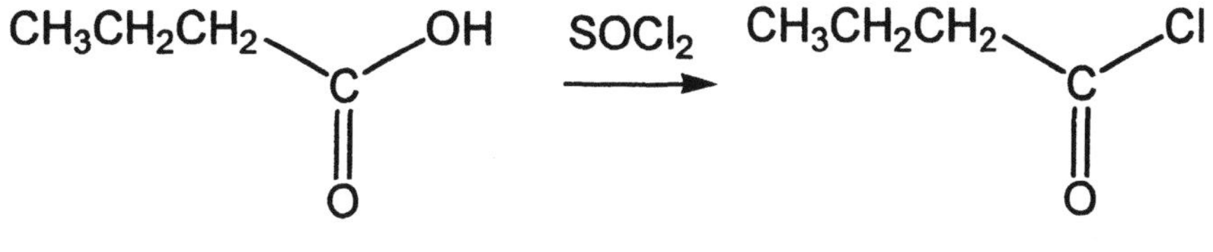

butanoic acid   butanoyl chloride

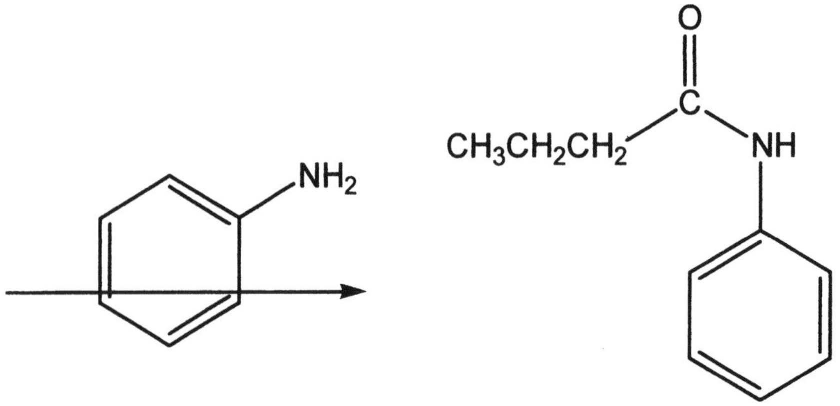

aniline

N-phenylbutyramide

c.

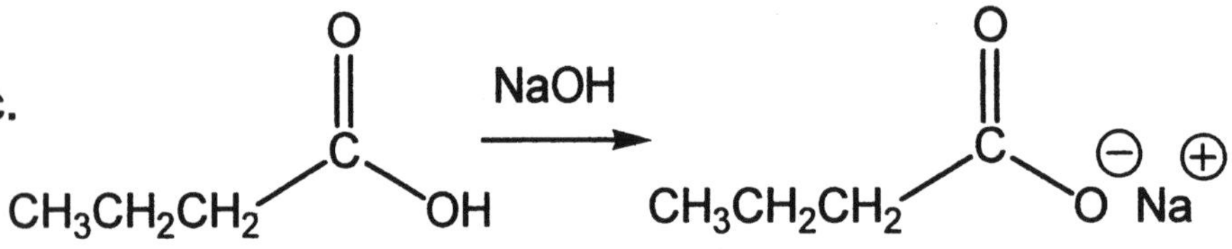

butanoic acid

sodium butanoate

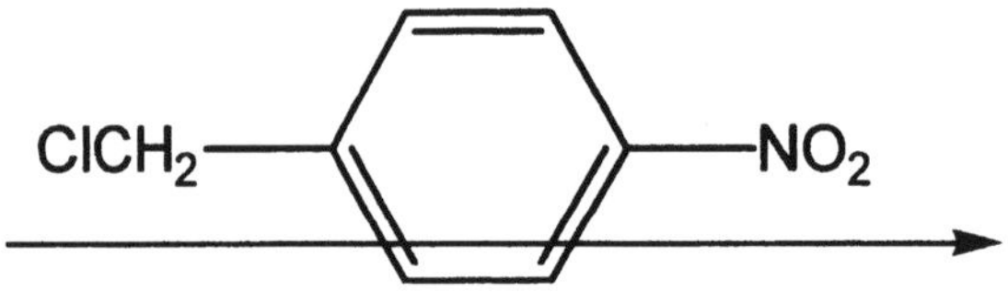

4-nitrobenzylchoride

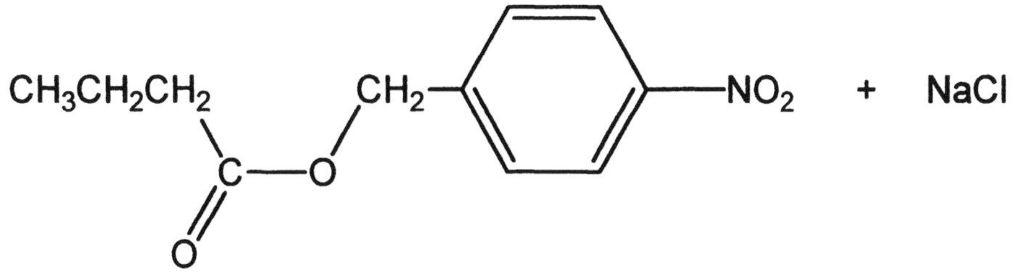

CH₃CH₂CH₂ —C—O—CH₂—⟨benzene ring⟩—NO₂   +   NaCl

(C=O)

4-nitrobenzyl butanoate

d.

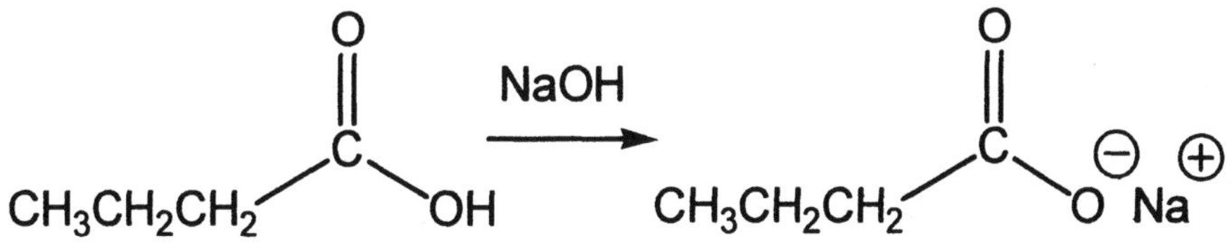

O
‖
CH₃CH₂CH₂—C—OH

butanoic acid

→ **NaOH** →

O
‖
CH₃CH₂CH₂—C—O⁻ Na⁺

sodium butanoate

O
‖
C—⟨benzene ring⟩—Br
|
BrCH₂

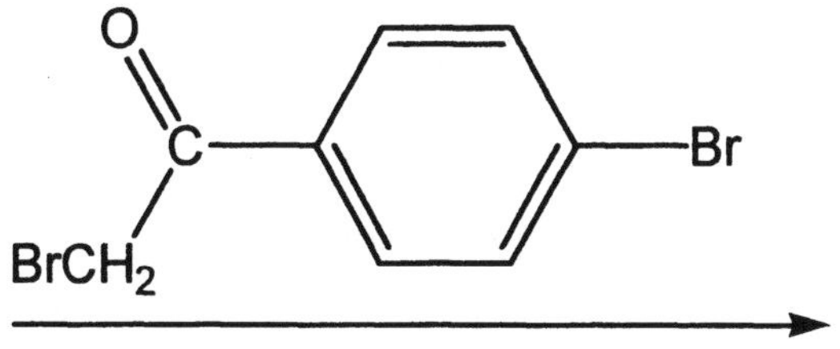

4-bromophenacyl bromide

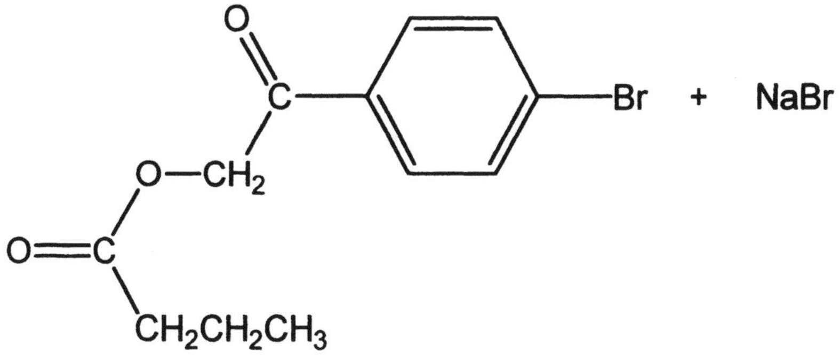

⟶

O
‖
C—⟨benzene ring⟩—Br   +   NaBr
|
O—CH₂
|
O=C
|
CH₂CH₂CH₃

2-(4-bromophenyl)-2-oxoethyl butanoate

e.

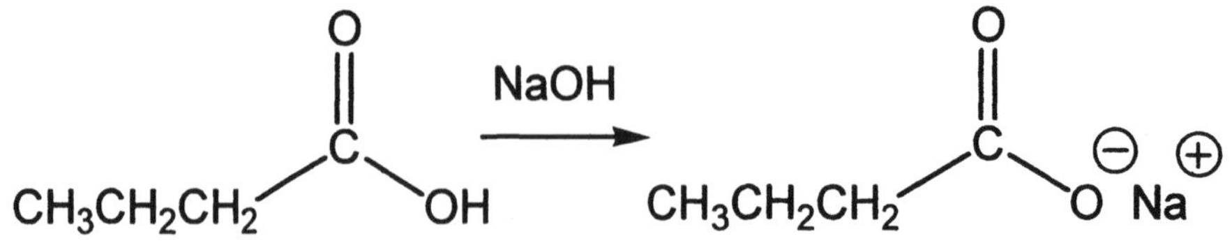

CH₃CH₂CH₂—C(=O)—OH   →[NaOH]   CH₃CH₂CH₂—C(=O)—O⁻ Na⁺

butanoic acid   sodium butanoate

→[C₆H₅CH₂SC(NH₂)₂⁺ Cl⁻]

S-benzylthiuronium salt

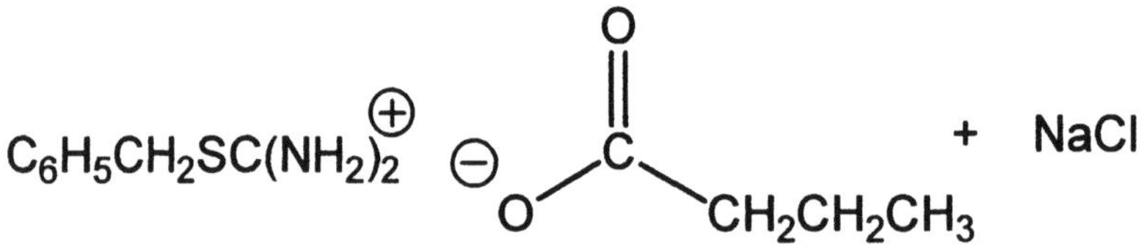

C₆H₅CH₂SC(NH₂)₂⁺ ⁻O—C(=O)—CH₂CH₂CH₃   + NaCl

S-benzylthiuronium salt of butanoic acid

f.

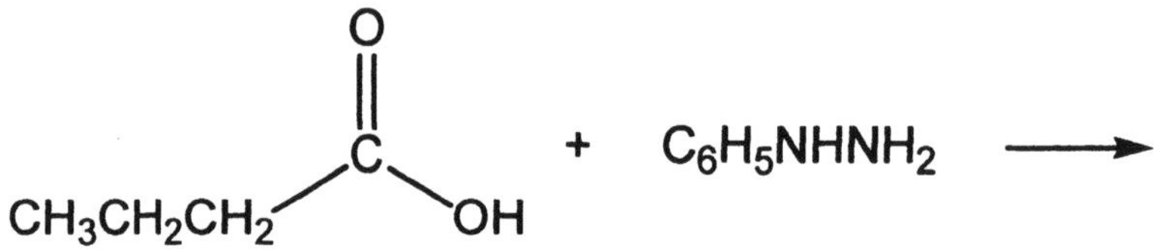

CH₃CH₂CH₂—C(=O)—OH + C₆H₅NHNH₂ →

butanoic acid   phenylhydrazine

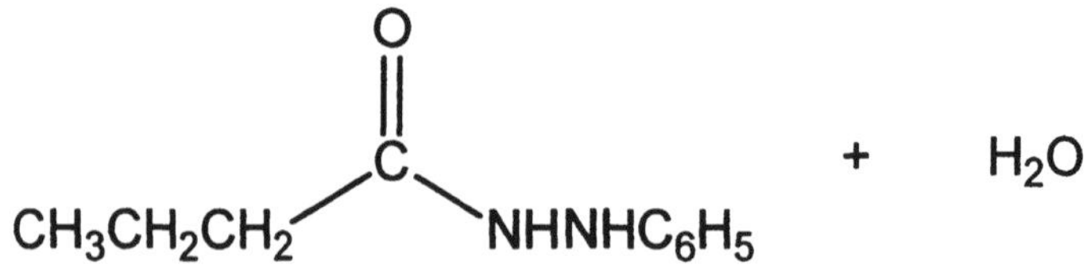

CH₃CH₂CH₂—C(=O)—NHNHC₆H₅   +   H₂O

N-phenylhydrazide of butanoic acid

7. **a.** CH₃CH₂CH₂CH₂OH + phenyl isocyanate →

1-butanol   phenyl isocyanate

butyl *N*-phenyl carbamate

b. CH₃CH₂CH₂CH₂OH + 1-naphthyl isocyanate →

1-butanol   1-naphthyl isocyanate

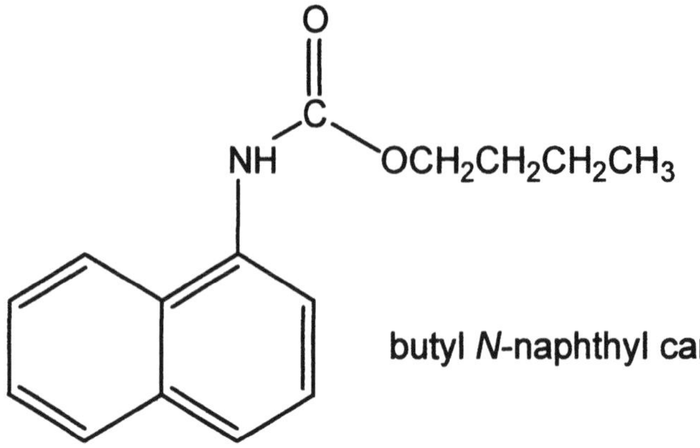

butyl *N*-naphthyl carbamate

c. 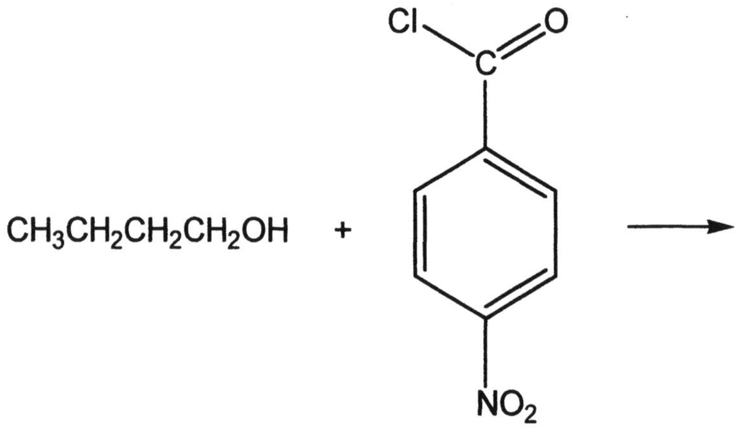 $CH_3CH_2CH_2CH_2OH$ +

1-butanol   4-nitrobenzoyl chloride

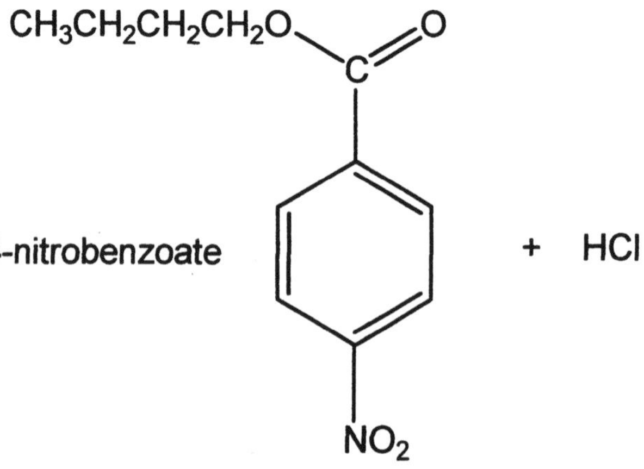

$CH_3CH_2CH_2CH_2O$

butyl 4-nitrobenzoate   +   HCl

d. CH₃CH₂CH₂CH₂OH +

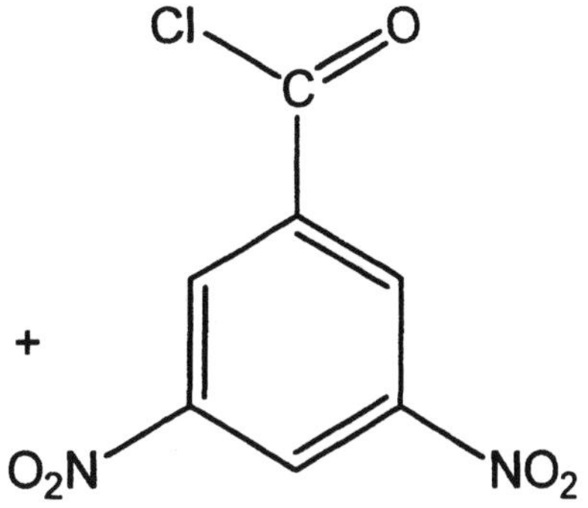

    1-butanol   3,5-dinitrobenzoyl chloride

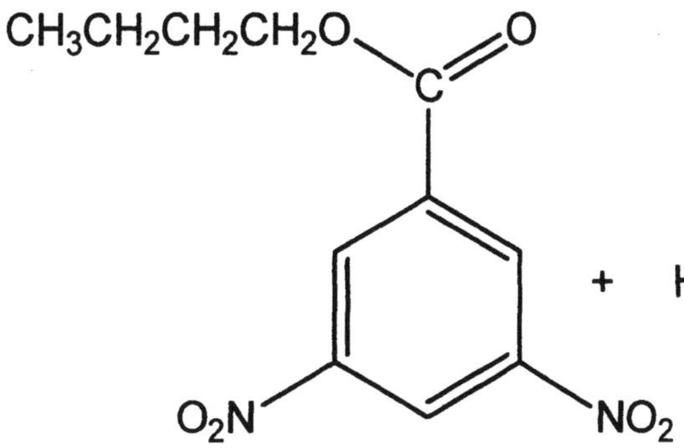

CH₃CH₂CH₂CH₂O

+   HCl

O₂N   NO₂

butyl 3,5-dinitrobenzoate

e. CH₃CH₂CH₂CH₂OH +

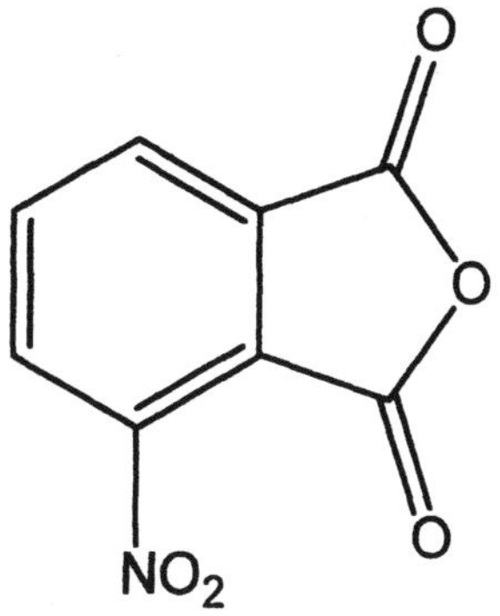

NO₂

    1-butanol   3-nitrophthalic anhydride

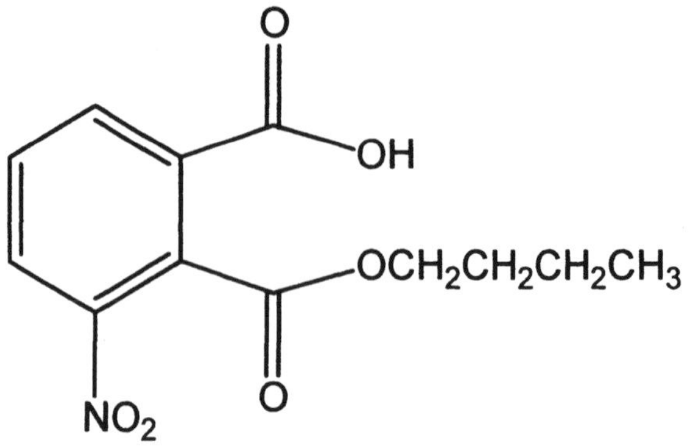

butyl 3-nitrophthalate

8. a.

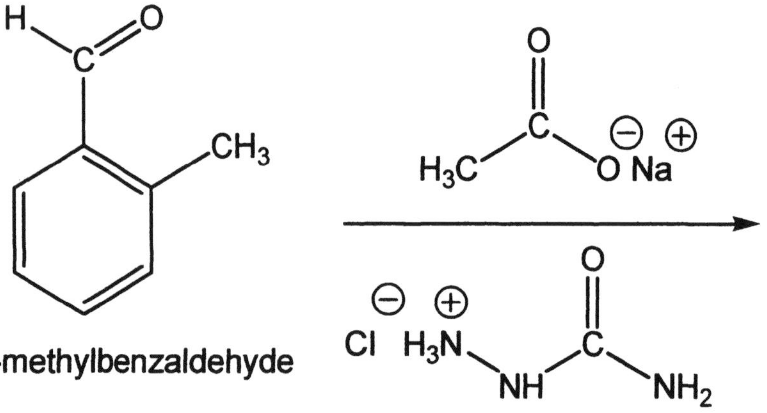

2-methylbenzaldehyde

semicarbazide
hydrochloride

semicarbazone of
2-methylbenzaldehyde

b.

2-methylbenzaldehyde   2,4-dinitrophenylhydrazine

2,4-dinitrophenylhydrazone
of 2-methylbenzaldehyde

c.

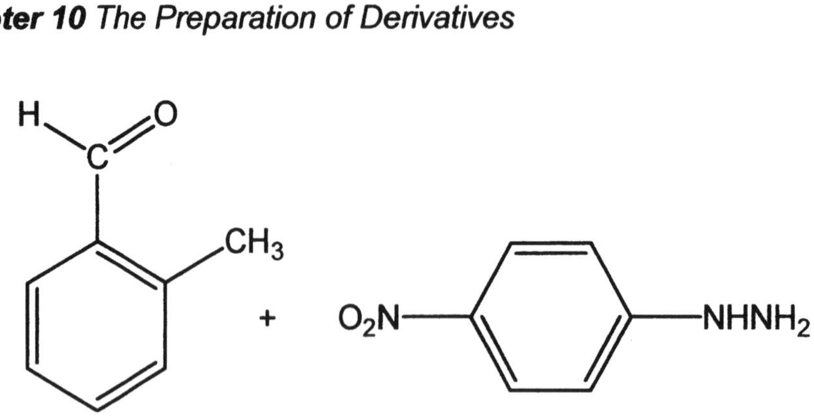

2-methylbenzaldehyde   4-nitrophenylhydrazine

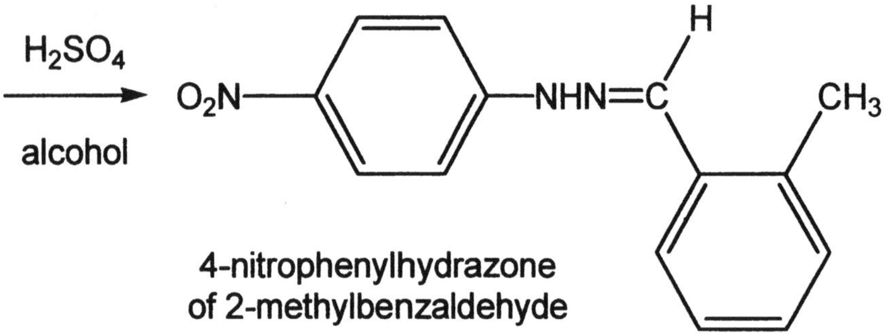

$\xrightarrow[\text{alcohol}]{H_2SO_4}$

4-nitrophenylhydrazone
of 2-methylbenzaldehyde

d.

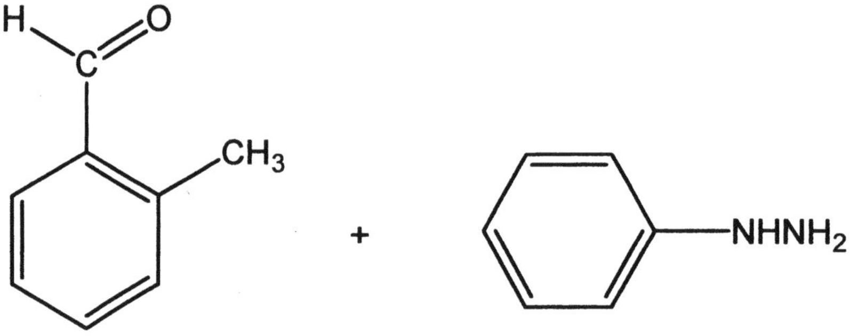

2-methylbenzaldehyde   phenylhydrazine

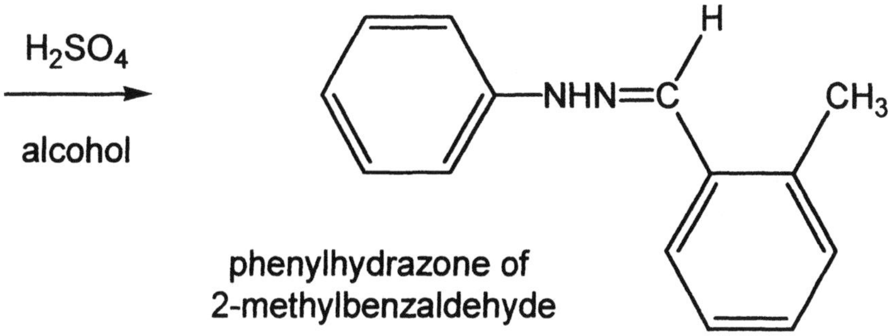

phenylhydrazone of
2-methylbenzaldehyde

e.

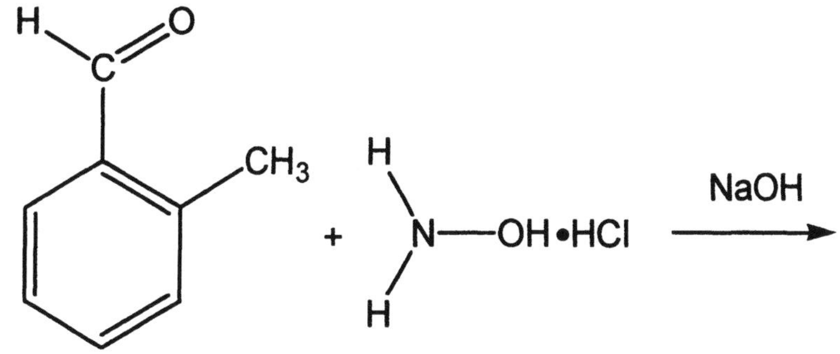

2-methylbenzaldehyde

hydroxylamine
hydrochloride

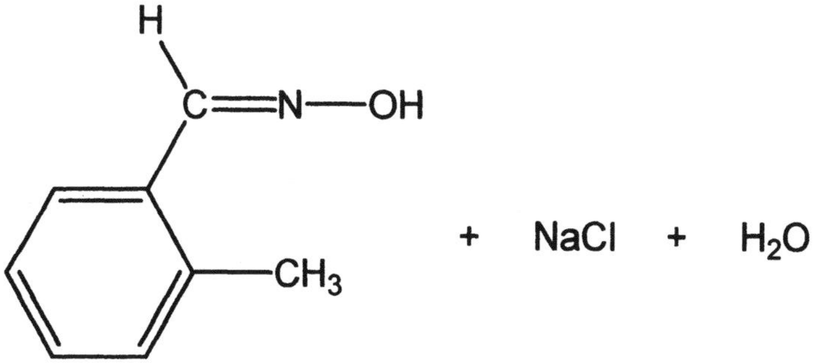

+ NaCl + H$_2$O

oxime of
2-methylbenzaldehyde

f.

2-methylbenzaldehyde   dimedon

methone derivative of 2-methylbenzaldehyde

9.

heptanamide   xanthydrol

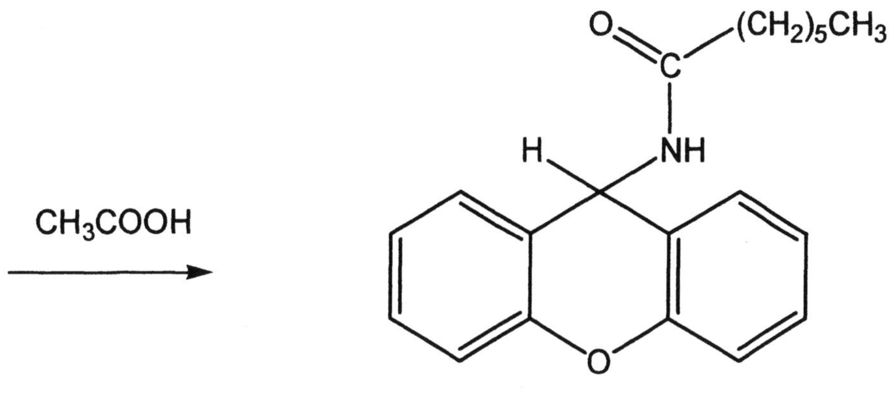

$$O=C-(CH_2)_5CH_3$$

9-hexanoylamidoxanthene

CH₃COOH →

10.

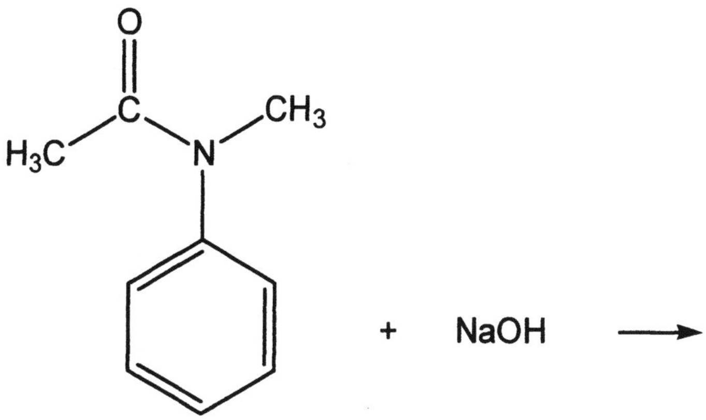

N-methyl-N-
phenylethanamide

+ NaOH →

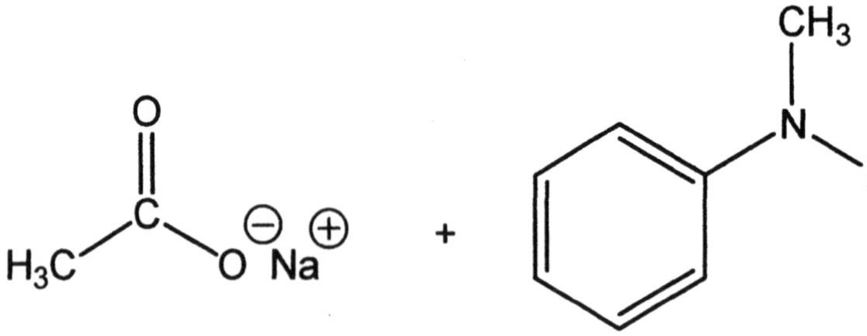

sodium acetate

+

N-methylaniline

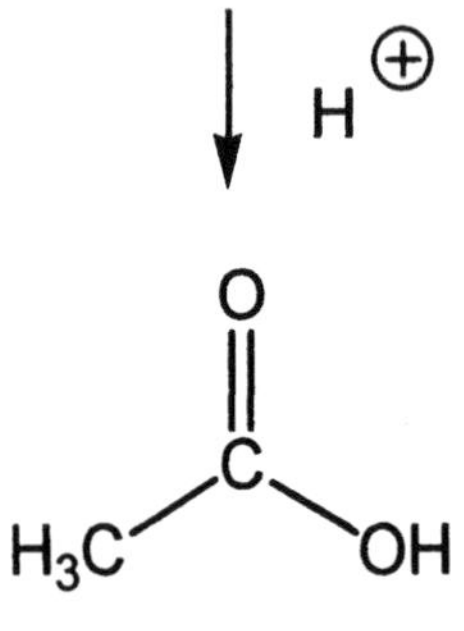

acetic acid

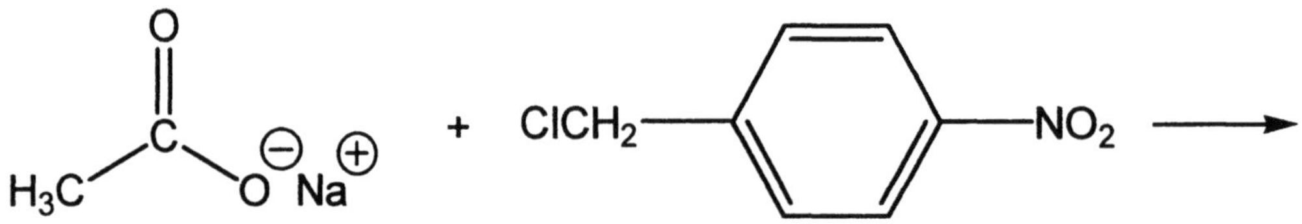

sodium acetate   4-nitrobenzyl chloride

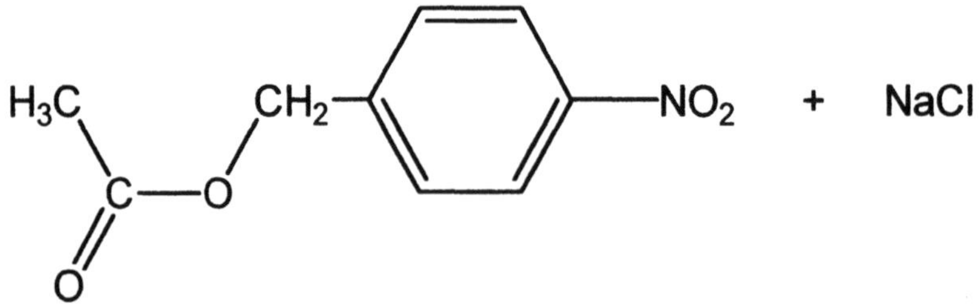

4-nitrobenzyl acetate

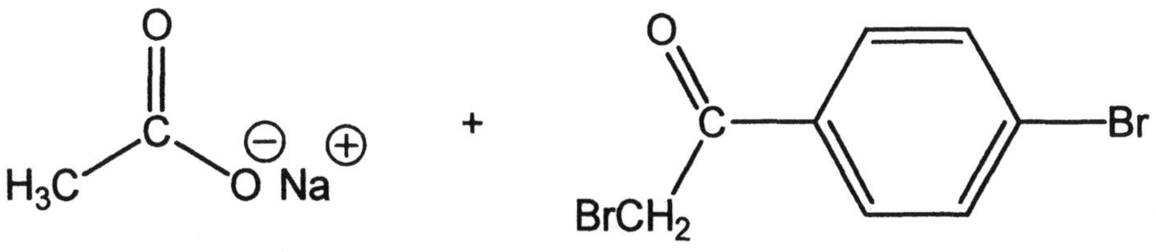

sodium acetate

4-bromophenacyl bromide

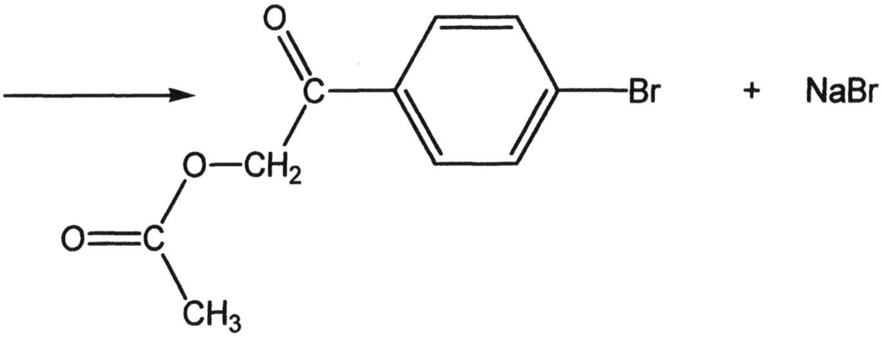

+ NaBr

2-(4-bromophenyl)-2-oxoethyl acetate

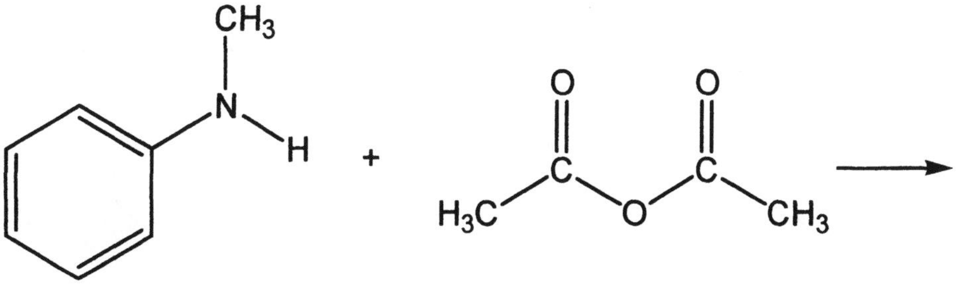

N-methylaniline

acetic anhydride

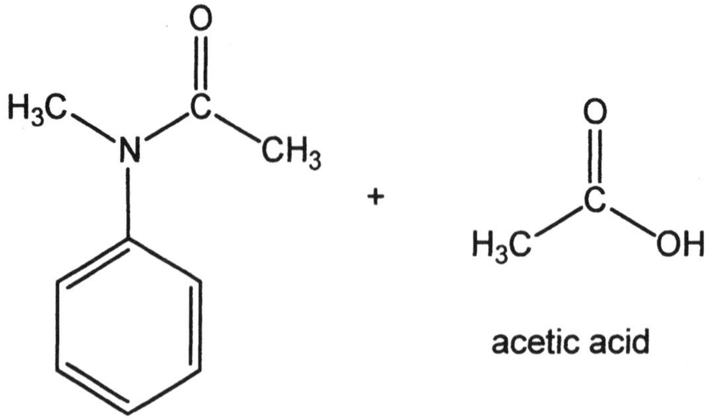

N-methyl-N-phenylacetamide

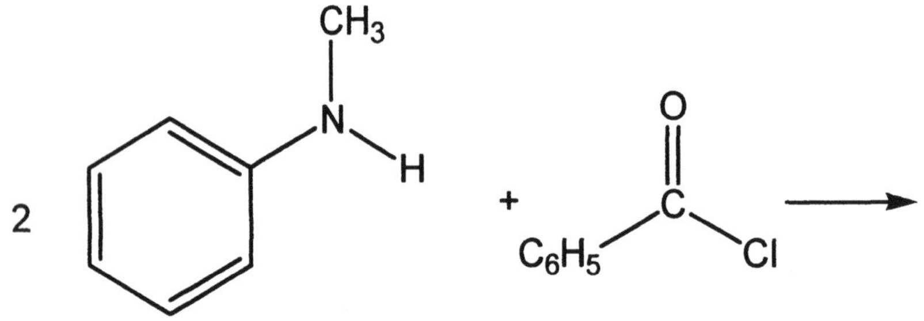

acetic acid

2   N-methylaniline   benzoyl chloride

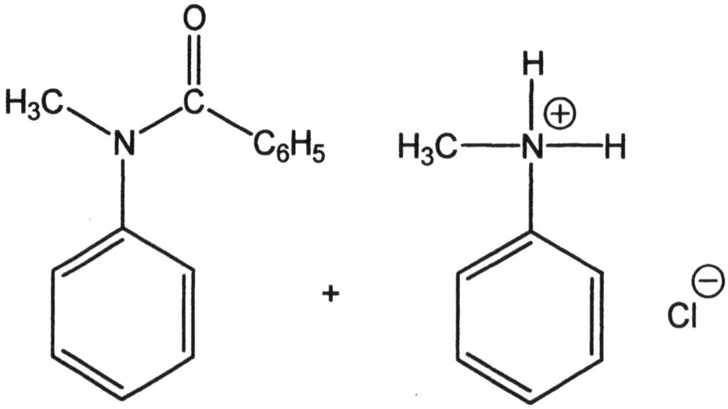

N-methyl-N-
phenylbenzamide

N-methylanilinium
chloride

11. **a.**

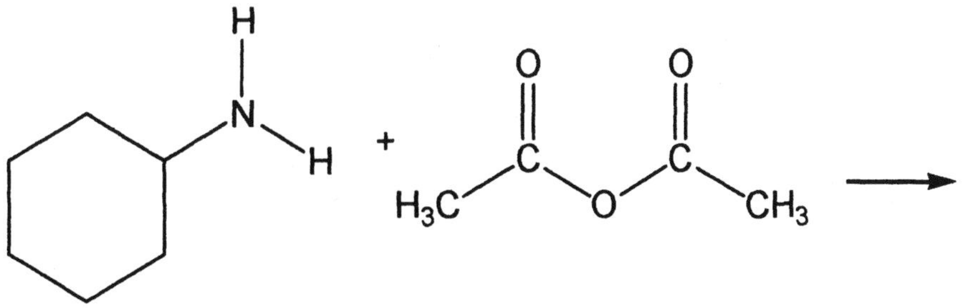

cyclohexylamine   acetic anhydride

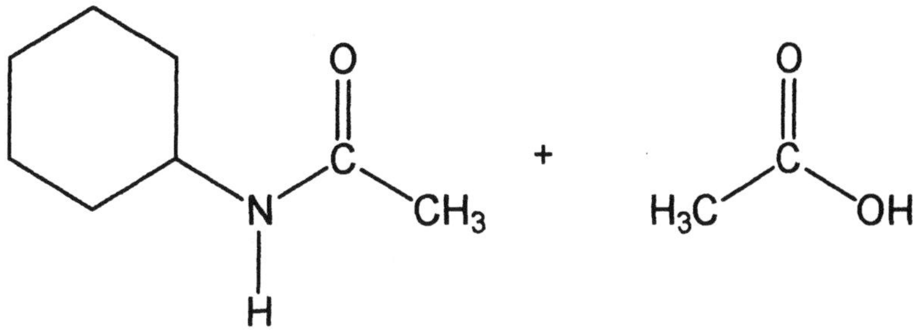

N-cyclohexylacetamide   acetic acid

b. 2

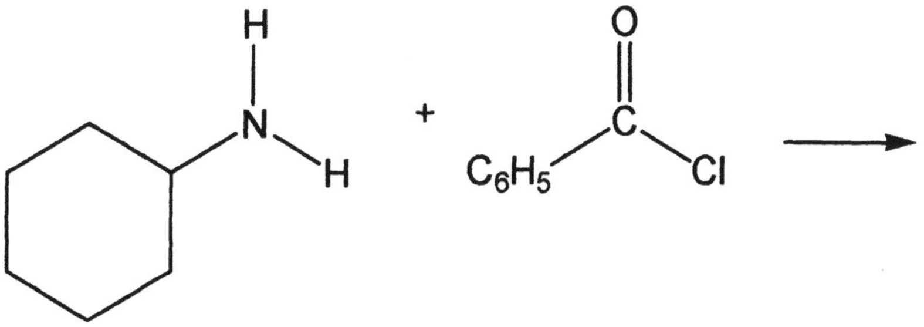

cyclohexylamine   benzoyl chloride

N-cyclohexylbenzamide   cyclohexyl ammonium
                                         chloride

c.

cyclohexylamine   benzenesulfonyl chloride

(soluble)   +   NaCl   +   2 H$_2$O

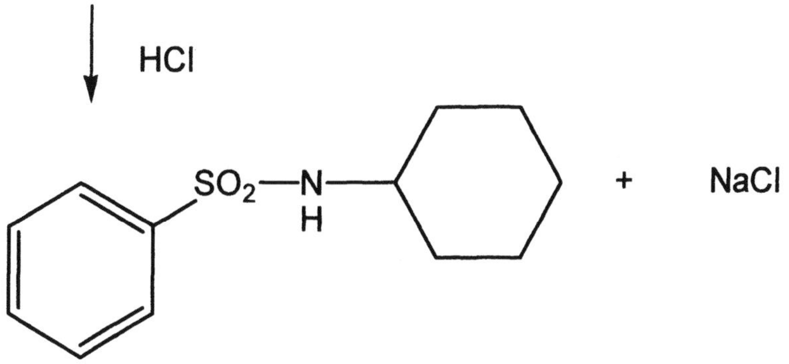

HCl

SO$_2$—N—cyclohexyl
H

+   NaCl

N-cyclohexyl benzenesulfonamide

d.

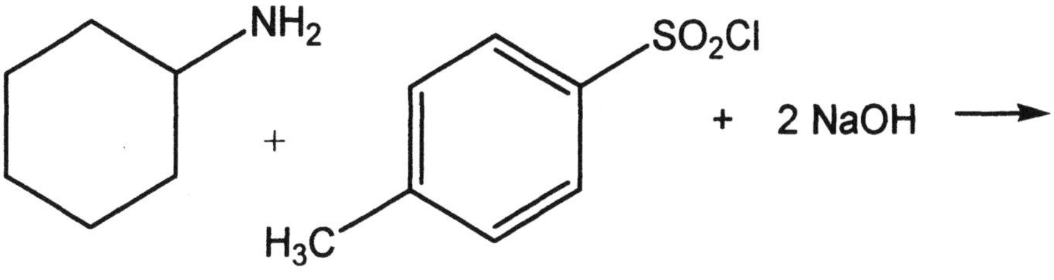

cyclohexylamine (with NH$_2$)   +   4-toluenesulfonyl chloride (H$_3$C—ring—SO$_2$Cl)   +   2 NaOH   →

H$_3$C—ring—SO$_2$—N$^{\ominus}$—cyclohexyl
Na$^{\oplus}$

cyclohexylamine   4-toluenesulfonyl chloride

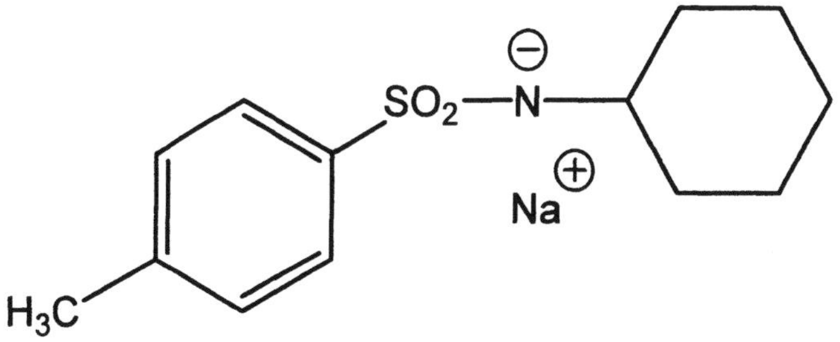

(soluble)   +   NaCl   +   2 H$_2$O

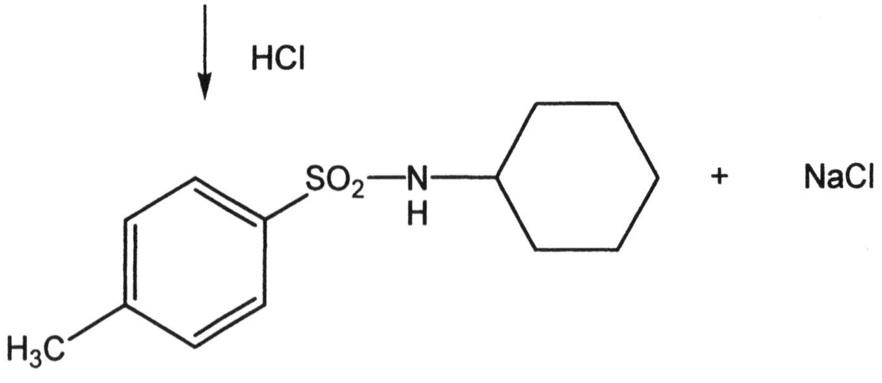

HCl

N-cyclohexyl 4-toluenesulfonamide + NaCl

e. cyclohexylamine

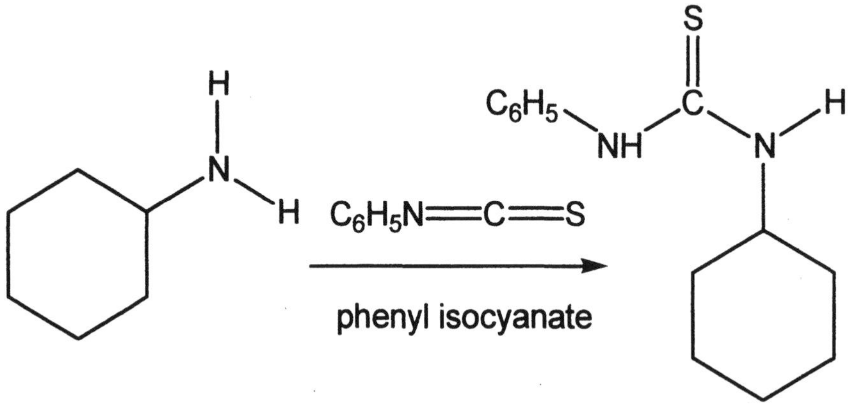

C₆H₅N=C=S

phenyl isocyanate

1-cyclohexyl-3-phenylthiourea

f. cyclohexylamine

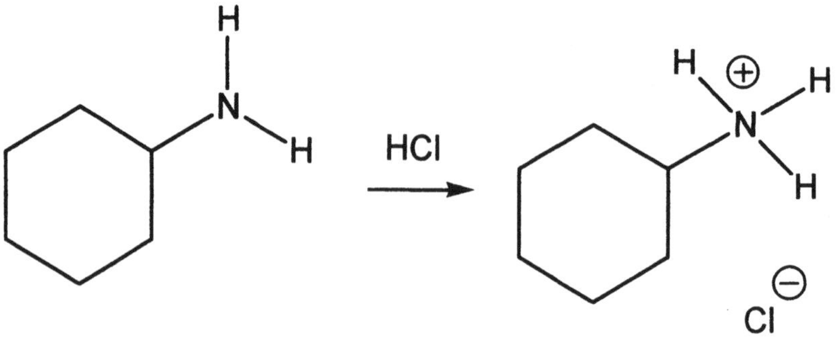

HCl

cyclohexylammonium hydrochloride

12. **a.**

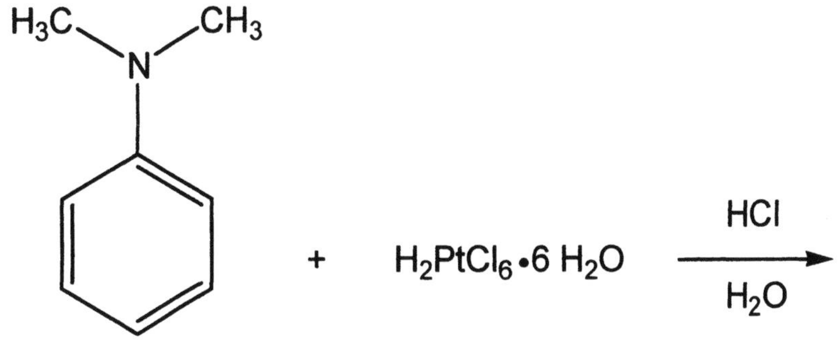

N,N-dimethylaniline   chloroplatinic acid

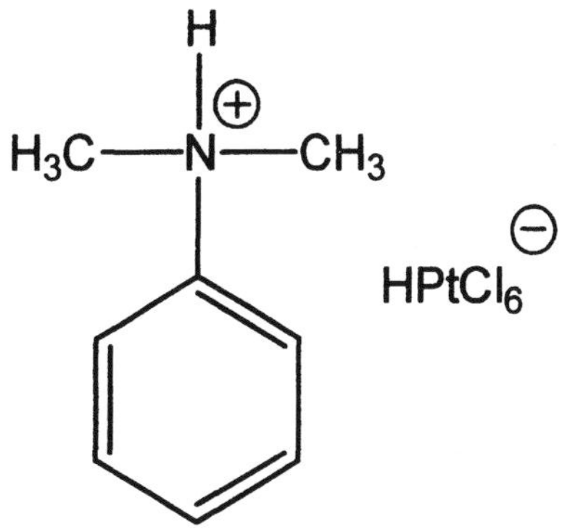

N,N-dimethylanilinium
platinate

b.

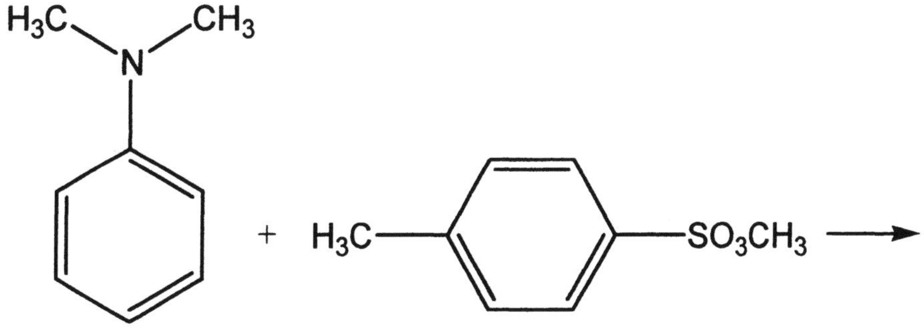

N,N-dimethylaniline   methyl 4-toluenesulfonate

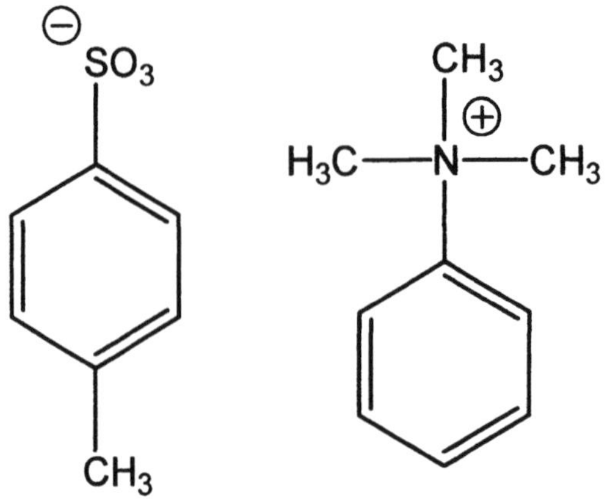

N,N,N-trimethylanilinium
4-toluenesulfonate

c.

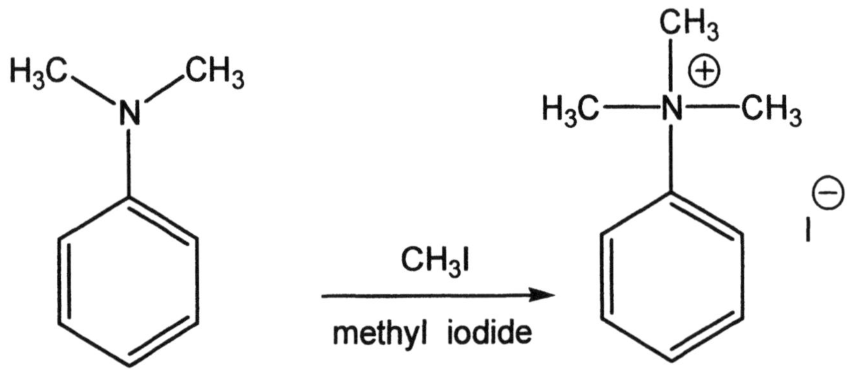

N,N-dimethylaniline

CH₃I
methyl iodide

N,N,N-trimethyl-
anilinium iodide

d.

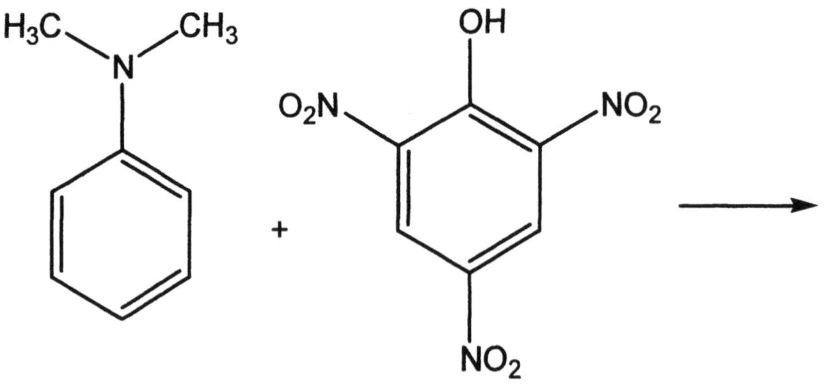

N,N-dimethylaniline   +   picric acid

N,N-dimethylanilinium picrate

e.

N,N-dimethylaniline   →(HCl)   *N,N*-dimethylanilinium chloride

13. a.

glycine   + NaOH →

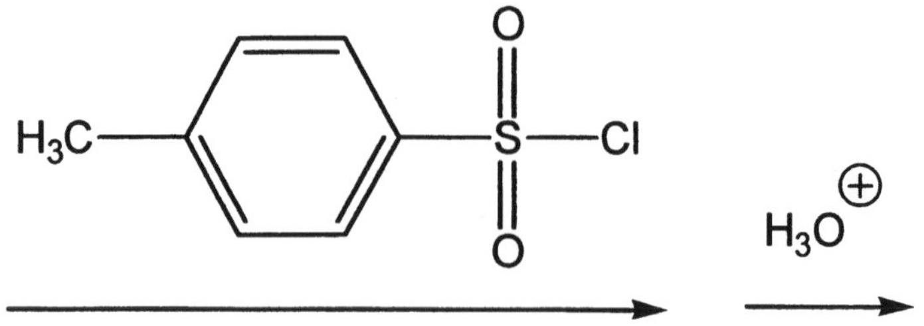

4-toluenesulfonyl chloride

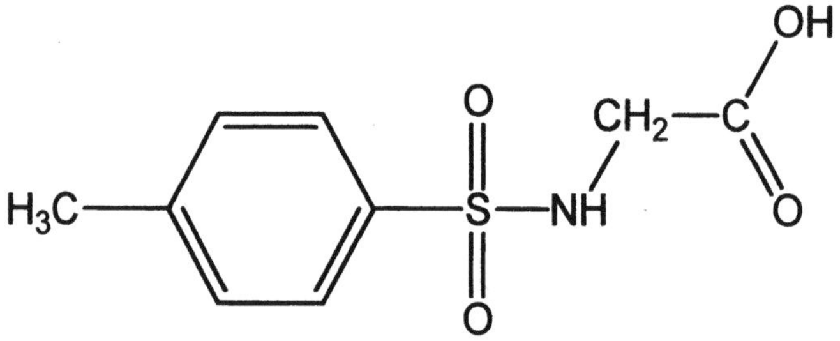

(4-toluenesulfonamido)acetic acid

b.

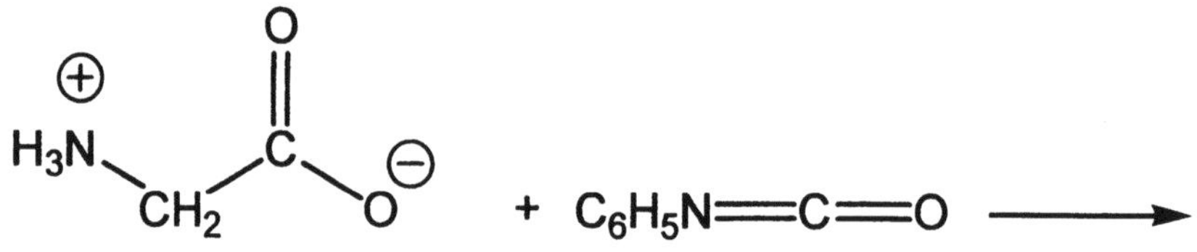

glycine   phenyl isocyanate

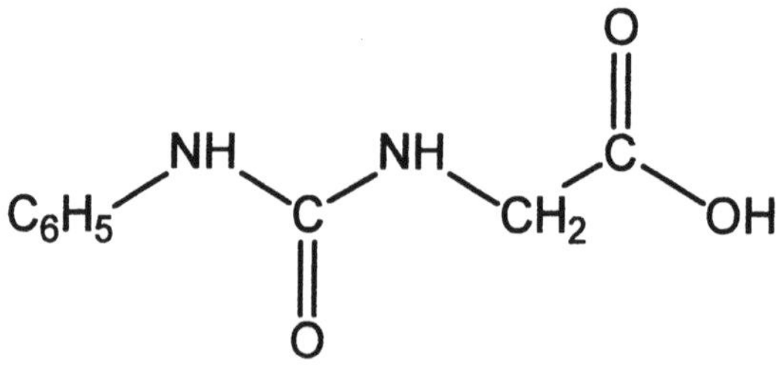

(3-phenylureido)-acetic acid

c.

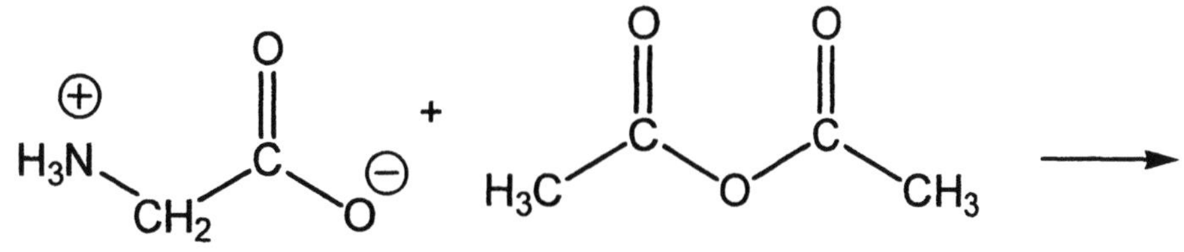

glycine   acetic anhydride

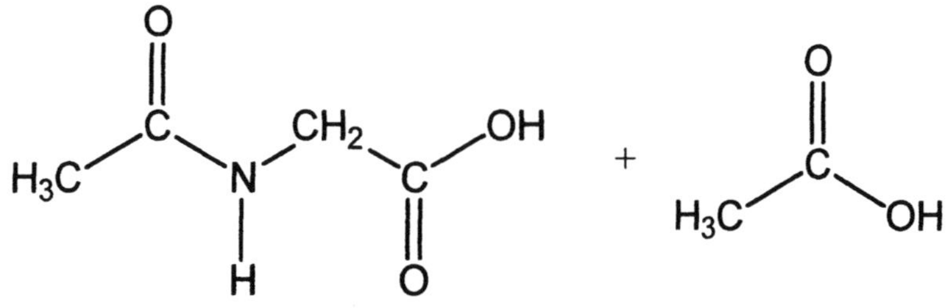

acetylaminoacetic acid   acetic acid

d.

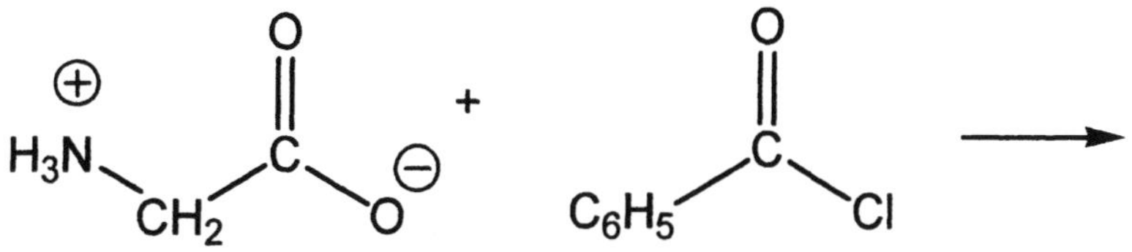

glycine   benzoyl chloride

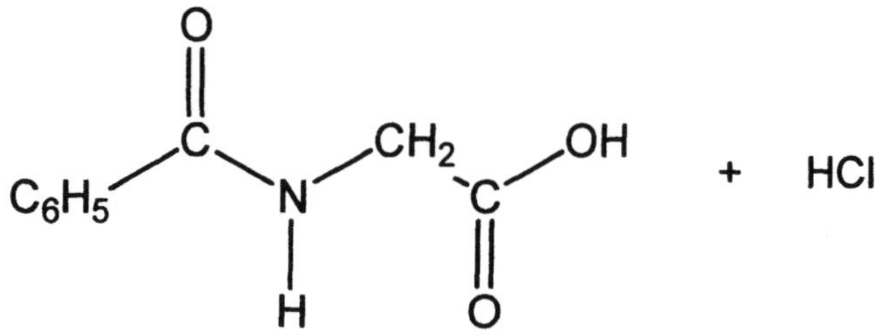

+ HCl

benzoylaminoacetic acid

e.

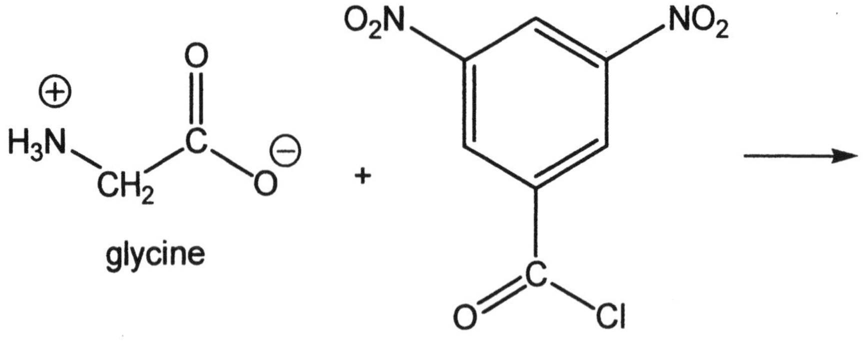

glycine

+

3,5-dinitrobenzoyl chloride

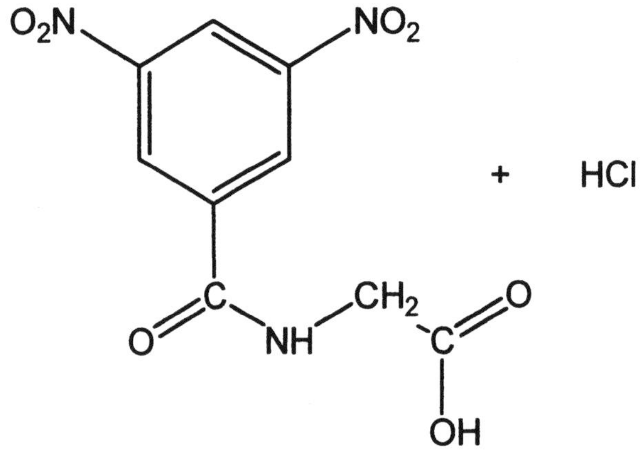

+ HCl

(3,5-dinitrobenzoylamino)acetic acid

f.

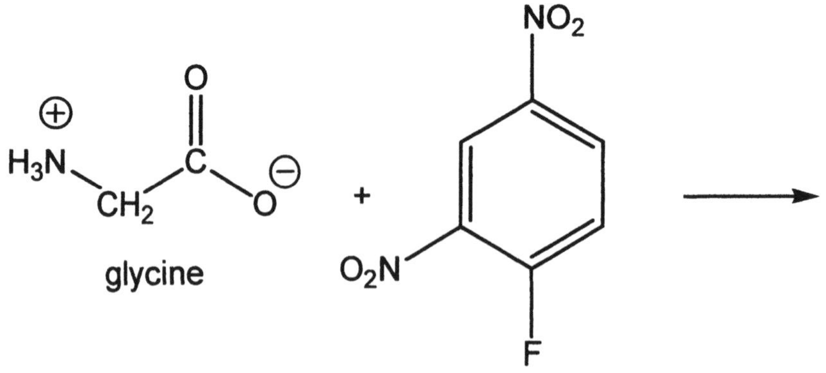

glycine

+

2,4-dinitrofluorobenzene

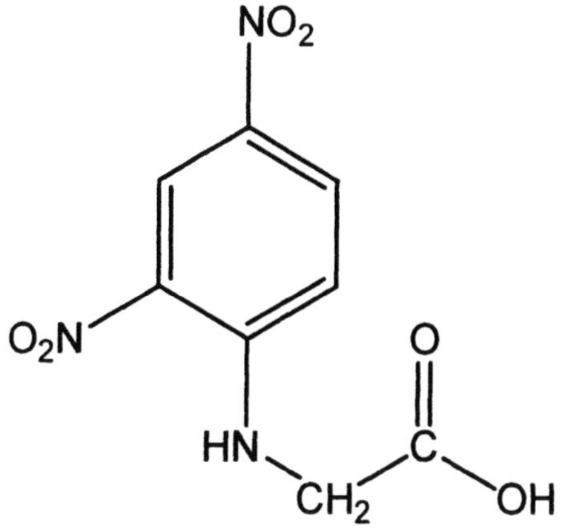

+ NaF

(2,4-dinitrophenylamino)acetic acid

14. a.

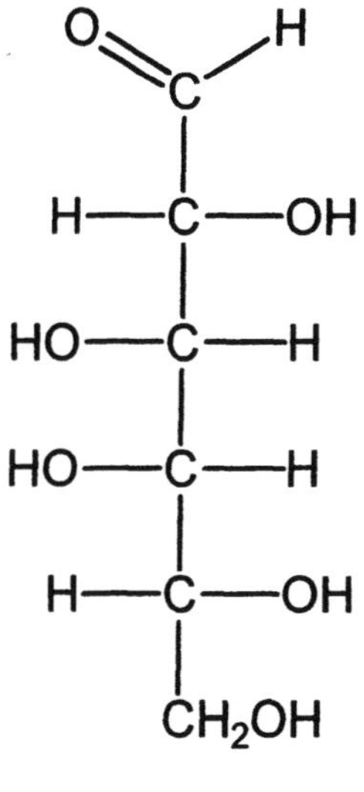

D-galactose

+ 3 $H_2NNHC_6H_5$ →

phenylhydrazine

$$
\begin{array}{c}
\text{H} \diagdown \quad \diagup \text{NNHC}_6\text{H}_5 \\
\text{C} \\
\| \\
\text{C}\!\!=\!\!\text{NNHC}_6\text{H}_5 \\
| \\
\text{HO}\!-\!\text{C}\!-\!\text{H} \\
| \\
\text{HO}\!-\!\text{C}\!-\!\text{H} \\
| \\
\text{H}\!-\!\text{C}\!-\!\text{OH} \\
| \\
\text{CH}_2\text{OH}
\end{array}
\quad \text{(s)} \quad + \quad C_6H_5NH_2 \quad + \quad NH_3 \quad + \quad 2\,H_2O
$$

aniline

osazone of *D*-galactose

b.

$$
\begin{array}{c}
\text{O} \diagdown \quad \diagup \text{H} \\
\text{C} \\
| \\
\text{H}\!-\!\text{C}\!-\!\text{OH} \\
| \\
\text{HO}\!-\!\text{C}\!-\!\text{H} \\
| \\
\text{HO}\!-\!\text{C}\!-\!\text{H} \\
| \\
\text{H}\!-\!\text{C}\!-\!\text{OH} \\
| \\
\text{CH}_2\text{OH}
\end{array}
\quad + \quad
$$

NHNH$_2$

NO$_2$

4-nitrophenylhydrazine

$\longrightarrow$

D-galactose

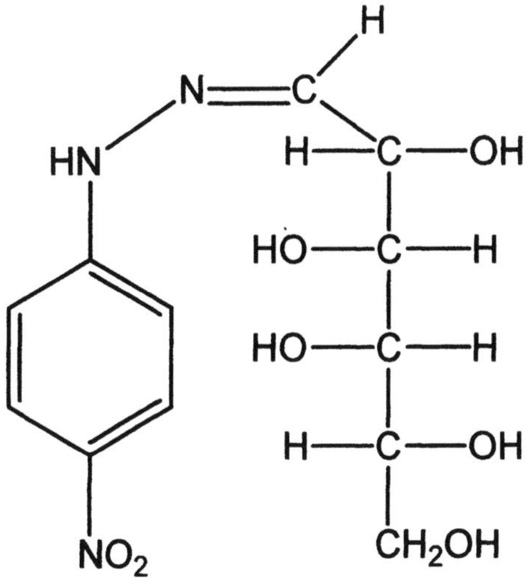

4-nitrophenylhydrazone
of *D*-galactose

c.

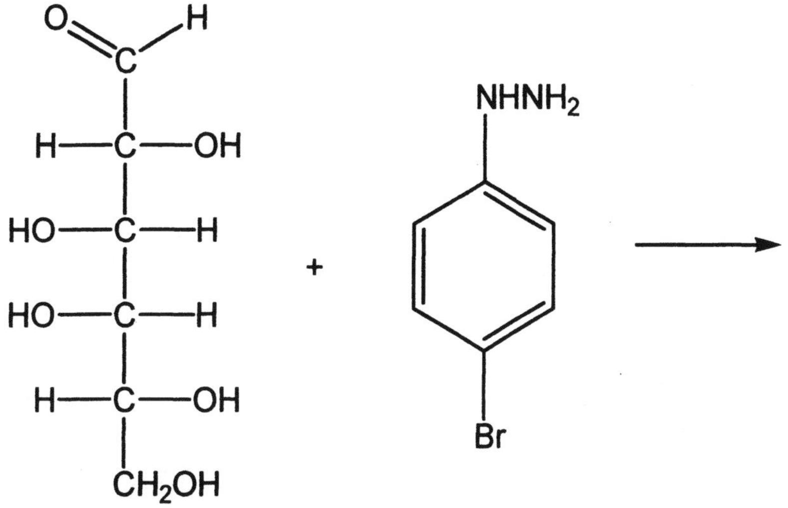

D-galactose

+

4-bromophenylhydrazine

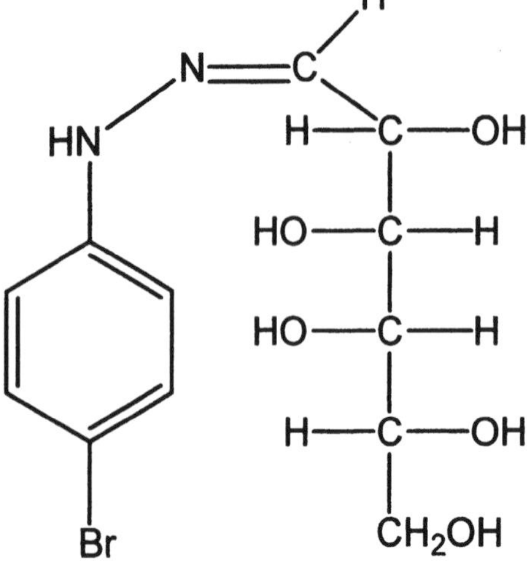

4-bromophenylhydrazone of *D*-galactose

d.

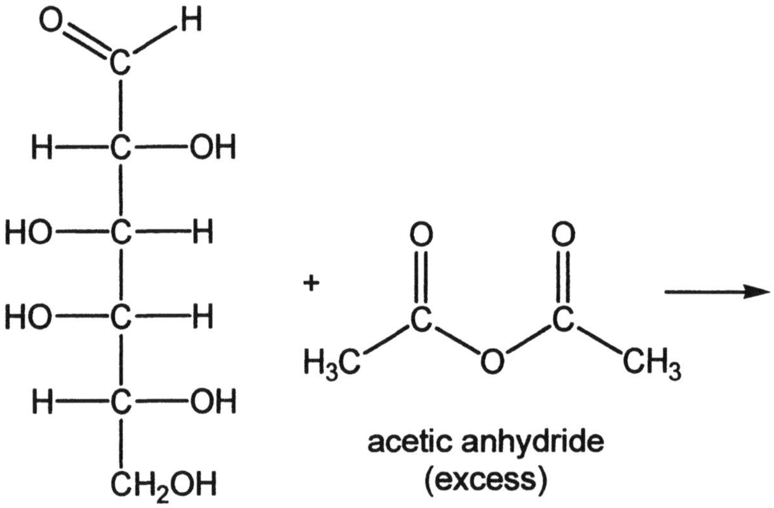

D-galactose

+

acetic anhydride
(excess)

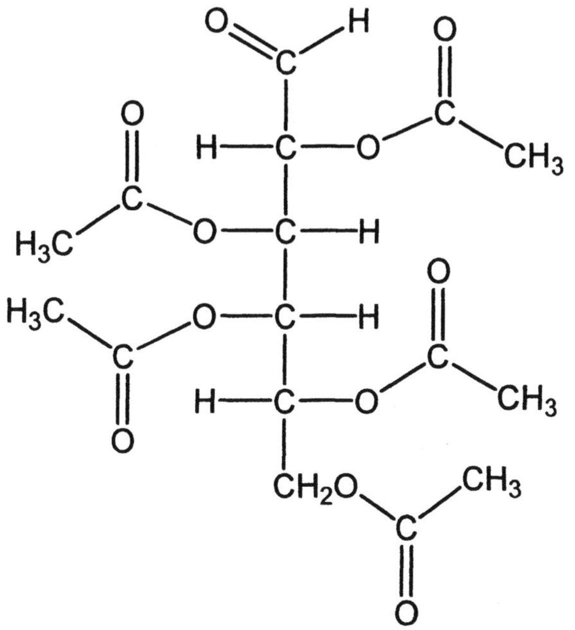

2,3,4,5,6-tetraacetoxy
derivative of
D-galactose

15. Simple esters, with bp of below 110°C, would be hydrolyzed easily by method **a**. Esters with bp of 110°C to 200°C would require a reflux time of 1–2 hours.

16. **a.**

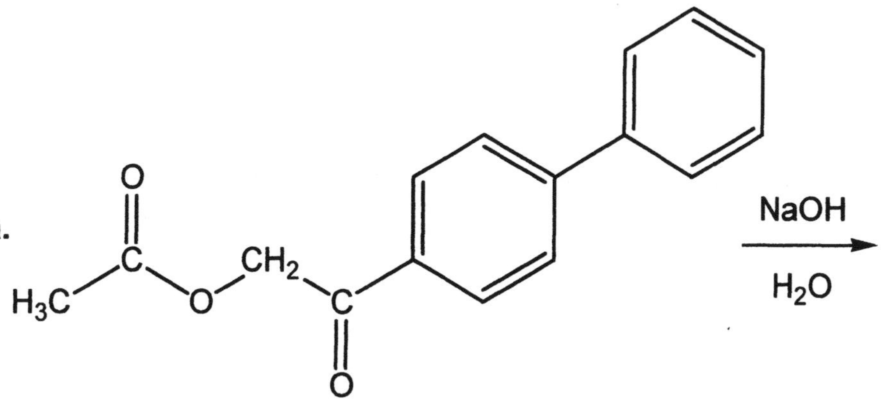

4-phenylphenacyl acetate

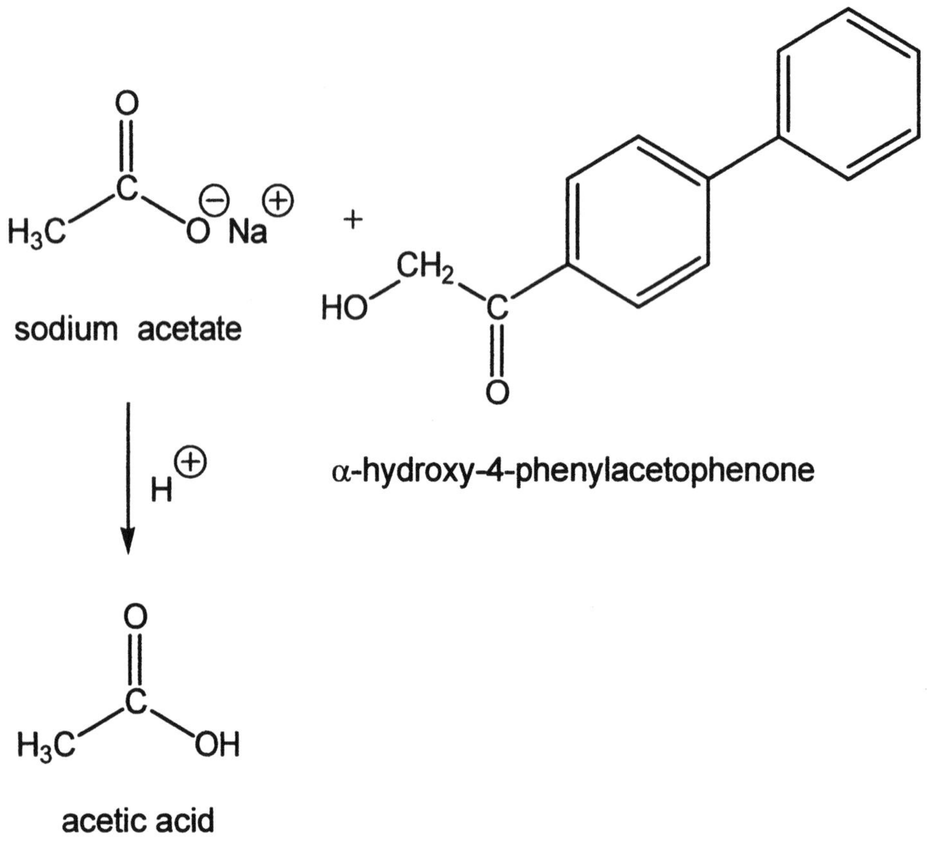

sodium acetate

+

α-hydroxy-4-phenylacetophenone

H⊕

acetic acid

b.

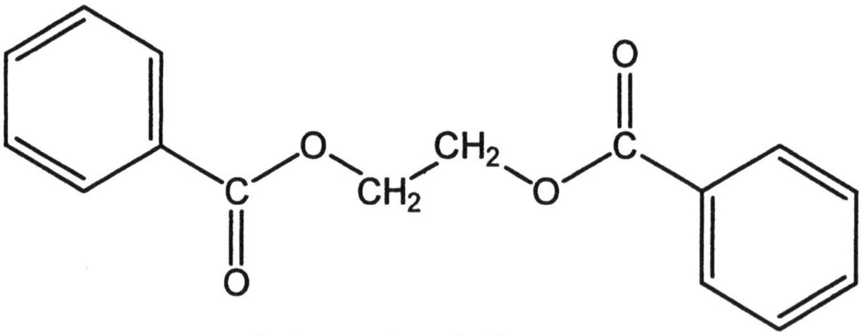

ethylene glycol dibenzoate

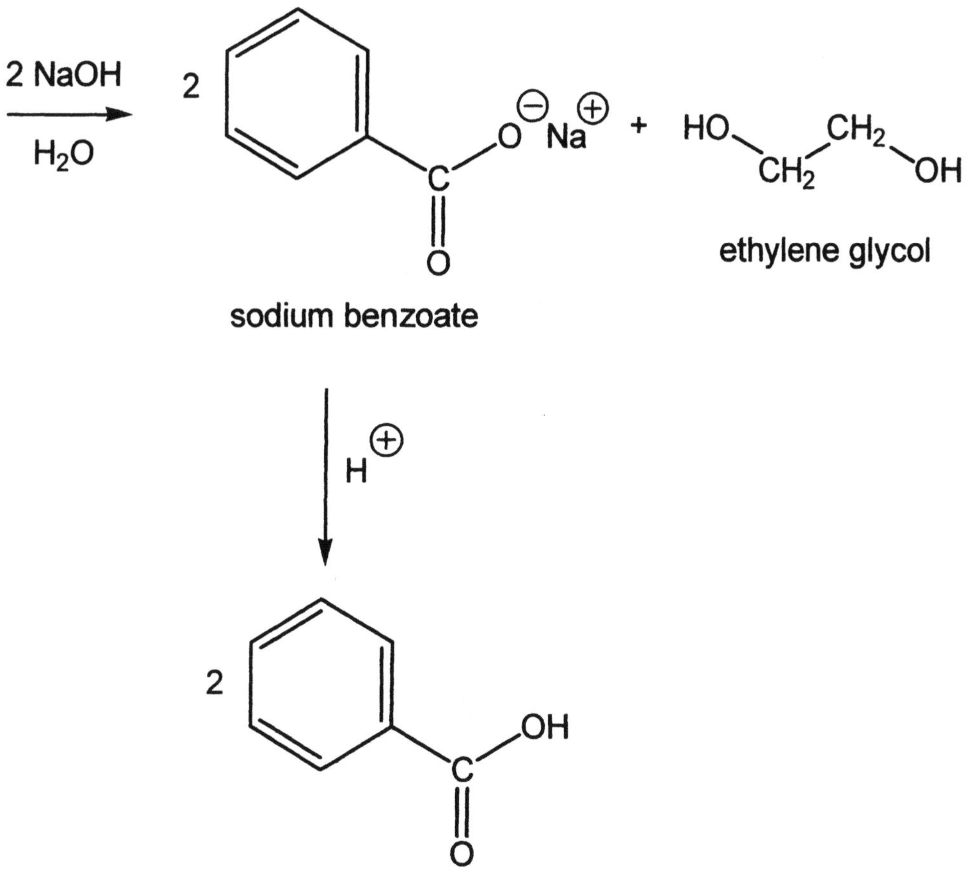

$$2 \text{ NaOH} \xrightarrow{\quad H_2O \quad}$$

2 [benzene ring]—C(=O)—O$^{\ominus}$ Na$^{\oplus}$ + HO—CH$_2$—CH$_2$—OH

sodium benzoate   ethylene glycol

↓ H$^{\oplus}$

2 [benzene ring]—C(=O)—OH

benzoic acid

c. CH$_3$CH$_2$CH$_2$CH$_2$—O—C(=O)—C(=O)—O—CH$_2$CH$_2$CH$_2$CH$_3$

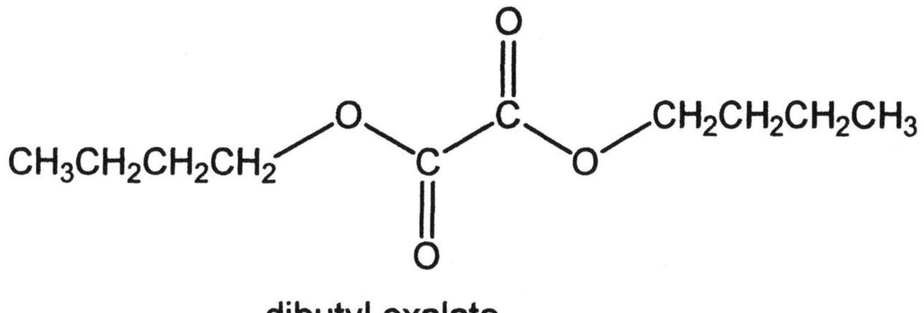

dibutyl oxalate

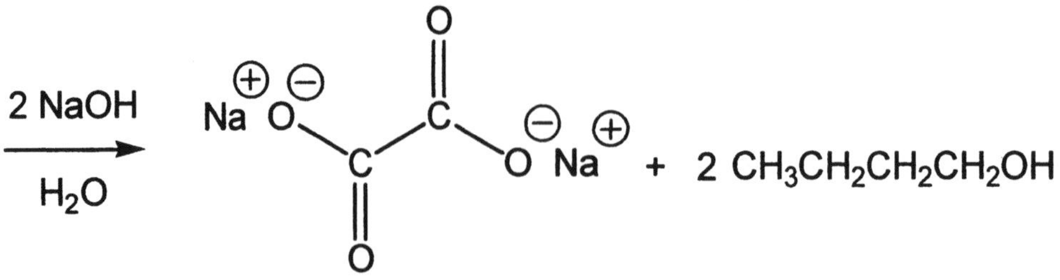

$$\xrightarrow[\text{H}_2\text{O}]{\text{2 NaOH}}$$

disodium oxalate   +   2 CH₃CH₂CH₂CH₂OH

1-butanol

$$\downarrow \text{H}^{\oplus}$$

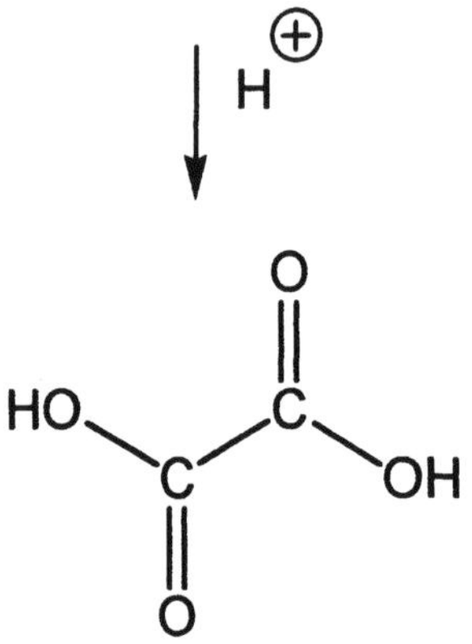

oxalic acid

d.

$$\xrightarrow[\text{H}_2\text{O}]{\text{3 NaOH}}$$

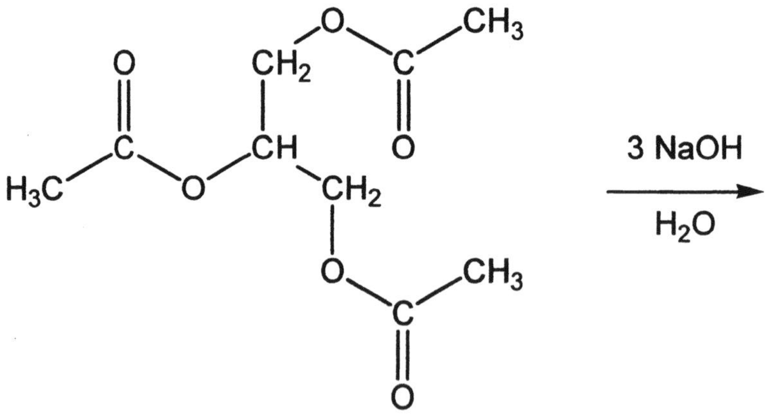

glycerol triacetate

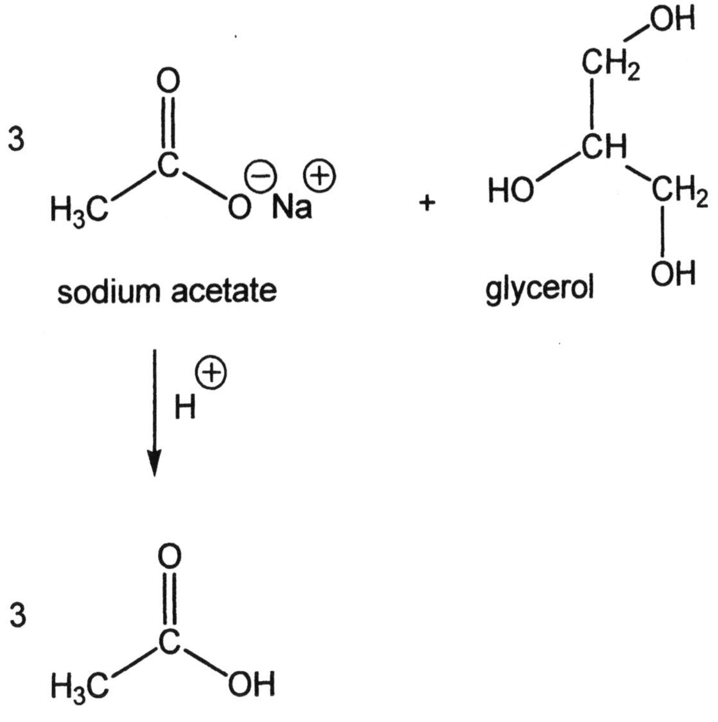

3 sodium acetate + glycerol

$\xrightarrow{H^{\oplus}}$

3 acetic acid

e. diethyl phthalate $\xrightarrow[\text{H}_2\text{O}]{\text{2 NaOH}}$

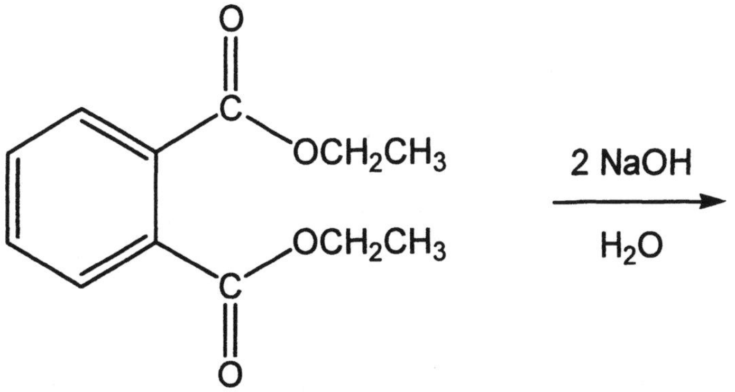

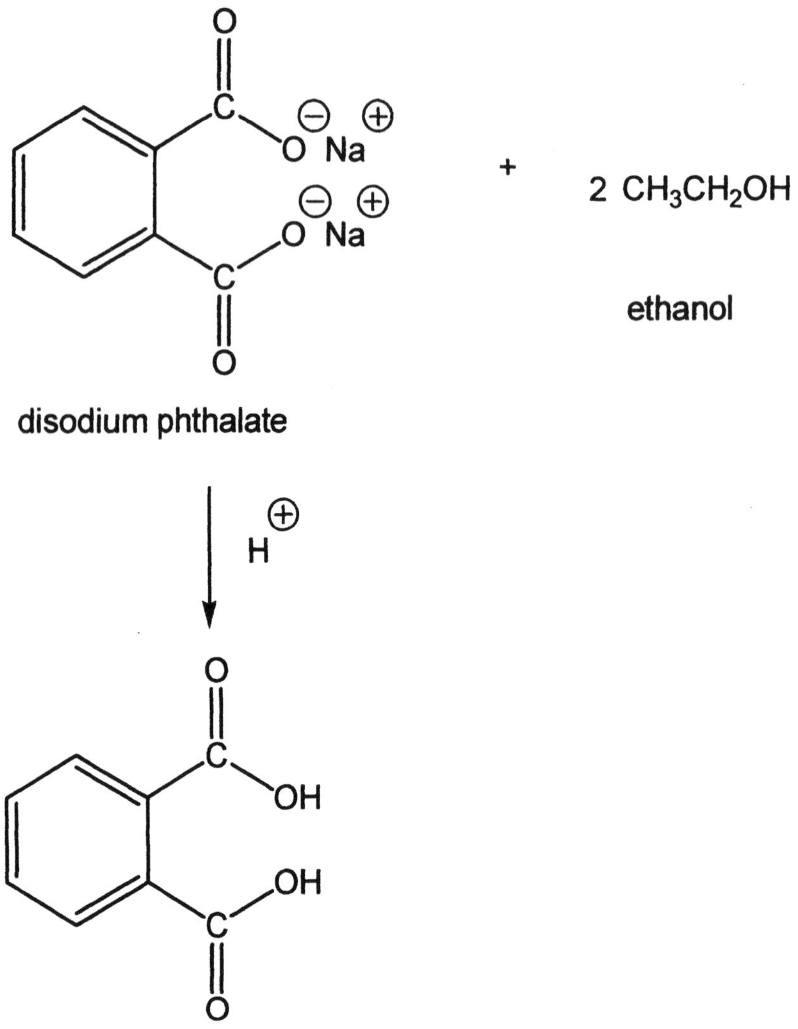

disodium phthalate

+   2 CH$_3$CH$_2$OH

ethanol

phthalic acid

For **a**, the semicarbazone derivative (mp 146°C) of α-hydroxy-4-phenylacetophenone (mp 86°C) and the 4-toluidide derivative (mp 153°C) of acetic acid (bp 118°C) could be prepared.

For **b**, the phenylurethane derivative (mp 157°C) of ethylene glycol (bp 197°C) and the 4-toluidide derivative (mp 158°C) of benzoic acid (mp 122°C) could be prepared.

For **c**, the phenylurethane derivative (mp 63°C) of 1-butanol (bp 118°C) and the di-4-toluidide derivative (mp 268°C) of oxalic acid [mp 101°C (dihydrate), mp 188°C (anhydrous)] could be prepared.

For **d**, the phenylurethane derivative (mp 180°C) of glycerol (bp 290°C) and the 4-toluidide derivative (mp 153°C) of acetic acid (bp 118°C) could be prepared.

For **e**, the phenylurethane derivative (mp 52°C) of ethanol (bp 78°C) and the dianilide derivative (mp 255°C) of phthalic acid (mp 208°C) could be prepared.

17.   **a.**   The saponification equivalent of ethyl acetoacetate is 65.

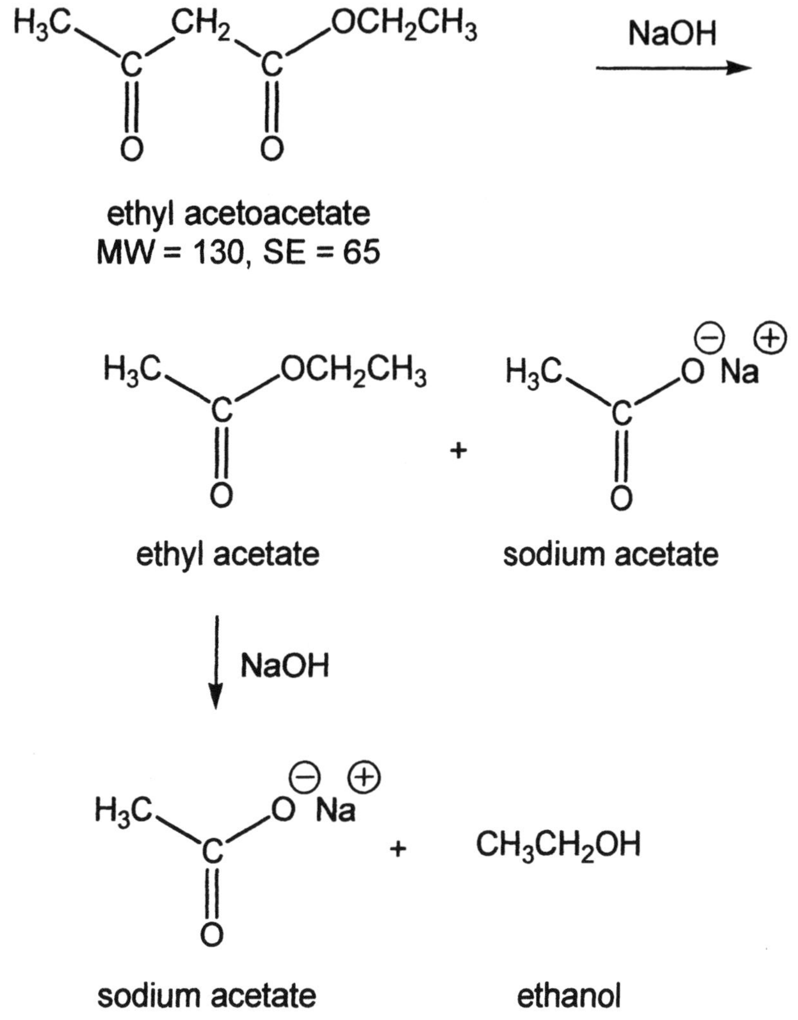

ethyl acetoacetate
MW = 130, SE = 65

ethyl acetate

+

sodium acetate

↓ NaOH

sodium acetate   + CH_3CH_2OH

ethanol

sodium acetate   acetic acid

b. The saponification equivalent of ethyl hydrogen phthalate is 97.

ethyl hydrogen phthalate
MW = 194, SE = 197

disodium phthalate

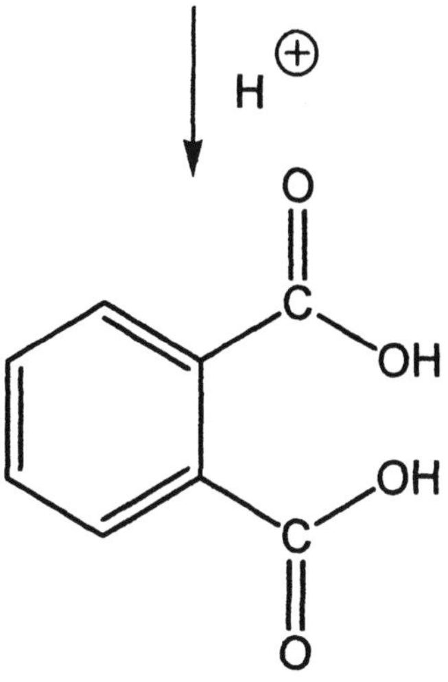

phthalic acid

c.   The saponification equivalent of diethyl propanedioate is 80.

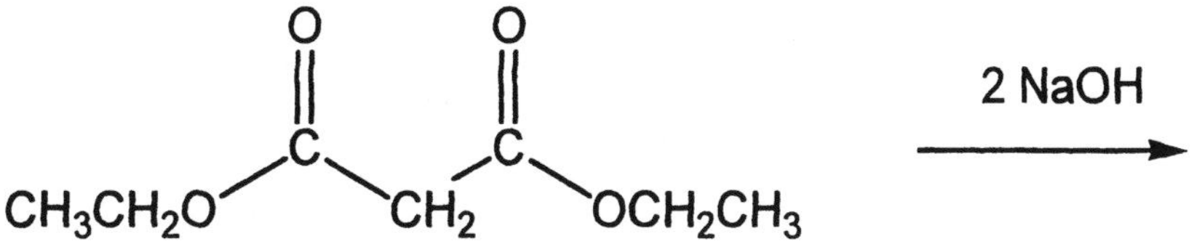

diethyl propanedioate
MW = 160, SE = 80

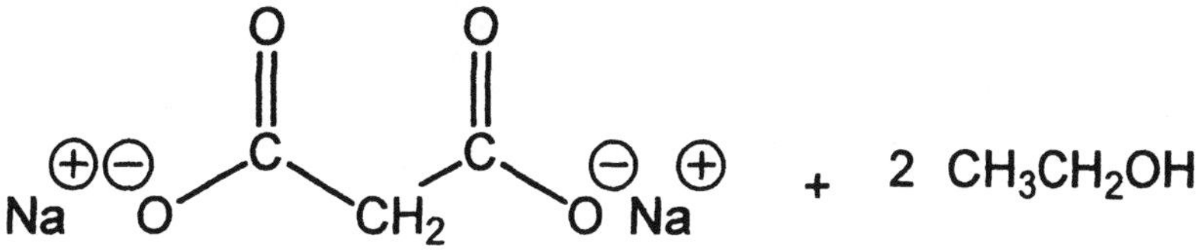

2 NaOH

disodium propanedioate

+   2 CH₃CH₂OH

ethanol

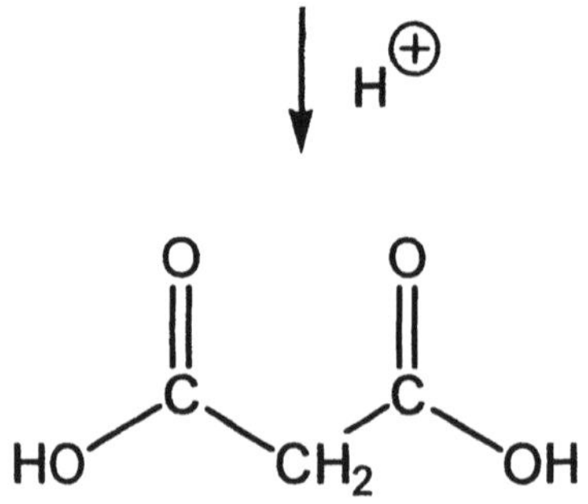

propanedioic acid

d.   The saponification equivalent of ethyl cyanoacetate is 56.5.

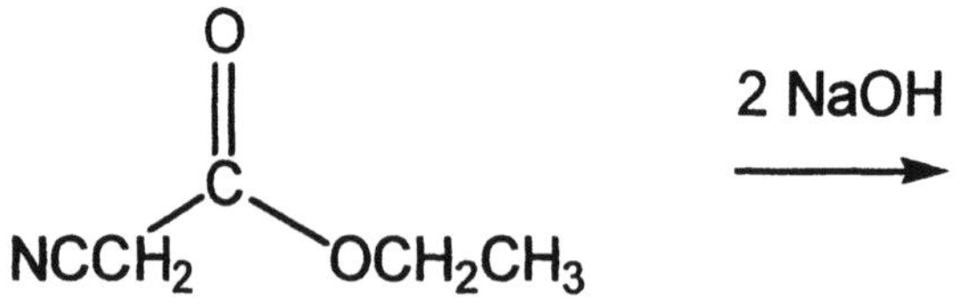

ethyl cyanoacetate
MW = 113, SE = 56.5

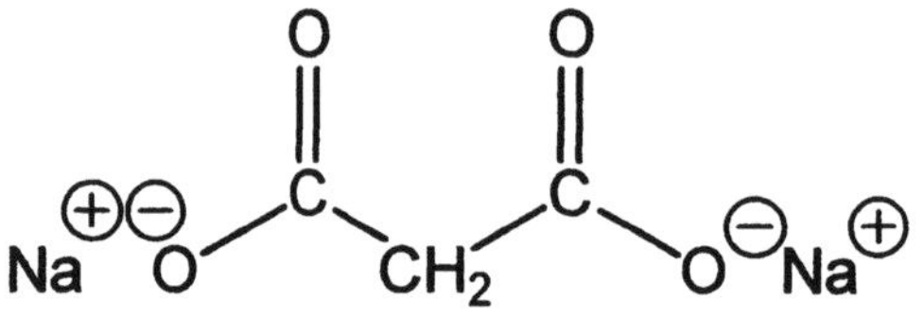

$2\ NaOH$ →

disodium propanedioate

$+\ CH_3CH_2OH\ +\ NH_3$

ethanol

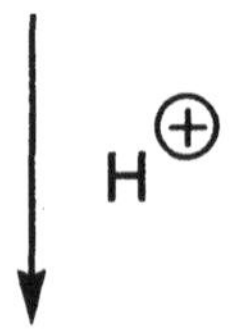

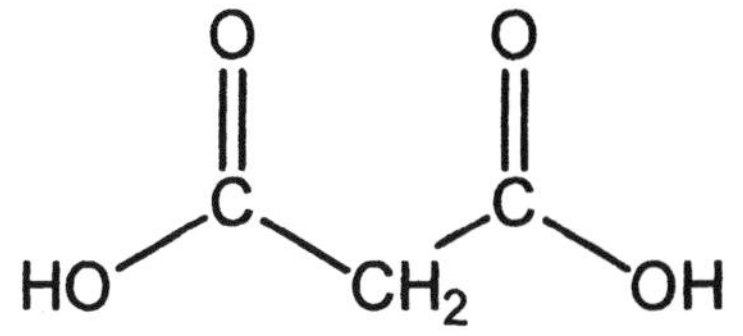

propanedioic acid

e. The saponification equivalent of dibutyl phthalate is 139.

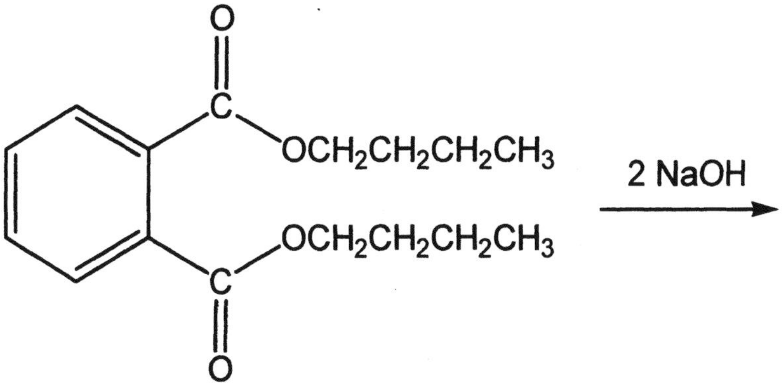

dibutyl phthalate
MW = 278, SE = 139

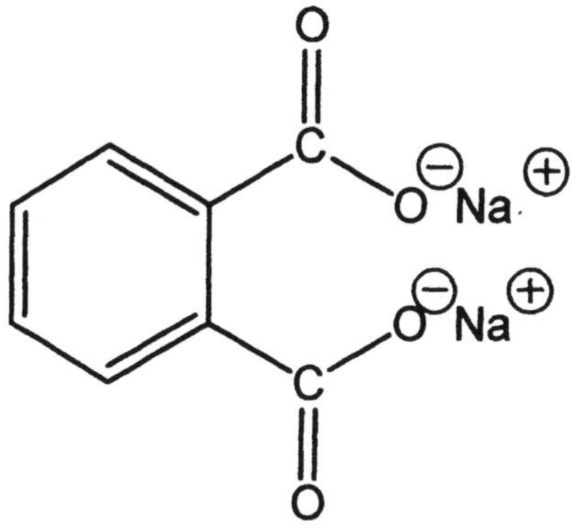

disodium phthalate

+ 2 $CH_3CH_2CH_2CH_2OH$

1-butanol

phthalic acid

18. When benzaldehyde is treated with sodium hydroxide, a Cannizzaro reaction occurs to form sodium benzoate and benzyl alcohol. Since only half of the benzaldehyde molecules become a carboxylic acid, the saponification equivalent is half of the molecular weight.

benzaldehyde
MW = 106, SE = 53

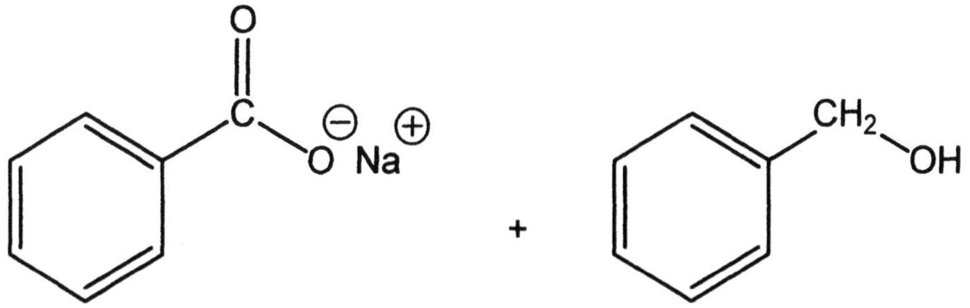

sodium benzoate   benzyl alcohol

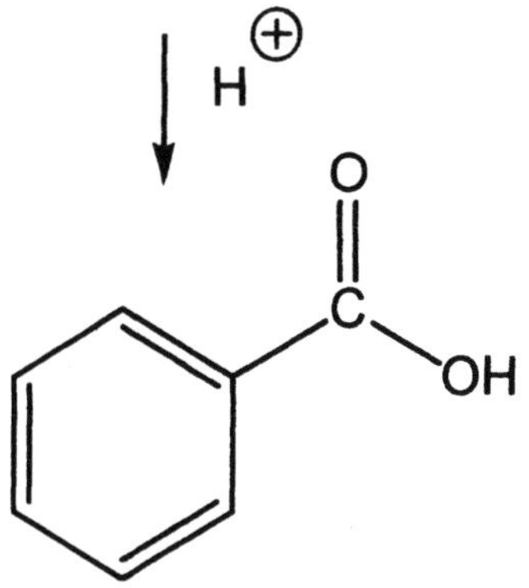

benzoic acid

19. If an ester is partially hydrolyzed, the saponification equivalent would be unreliable, since the value would be between the molecular weight of the carboxylic acid and the molecular weight of the ester.

20. The calculations are shown below.

$$\text{moles of ethyl bromide} = \frac{2.5 \text{ g}}{109 \text{ g/mole}} = 0.023 \text{ moles of ethyl bromide}$$

$$\text{moles of magnesium} = \frac{0.5 \text{ g}}{24.3 \text{ g/mole}} = 0.021 \text{ moles of magnesium}$$

The magnesium is the limiting reagent.

21.   **a.**

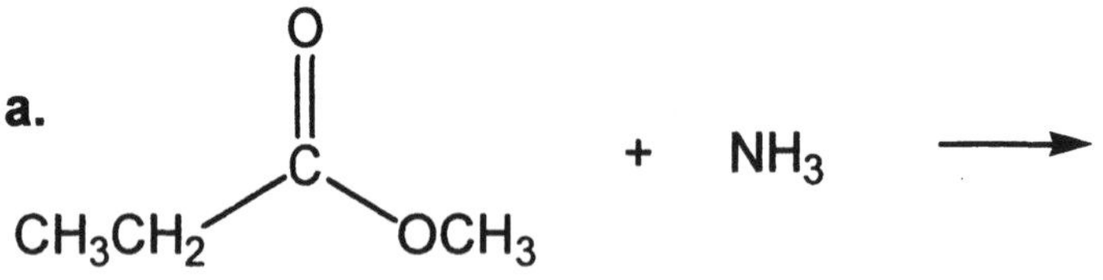

methyl propanoate   ammonia

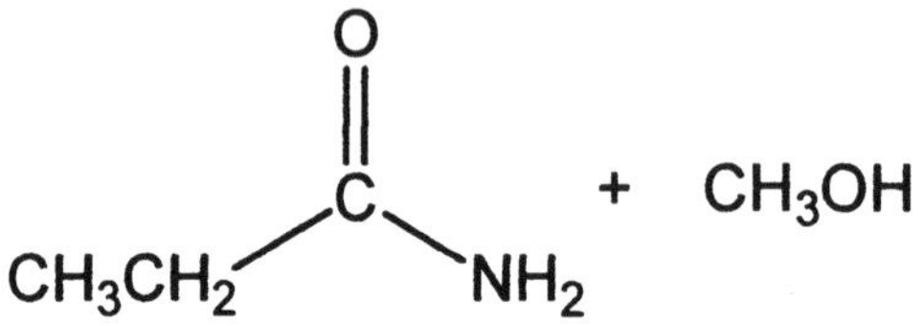

propanamide   methanol

b. CH_3CH_2Br +   Mg $\longrightarrow$ CH_3CH_2MgBr

ethyl bromide   magnesium   ethylmagnesium bromide

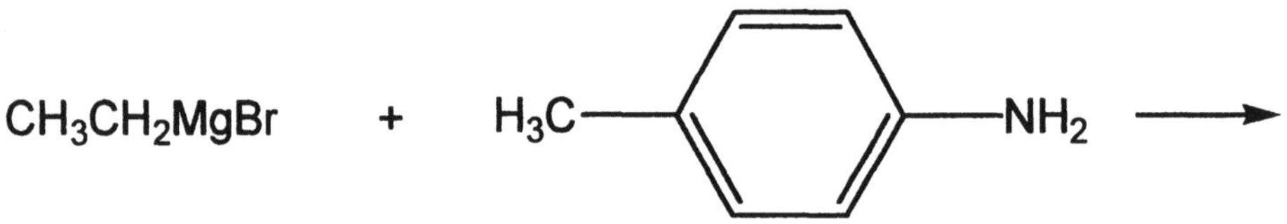

CH_3CH_2MgBr + $\longrightarrow$

ethylmagnesium bromide   4-toluidine (4-aminotoluene)

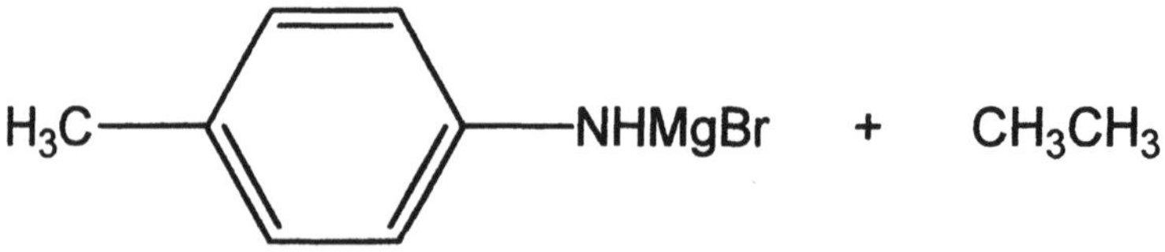

+ CH_3CH_3

4-toluidino magnesium bromide

$$
\underset{\substack{\text{methyl propanoate}}}{\text{CH}_3\text{CH}_2-\overset{\displaystyle \overset{O}{\|}}{\text{C}}-\text{OCH}_3}
\;+\; 2\;\; \underset{\substack{\text{4-toluidino magnesium bromide}}}{\text{H}_3\text{C}-\!\!\!\bigcirc\!\!\!-\text{NHMgBr}} \longrightarrow
$$

$$
\underset{\substack{\text{1,1-di-(4-toluidino)-1-propoxy-}\\\text{magnesium bromide}}}{(\text{H}_3\text{C}-\!\!\!\bigcirc\!\!\!-\text{NH})_2\;\underset{\displaystyle\overset{|}{\text{C}}-\text{OMgBr}}{\overset{\displaystyle \text{CH}_3\text{CH}_2}{}}}
\;\;+\;\; \underset{\substack{\text{methoxymagnesium}\\\text{bromide}}}{\text{CH}_3\text{OMgBr}}
$$

$$
\underset{\substack{\text{1,1-di-(4-toluidino)-1-propoxy-}\\\text{magnesium bromide}}}{(\text{H}_3\text{C}-\!\!\!\bigcirc\!\!\!-\text{NH})_2\;\underset{\displaystyle\overset{|}{\text{C}}-\text{OMgBr}}{\overset{\displaystyle \text{CH}_3\text{CH}_2}{}}}
\;\;+\;\; 2\;\text{HCl} \longrightarrow
$$

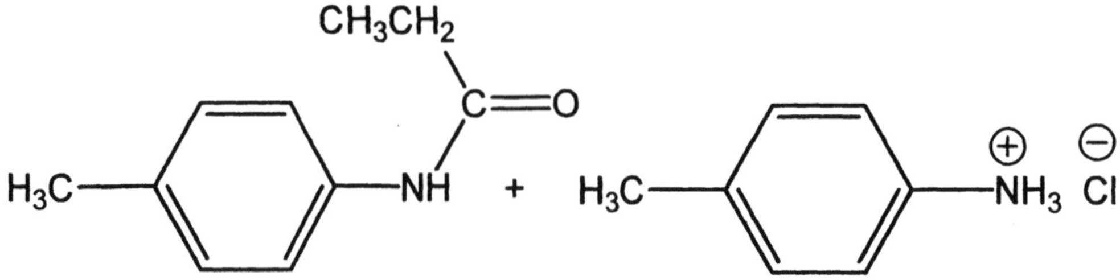

N-*p*-tolylpropionamide

4-toluidinium chloride

+   BrMgCl

c.

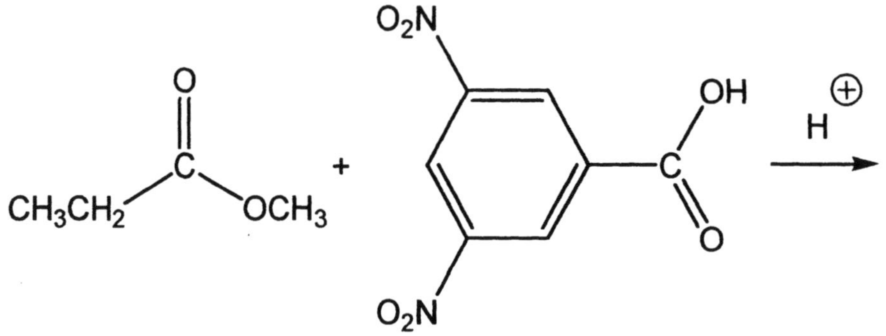

methyl propanoate   3,5-dinitrobenzoic acid

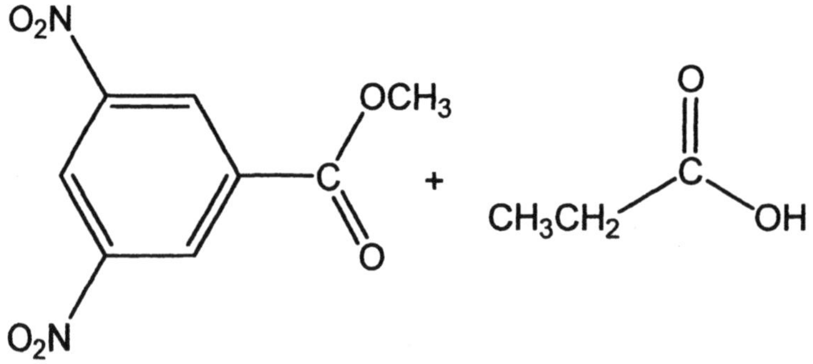

methyl 3,5-dinitrobenzoate   propanoic acid

d.

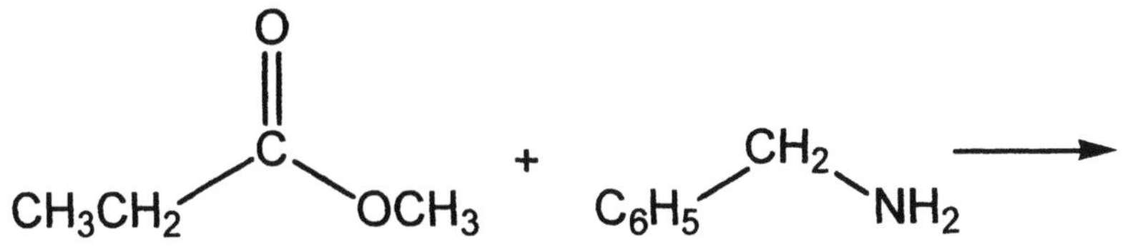

methyl propanoate   benzylamine

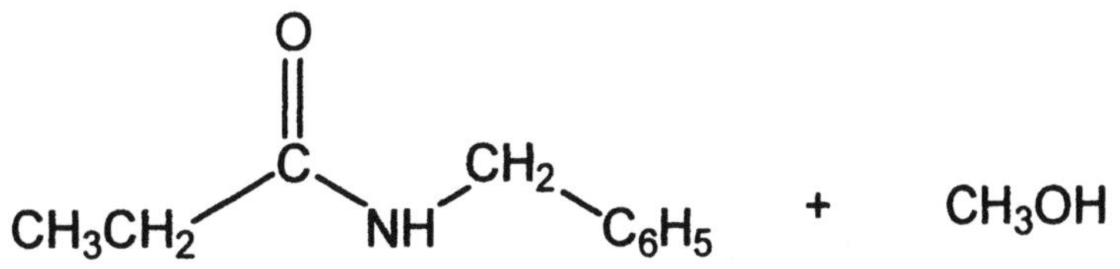

$+$ CH_3OH

N-benzylpropanamide   methanol

e.

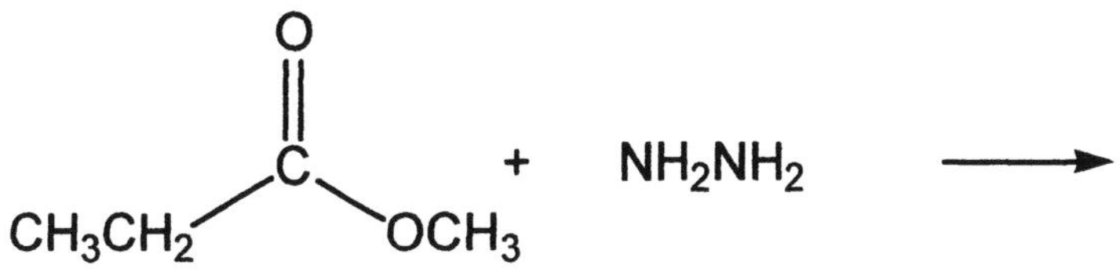

$+$ NH_2NH_2 $\longrightarrow$

methyl propanoate   hydrazine

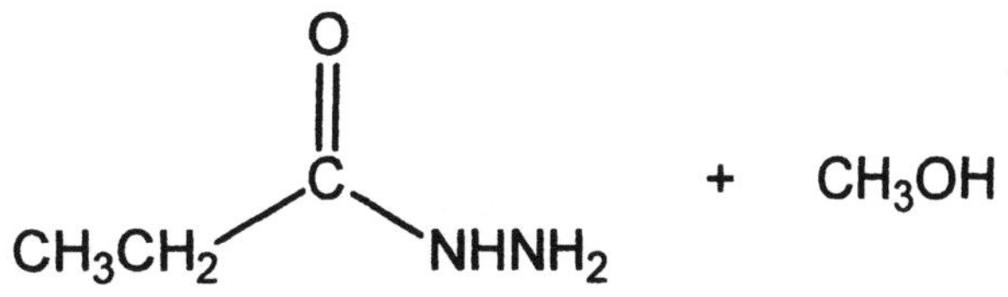

$+$ CH_3OH

propanoic acid hydrazide   methanol

22. If an unsymmetrical ether was treated with 3,5-dinitrobenzoyl chloride, then two 3,5-dinitrobenzoates would be formed. An example is shown below with methyl ethyl ether.

2 CH$_3$OCH$_2$CH$_3$ + 2 [3,5-dinitrobenzoyl chloride structure] $\longrightarrow$

ethyl methyl ether   3,5-dinitrobenzoyl chloride

[methyl 3,5-dinitrobenzoate structure] + [ethyl 3,5-dinitrobenzoate structure]

methyl 3,5-dinitrobenzoate   ethyl 3,5-dinitrobenzoate

+   CH$_3$CH$_2$Cl   +   CH$_3$Cl

chloroethane   chloromethane

23. 2 CH₃CH₂CH₂OCH₂CH₂CH₃ + 2

$$2\ CH_3CH_2CH_2OCH_2CH_2CH_3\ +\ 2$$

3,5-dinitrobenzoyl chloride

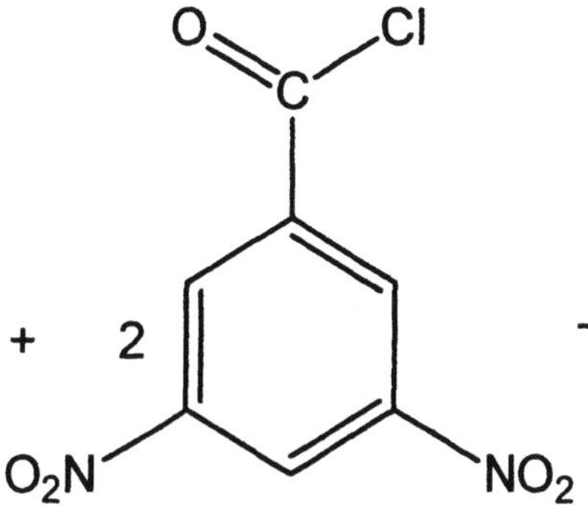

dipropyl ether   3,5-dinitrobenzoyl chloride

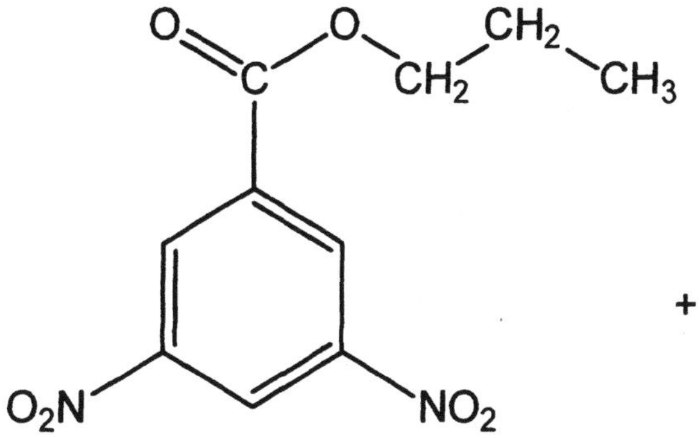

propyl 3,5-dinitrobenzoate   +   CH₃CH₂CH₂Cl

propyl 3,5-dinitrobenzoate   1-chloropropane

24.   **a.**

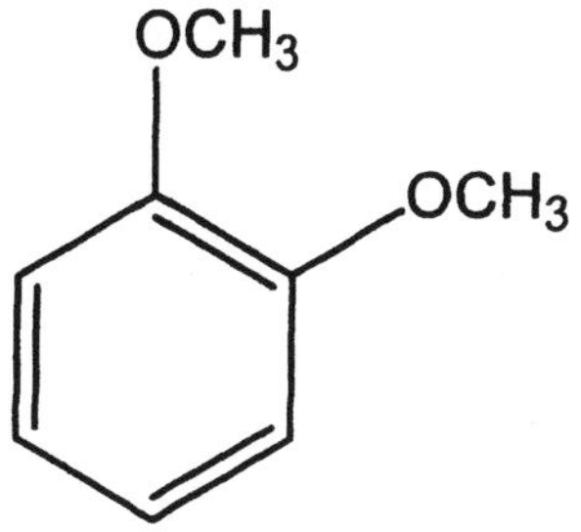

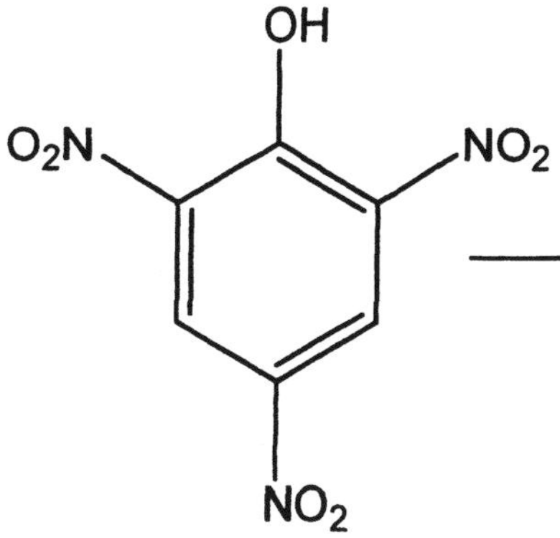

1,2-dimethoxybenzene   picric acid

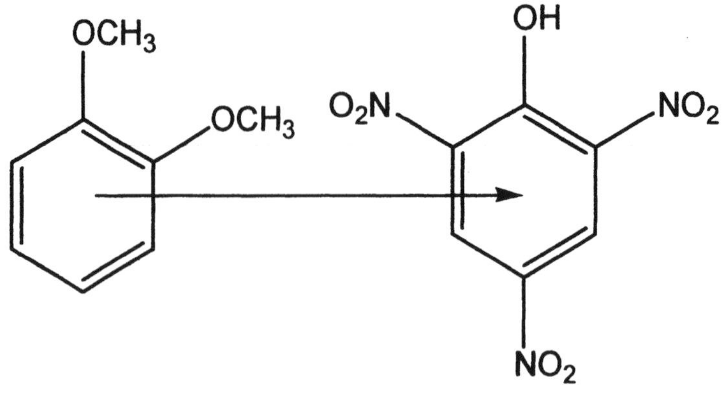

1,2-dimethoxybenzene picrate

b.

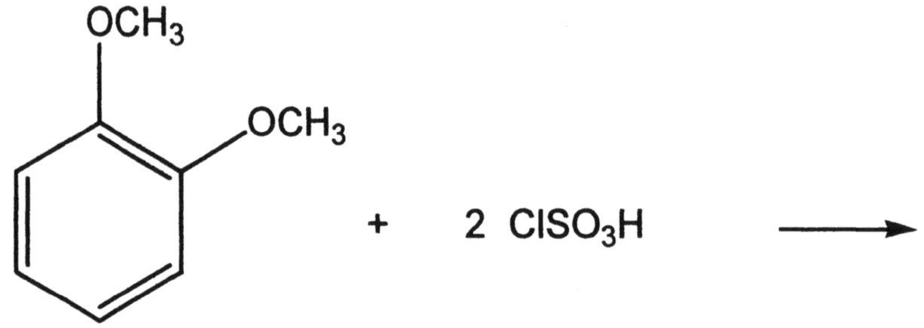

+   2 ClSO$_3$H   ⟶

1,2-dimethoxybenzene   chlorosulfonic acid

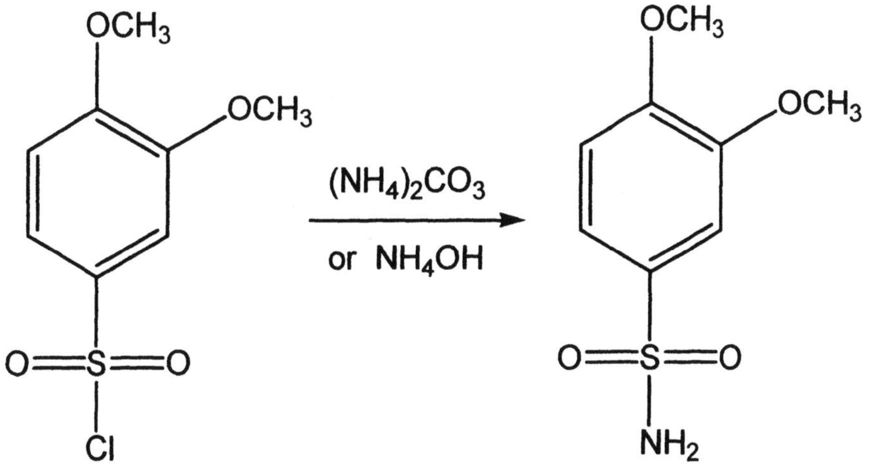

1,2-dimethoxybenzenesulfonyl
chloride

1,2-dimethoxybenzene-
sulfonamide

+ HCl + H_2SO_4

c.

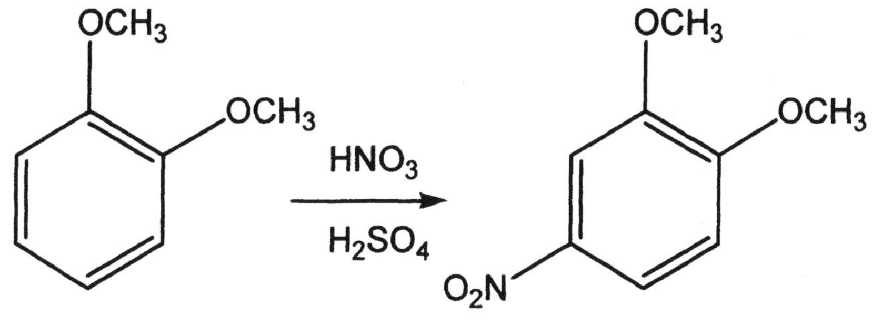

1,2-dimethoxybenzene

1,2-dimethoxy-4-
nitrobenzene

d.

1,2-dimethoxybenzene   4,5-dibromo-1,2-
                                dimethoxybenzene

25.   Potassium iodide is added to alkyl chlorides to speed up the
      reaction. This is not needed for vicinal dihalides. A base added to a
      vicinal dihalide may result in the formation of an alkyne.

vicinal dihalide   alkene   alkyne

26.   **a.** $BrCH_2$—$\overset{\overset{\textstyle CH_3}{|}}{CH}$—$CH_3$ +   Mg $\longrightarrow$

            1-bromo-2-   magnesium
            methylpropane

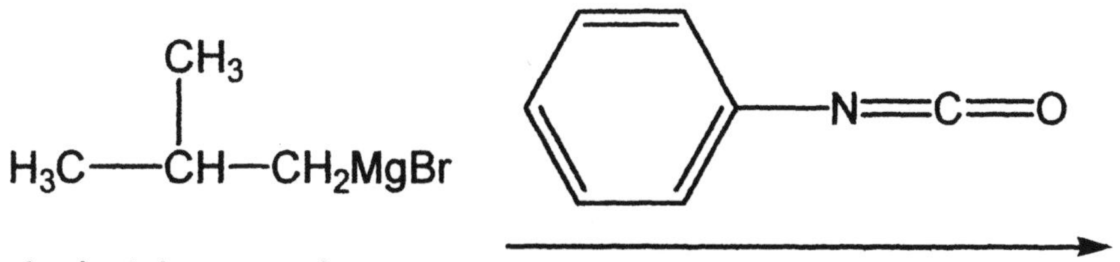

isobutylmagnesium
bromide

phenyl isocyanate

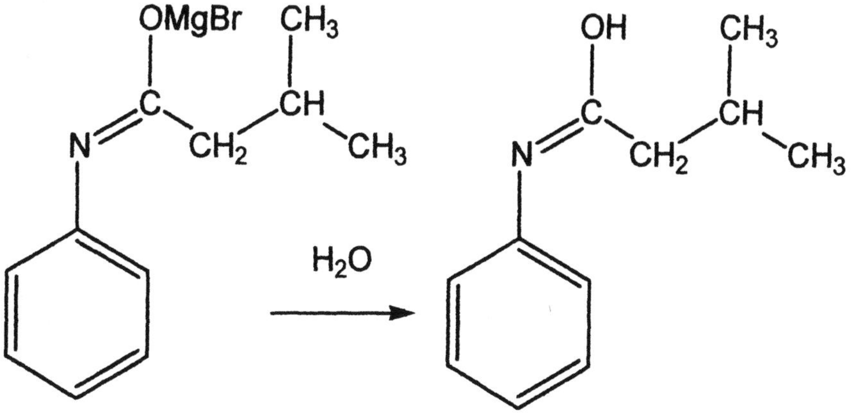

H_2O

3-methyl-*N*-phenyl-
butanimidic acid

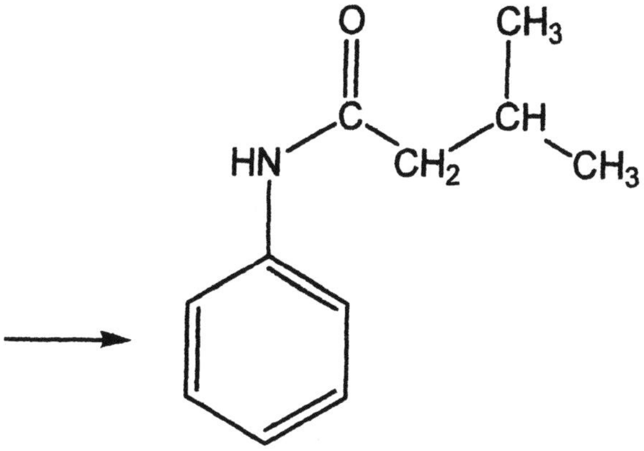

3-methyl-*N*-phenylbutanamide

b.
$$BrCH_2\!\!-\!\!\underset{\underset{CH_3}{|}}{CH}\!\!-\!\!CH_3 \quad + \quad Mg \quad \longrightarrow$$

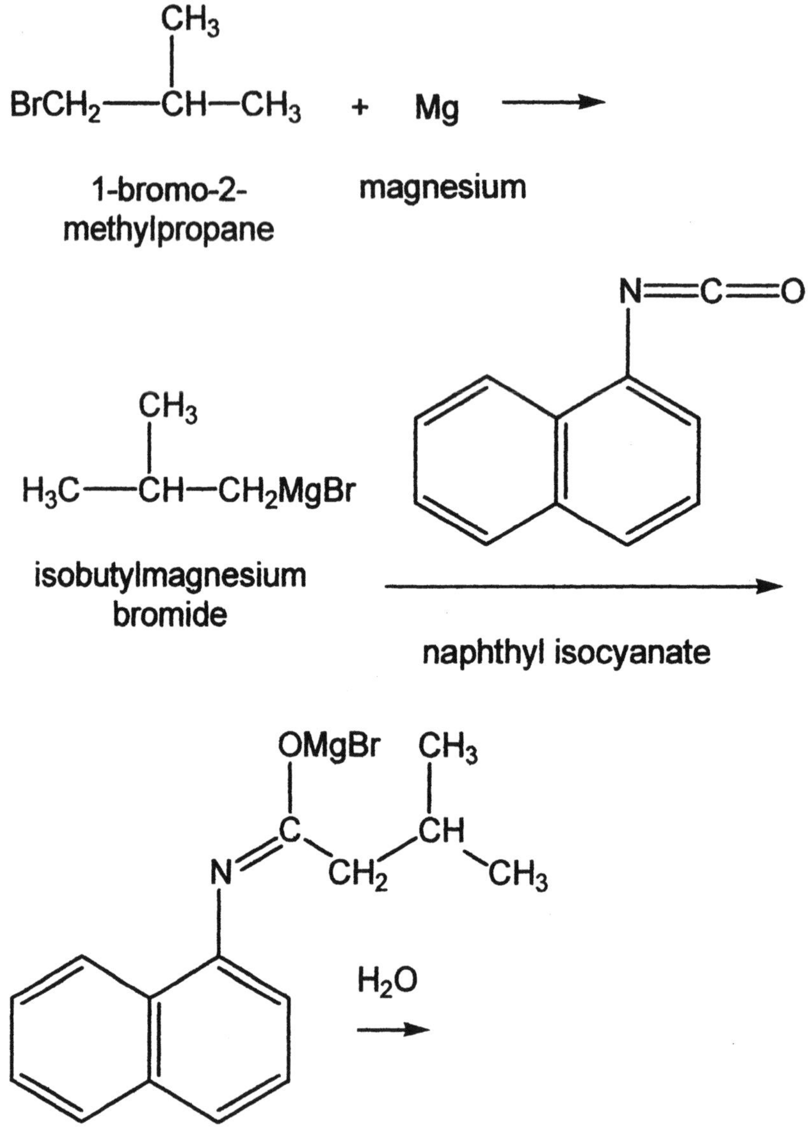

1-bromo-2-
methylpropane

magnesium

isobutylmagnesium
bromide

naphthyl isocyanate

H_2O

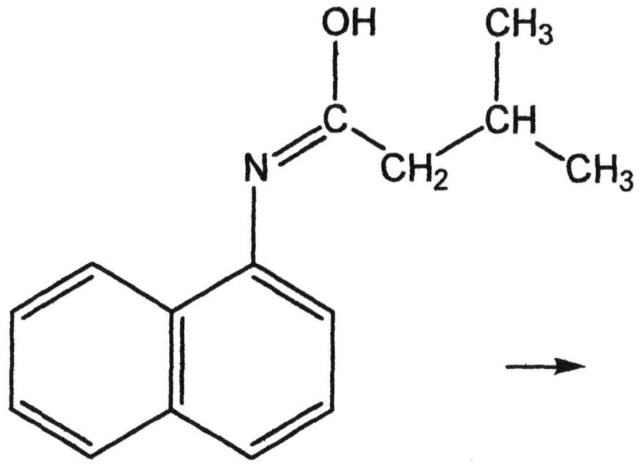

3-methyl-*N*-naphthalen-1-yl-
butanamidic acid

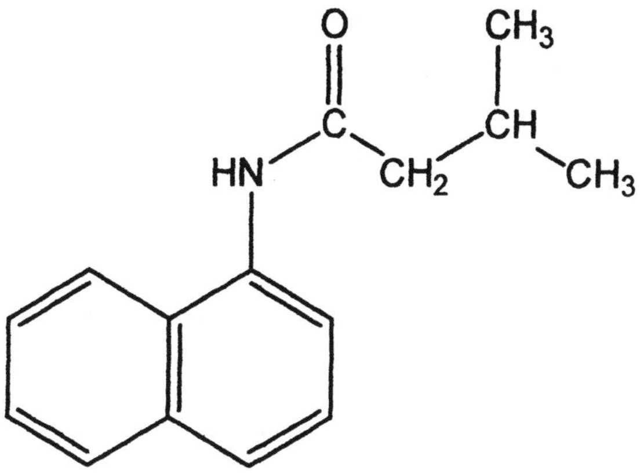

3-methyl-*N*-naphthalen-1-yl-
butanamide

c.  BrCH$_2$—CH—CH$_3$ + Mg $\longrightarrow$

CH$_3$ (above CH)

1-bromo-2-
methylpropane

magnesium

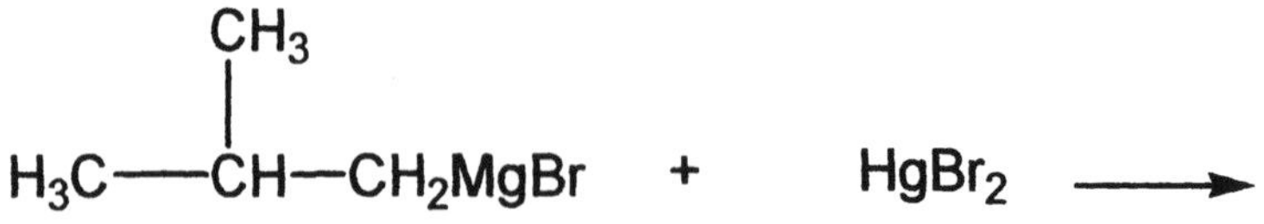

$$
\begin{array}{c}
\overset{\displaystyle CH_3}{\underset{\displaystyle |}{}} \\
H_3C\!-\!\!-\!CH\!-\!CH_2MgBr \quad + \quad HgBr_2 \longrightarrow
\end{array}
$$

isobutylmagnesium
bromide

mercuric bromide

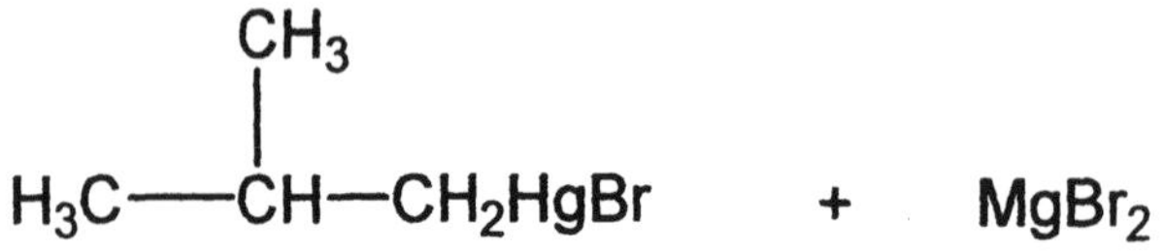

$$
\begin{array}{c}
\overset{\displaystyle CH_3}{\underset{\displaystyle |}{}} \\
H_3C\!-\!\!-\!CH\!-\!CH_2HgBr \quad + \quad MgBr_2
\end{array}
$$

isobutylmercuric
bromide

d.

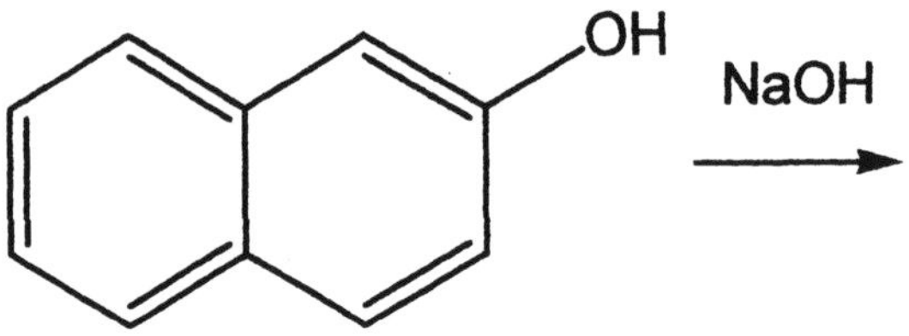

2-naphthol

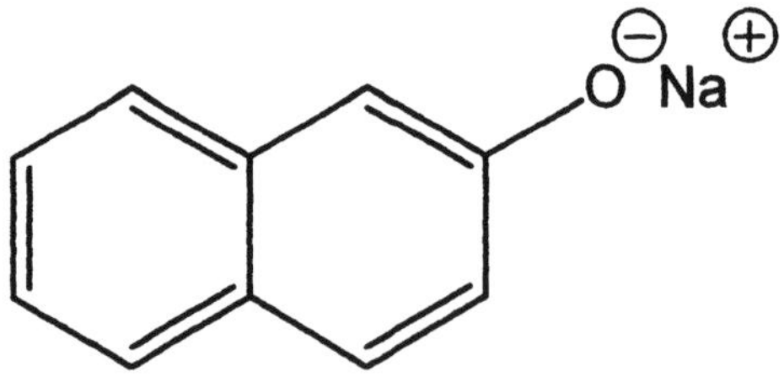

$\xrightarrow{\text{NaOH}}$

sodium naphthoxide

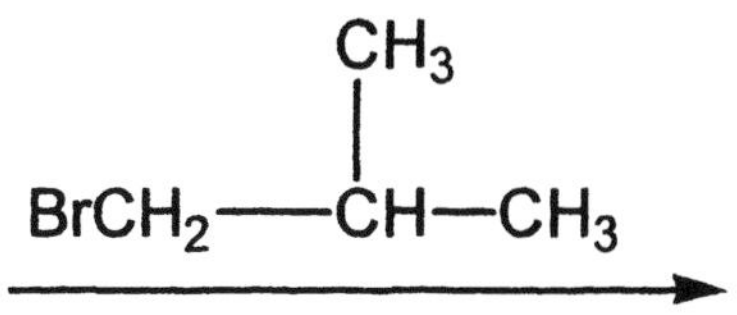

1-bromo-2-methylpropane

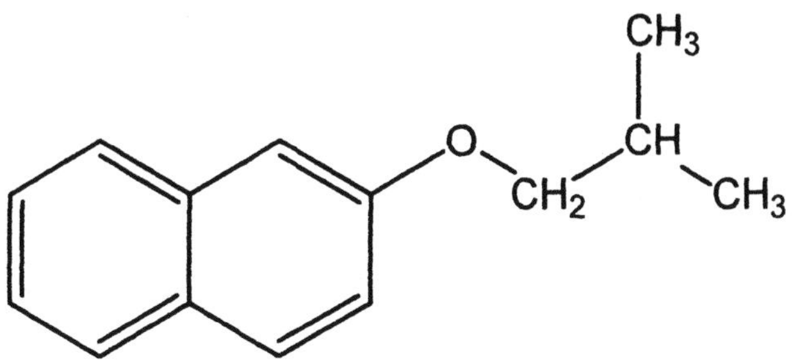

2-methylpropoxy naphthalene

e.

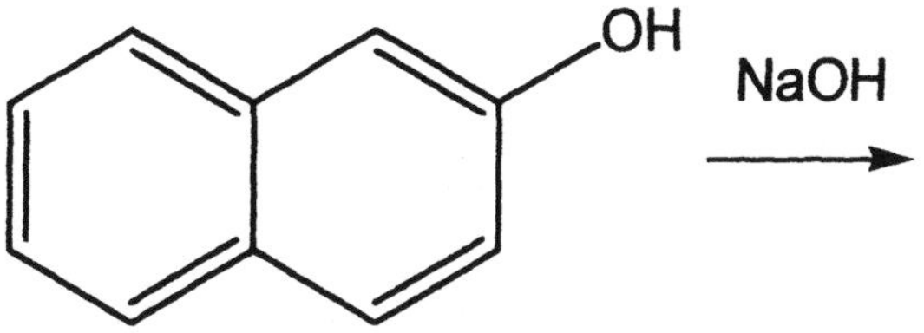

2-naphthol

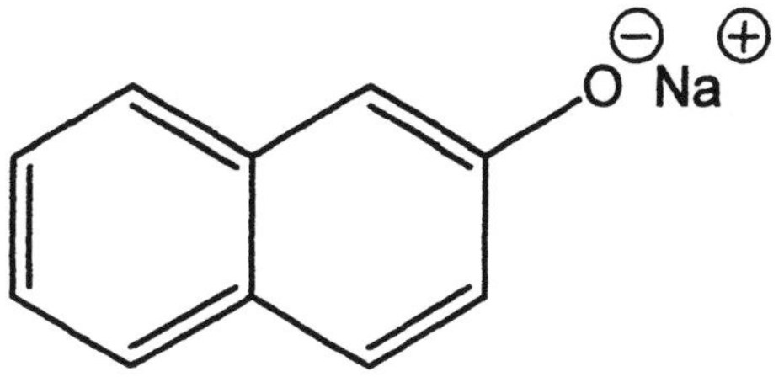

NaOH

sodium naphthoxide

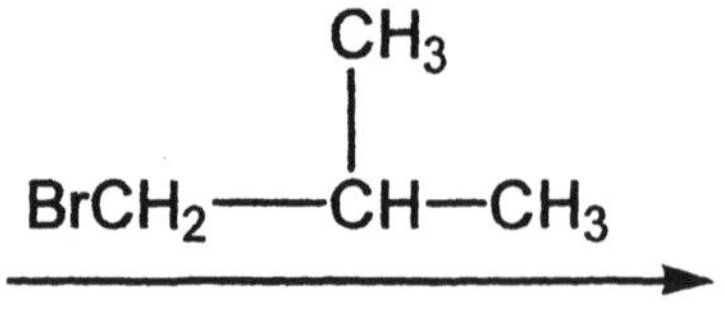

$$BrCH_2-\overset{\overset{\displaystyle CH_3}{|}}{CH}-CH_3 \longrightarrow$$

1-bromo-2-methylpropane

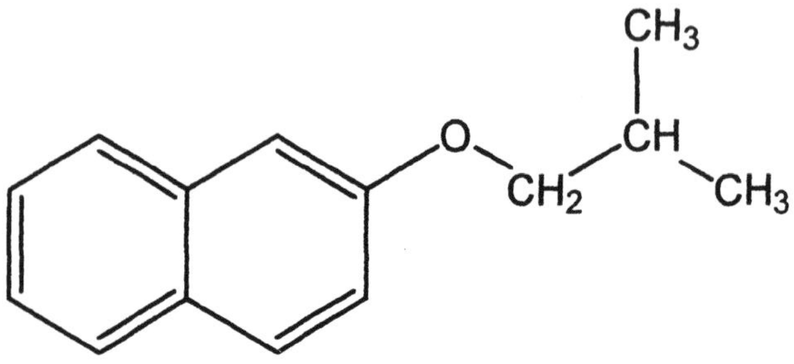

2-methylpropoxy naphthalene

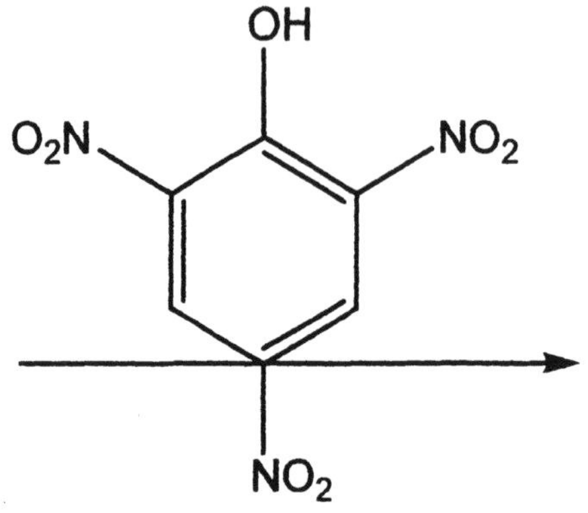

picric acid

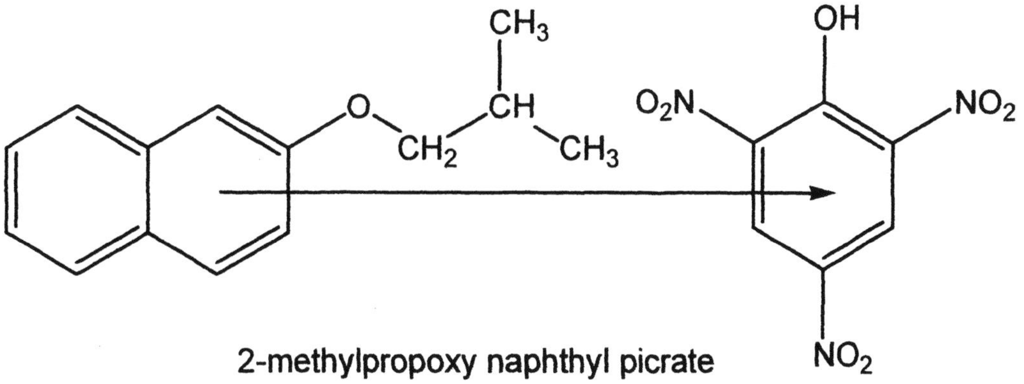

2-methylpropoxy naphthyl picrate

f.

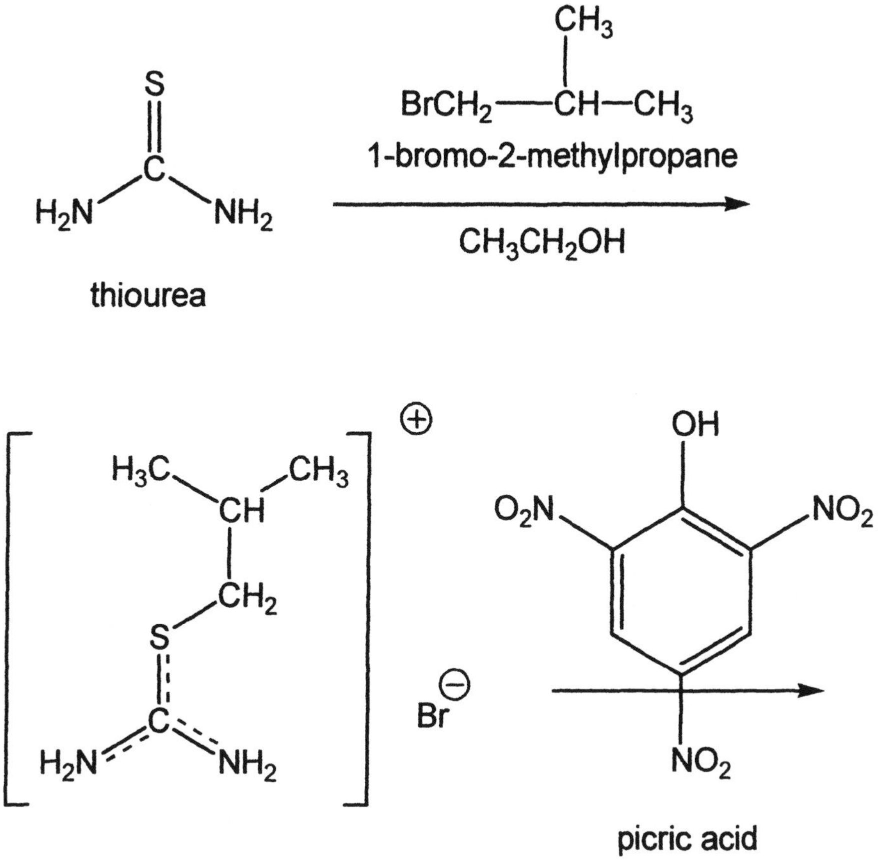

thiourea

1-bromo-2-methylpropane

CH_3CH_2OH

picric acid

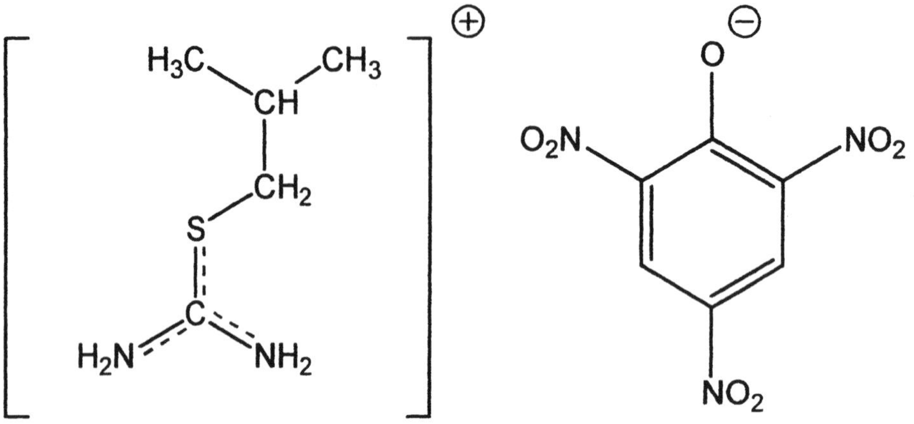

S-2-methylpropylthiuronium picrate

27.   a.

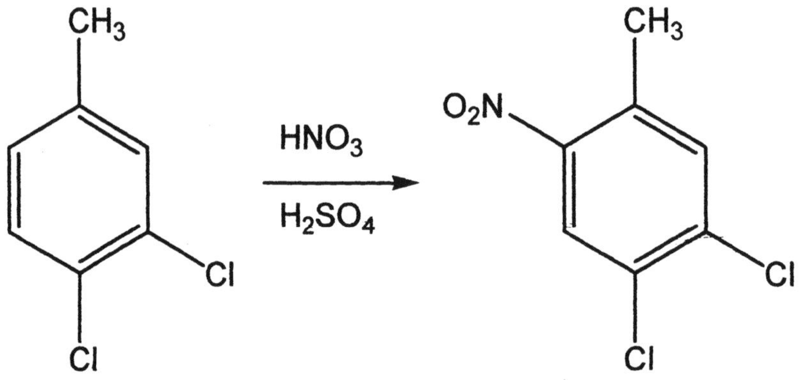

3,4-dichlorotoluene

3,4-dichloro-6-
nitrotoluene

b.

3,4-dichlorotoluene → (2 ClSO$_3$H) → 3,4-dichloro-2-methyl-benzenesulfonyl chloride

→ (NH$_4$OH) → 3,4-dichloro-2-methyl-benzenesulfonamide

c.

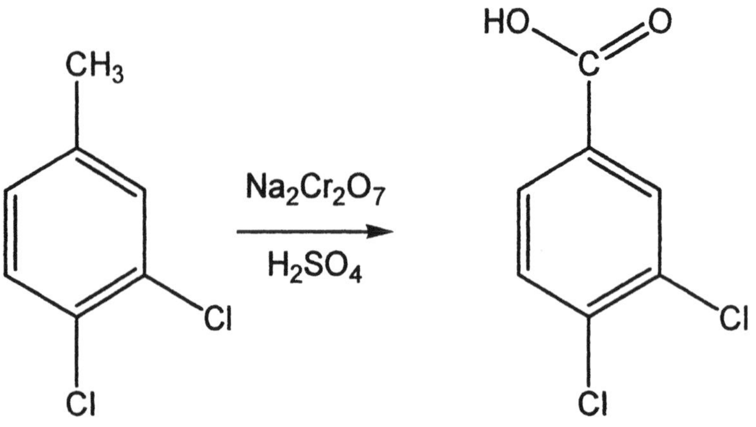

3,4-dichlorotoluene   3,4-dichlorobenzoic acid

28. The value of 148 corresponds to the disubstituted aromatic ring, the two carbonyl groups, and the hydroxy group minus the –H on the aromatic compound.

29. **a.**

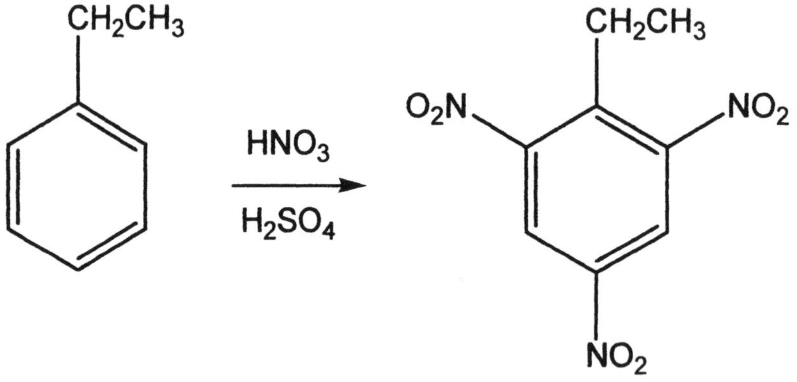

ethylbenzene   1-ethyl-2,4,6-
trinitrobenzene

b.

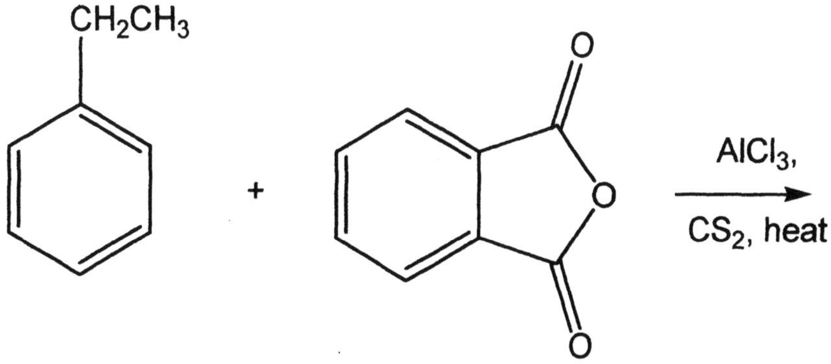

ethylbenzene + phthalic anhydride

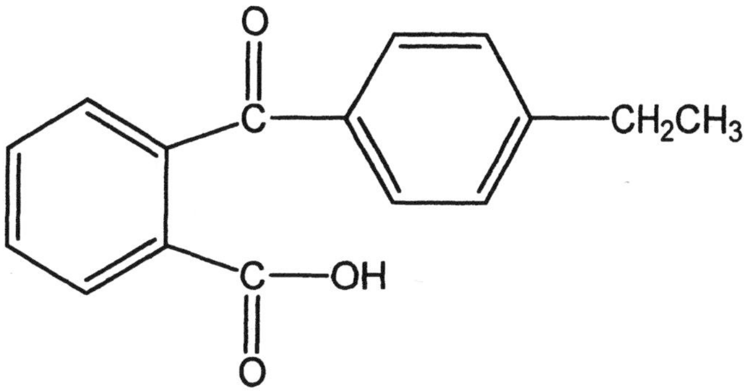

$\xrightarrow[\text{CS}_2,\text{ heat}]{\text{AlCl}_3,}$

2-(4-ethylbenzoyl)benzoic acid

c.

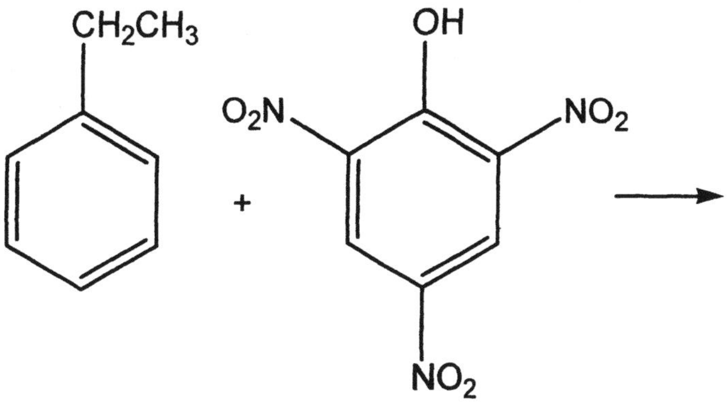

ethylbenzene + picric acid →

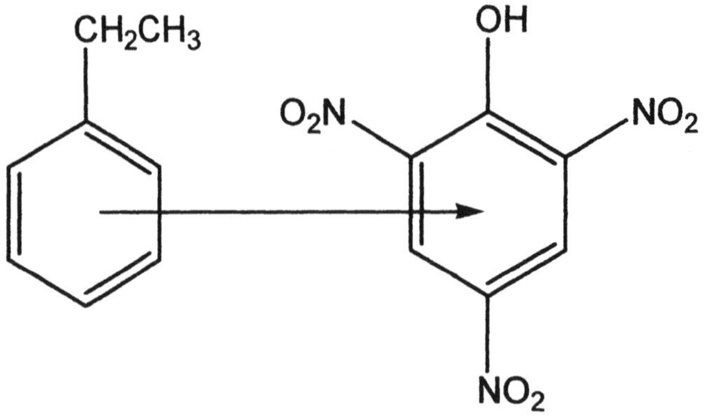

ethylbenzene picrate

30. a.

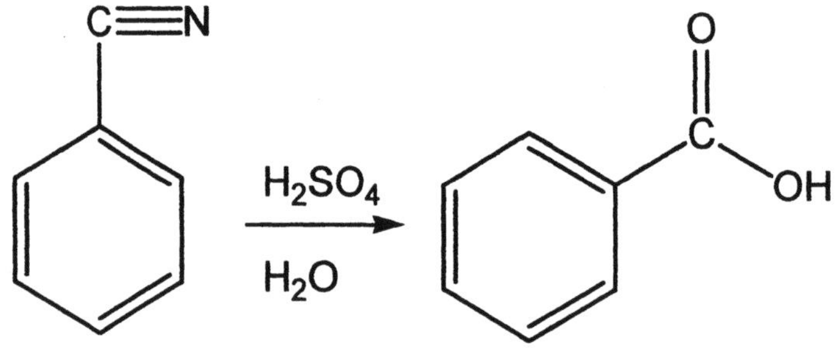

benzonitrile   benzoic acid

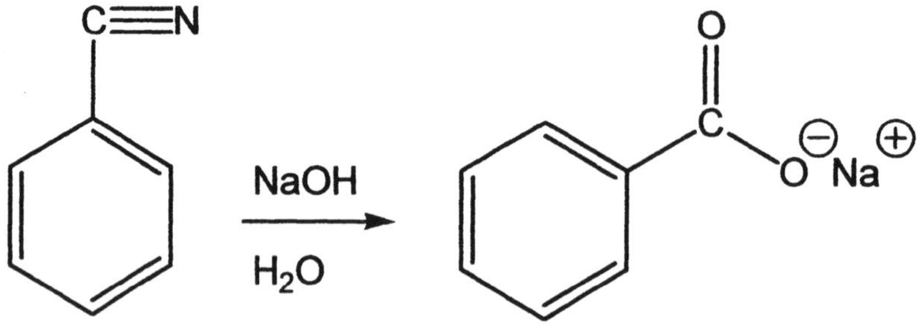

benzonitrile   sodium benzoate

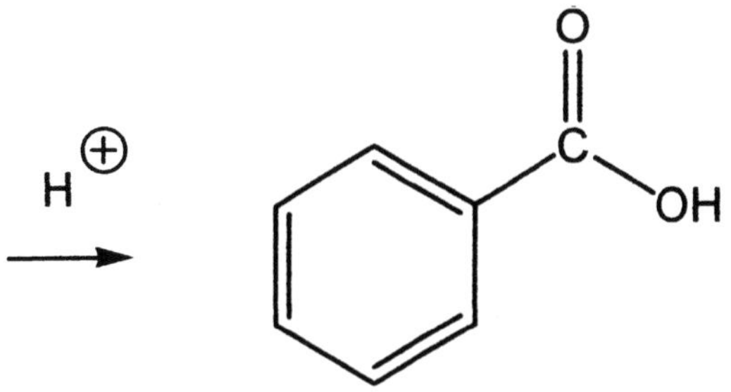

$H^{\oplus}$ →

benzoic acid

b.

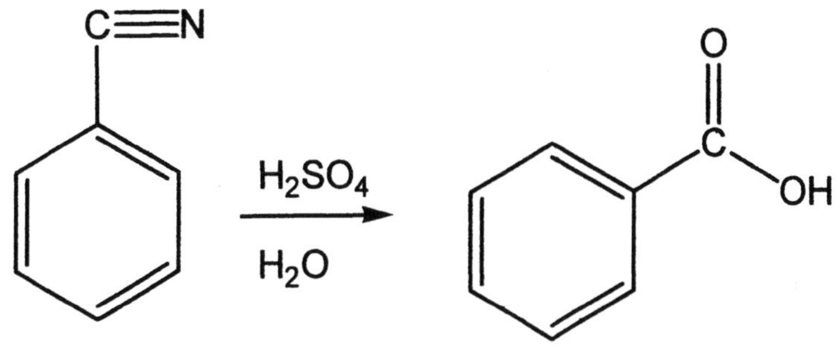

benzonitrile

H_2SO_4

H_2O

→

benzoic acid

$SOCl_2$

thionyl chloride

→

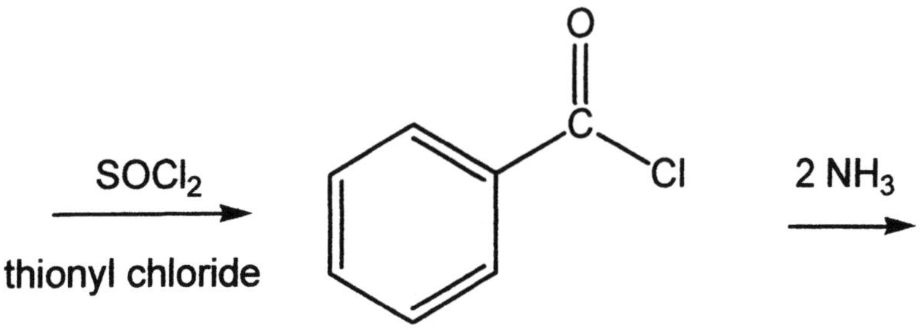

benzoyl chloride

2 NH_3 →

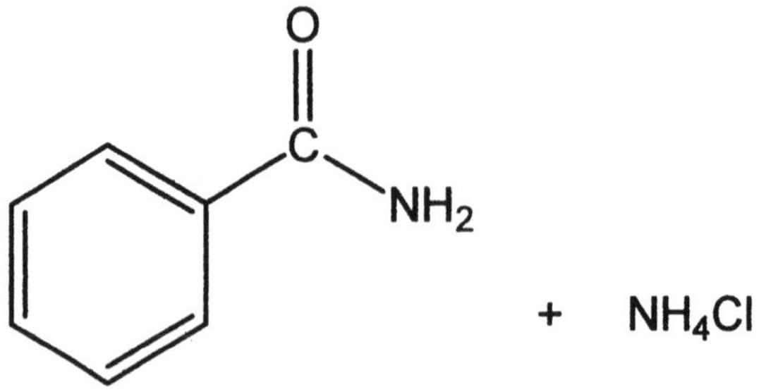

benzamide

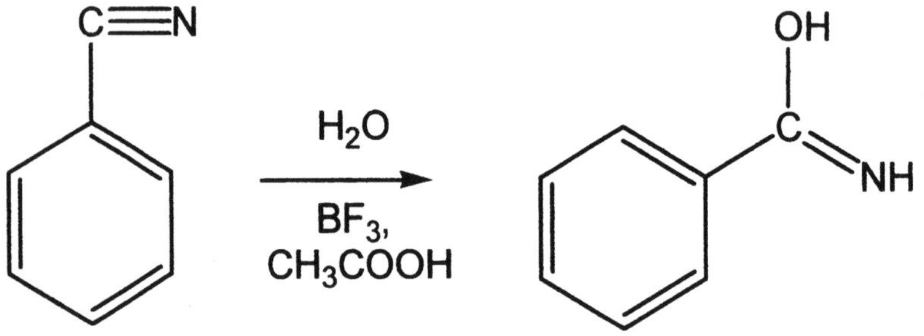

$+$ NH_4Cl

benzonitrile

$$\xrightarrow[\substack{BF_3, \\ CH_3COOH}]{H_2O}$$

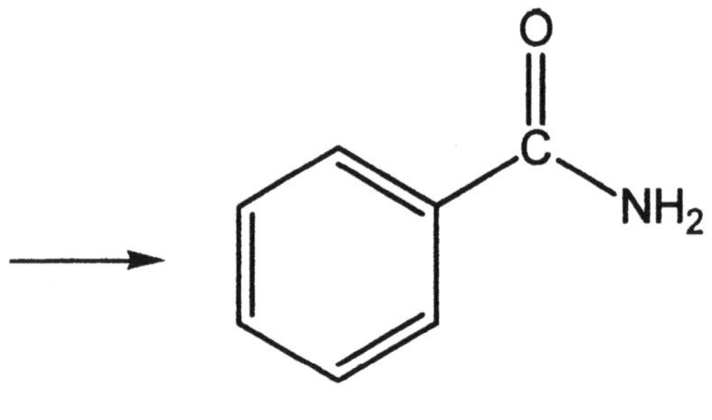

benzamide

c.

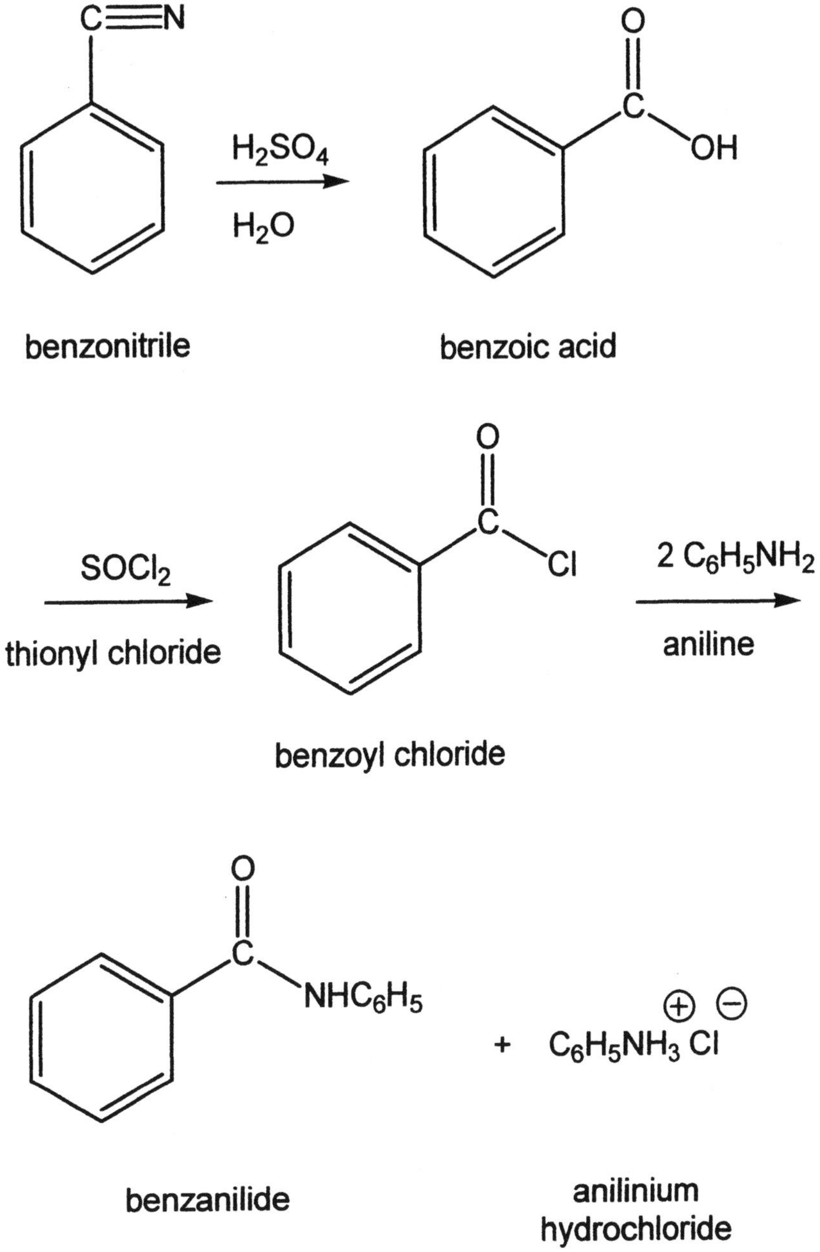

benzonitrile

$\xrightarrow[\text{H}_2\text{O}]{\text{H}_2\text{SO}_4}$

benzoic acid

$\xrightarrow[\text{thionyl chloride}]{\text{SOCl}_2}$

benzoyl chloride

$\xrightarrow[\text{aniline}]{2\ \text{C}_6\text{H}_5\text{NH}_2}$

benzanilide

$+\quad \text{C}_6\text{H}_5\overset{\oplus}{\text{NH}_3}\ \overset{\ominus}{\text{Cl}}$

anilinium
hydrochloride

d.

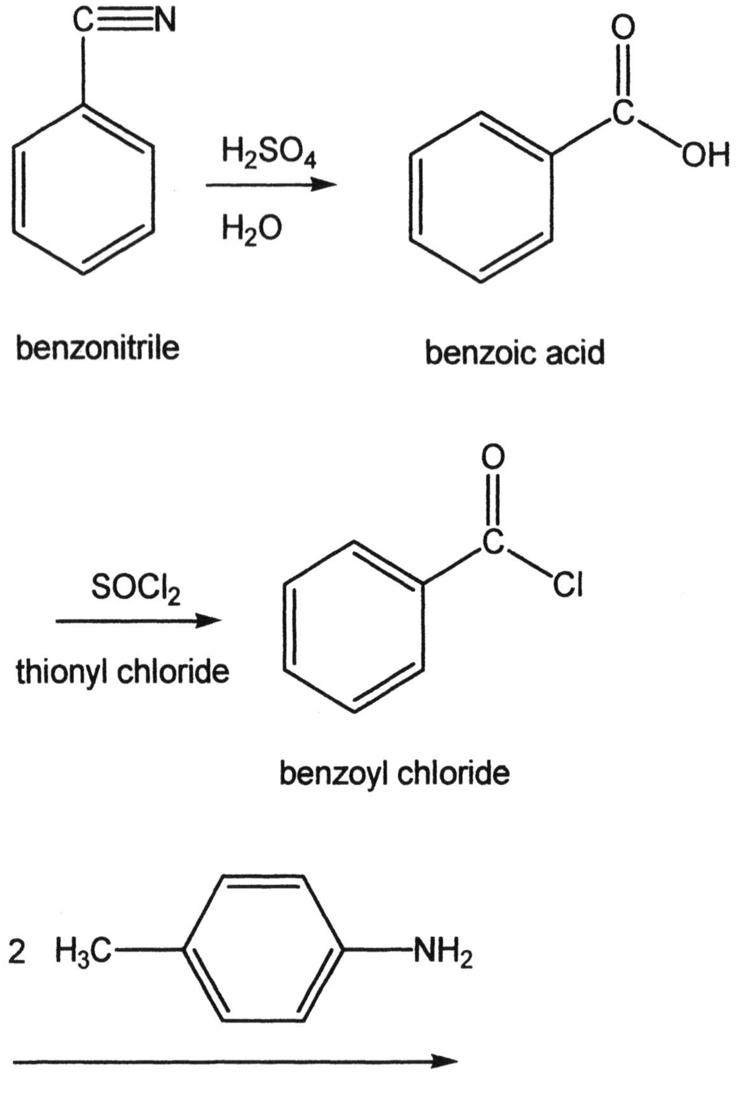
C≡N
H2SO4
H2O
O
C
OH
benzonitrile
benzoic acid
O
C
Cl
SOCl2
thionyl chloride
benzoyl chloride
2 H3C
NH2
4-toluidine

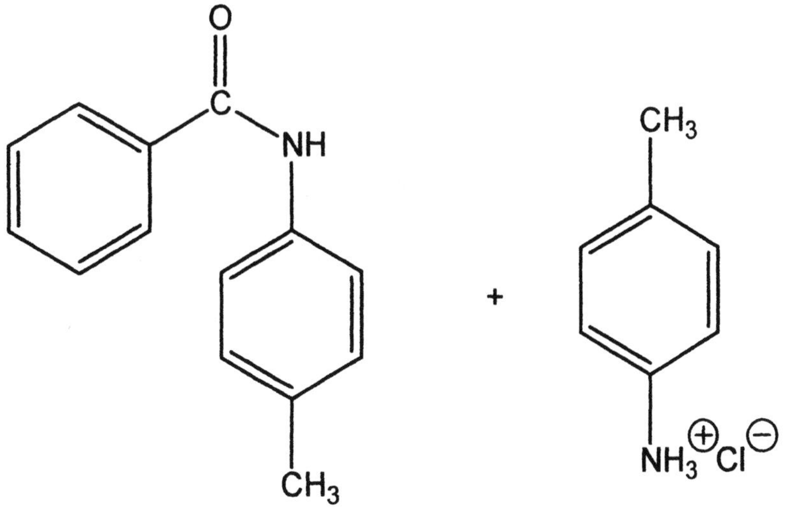
O
C
NH
CH3
+
CH3
NH3 Cl

4-phenyltoluidide

4-toluidinium hydrochloride

e.

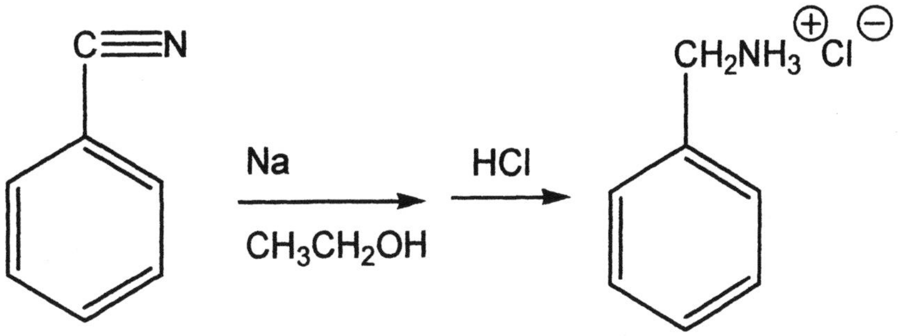
C≡N
Na
CH3CH2OH
HCl
CH2NH3 Cl

benzonitrile

benzylammonium chloride

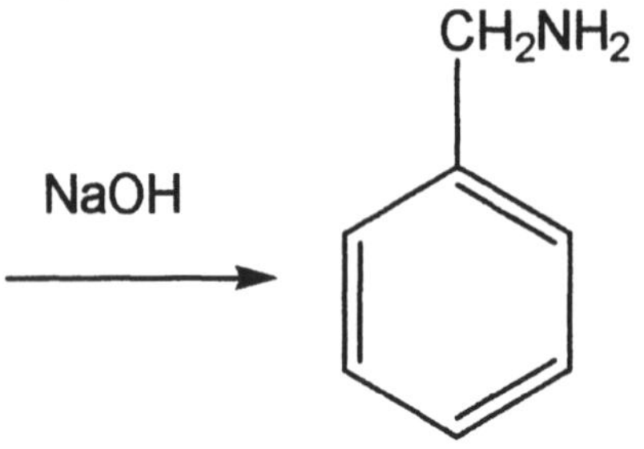

$$\text{NaOH} \longrightarrow$$

CH$_2$NH$_2$

benzylamine

f.

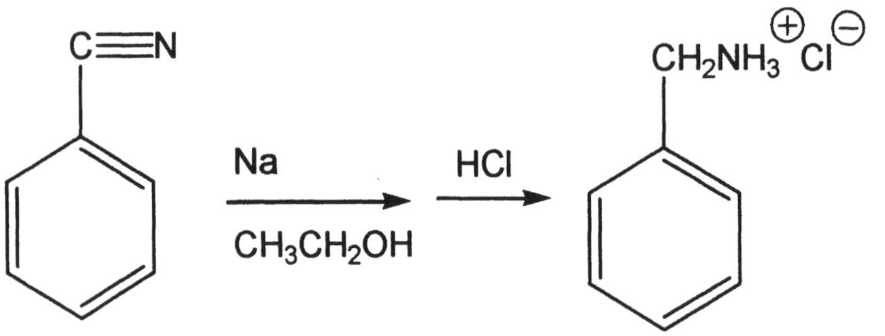

C≡N

benzonitrile

$$\xrightarrow[\text{CH}_3\text{CH}_2\text{OH}]{\text{Na}} \xrightarrow{\text{HCl}}$$

CH$_2$NH$_3^{\oplus}$ Cl$^{\ominus}$

benzylammonium
chloride

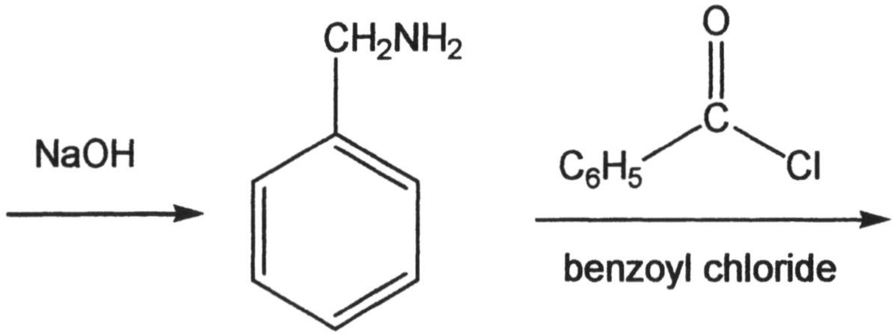

$$\xrightarrow{\text{NaOH}}$$

CH$_2$NH$_2$

benzylamine

$$C_6H_5 - \overset{\overset{\displaystyle O}{\|}}{C} - Cl$$

benzoyl chloride
$$\longrightarrow$$

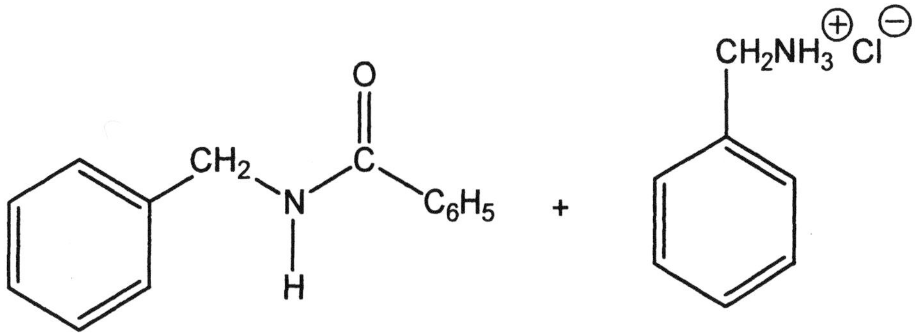
CH₂NH₃ Cl
O
C
CH₂
N
H
C₆H₅
+

N-benzylbenzamide benzylammonium hydrochloride

g.

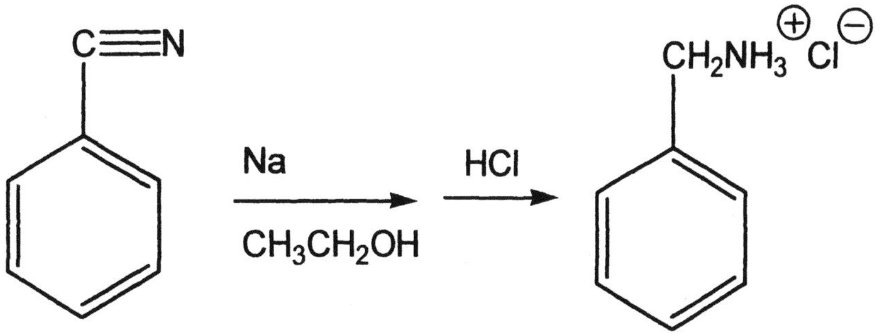
C≡N
Na
CH₃CH₂OH
HCl
CH₂NH₃ Cl

benzonitrile benzylammonium chloride

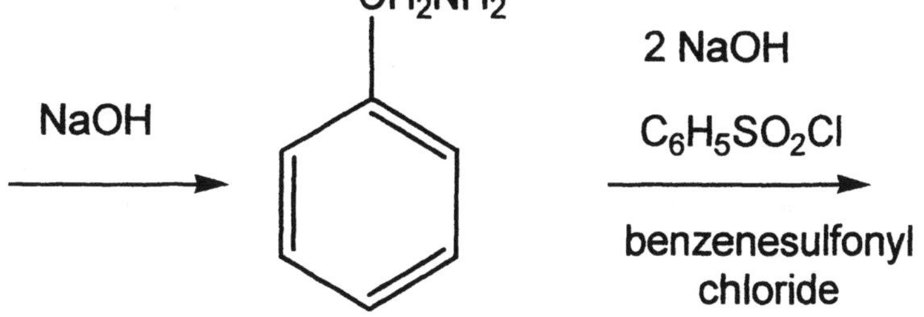
NaOH
CH₂NH₂
2 NaOH
C₆H₅SO₂Cl
benzenesulfonyl
chloride

benzylamine

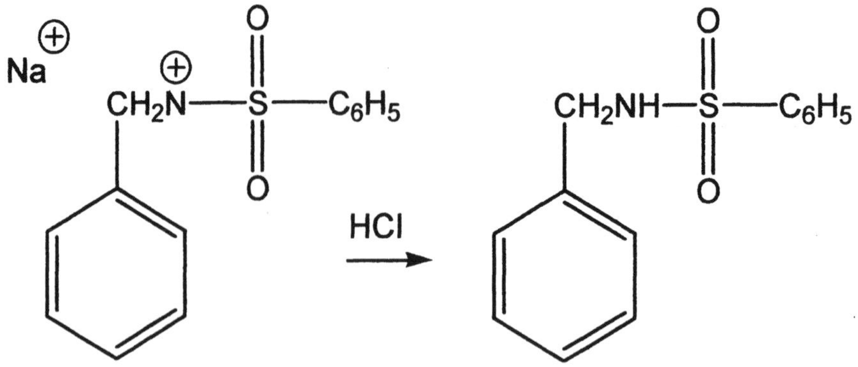

N-benzylbenzene-
sulfonamide

h.

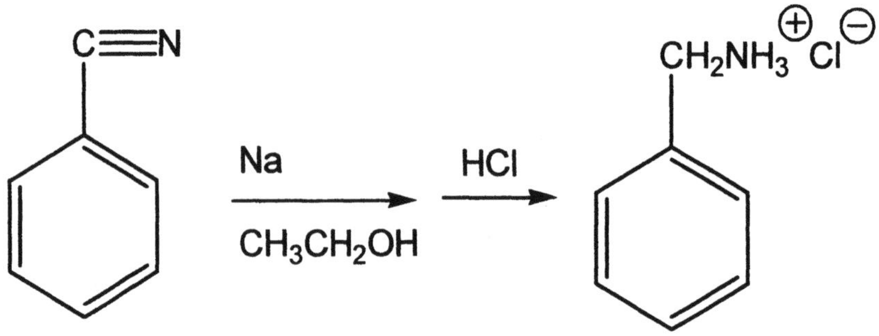

benzonitrile

benzylammonium
chloride

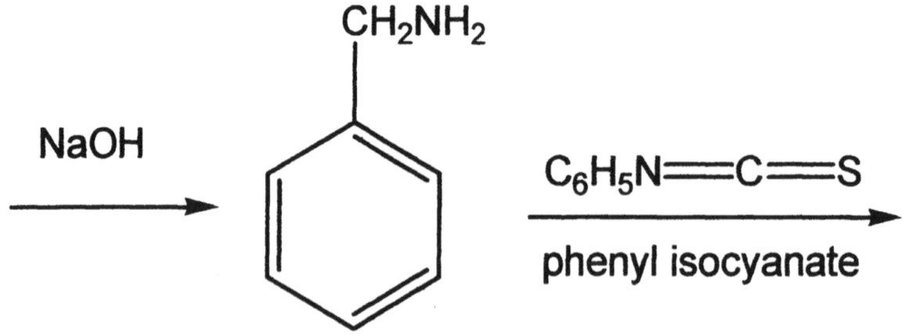

benzylamine

1-benzyl-3-phenylthiourea

i.

benzonitrile   thioglycolic acid

phenyl α-(imidioylthio)acetic acid hydrochloride

31.   **a.**

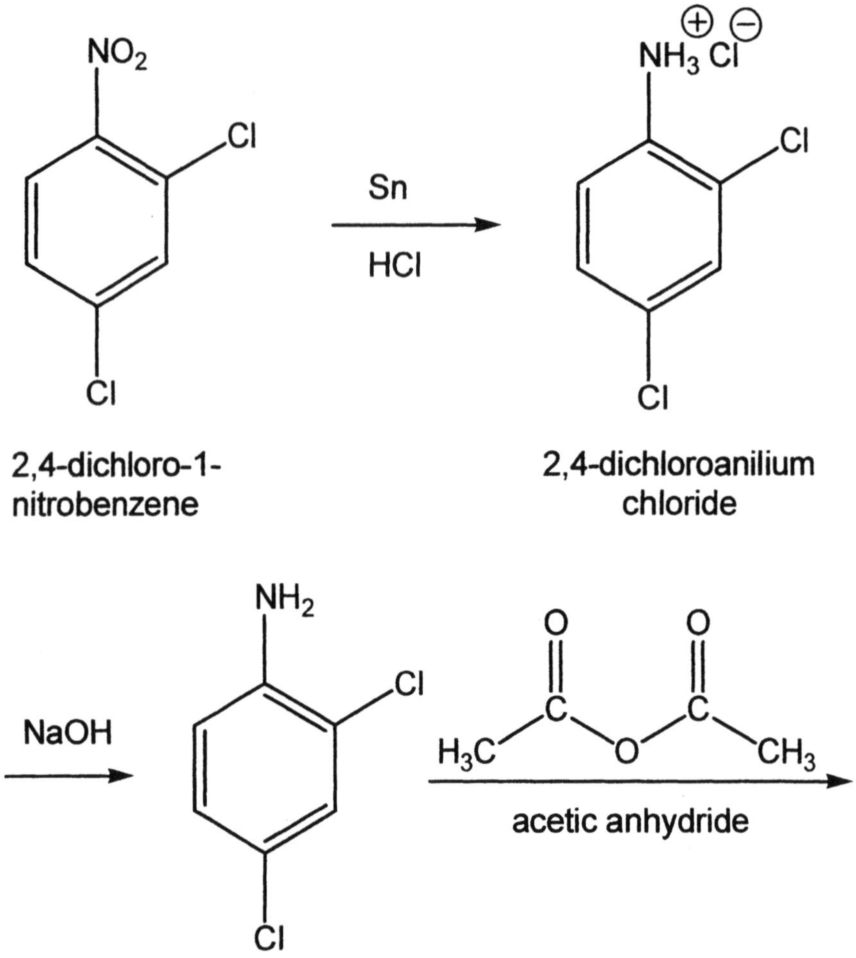

2,4-dichloro-1-
nitrobenzene

Sn

HCl

2,4-dichloroanilium
chloride

NaOH

2,4-dichloroaniline

acetic anhydride

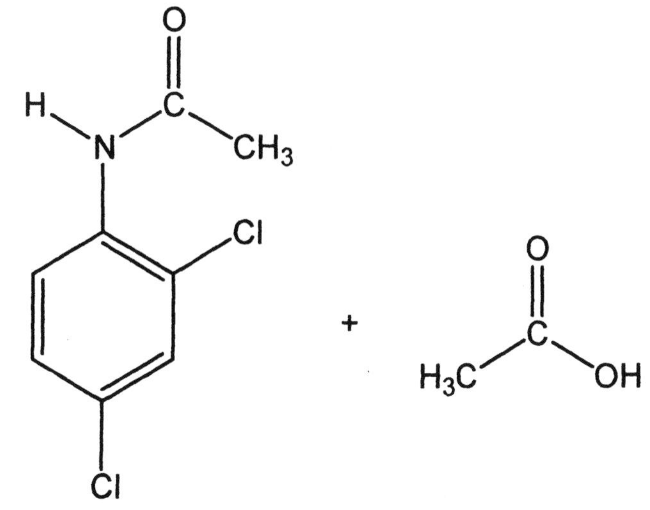

N-(2,4-dichlorophenyl)-
acetamide

acetic acid

b.

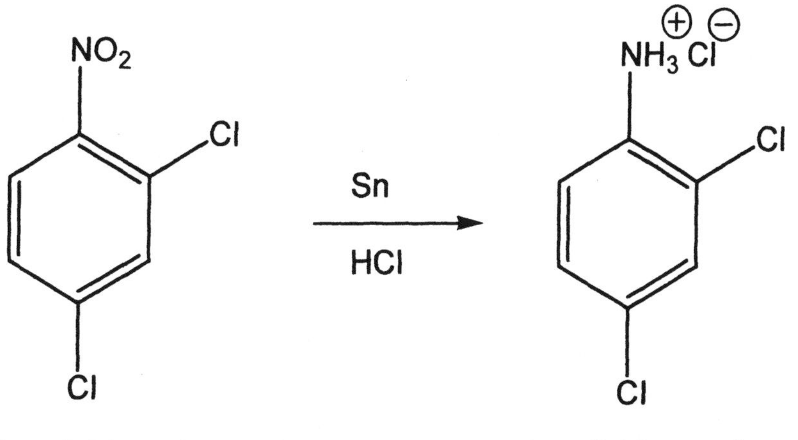

Sn

HCl

2,4-dichloro-1-
nitrobenzene

2,4-dichloroanilium
chloride

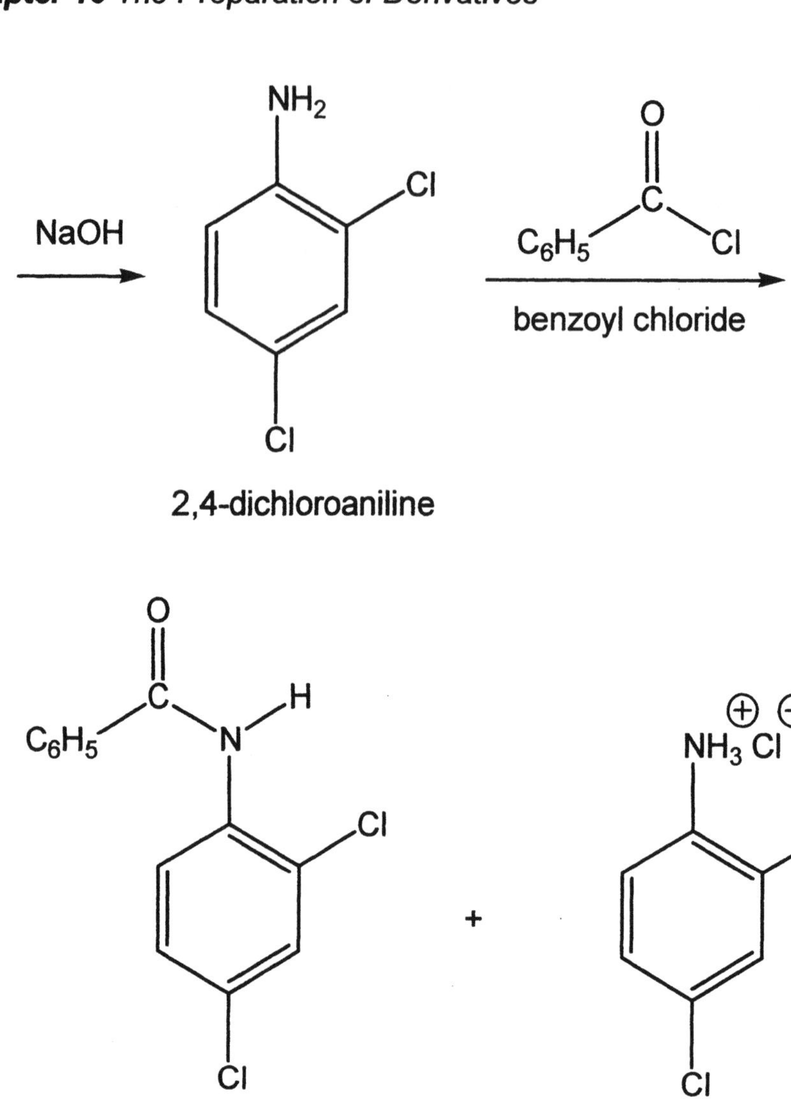

2,4-dichloroaniline

benzoyl chloride

N-(2,4-dichlorophenyl)–
benzamide

+

2,4-dichloroanilium
chloride

c.

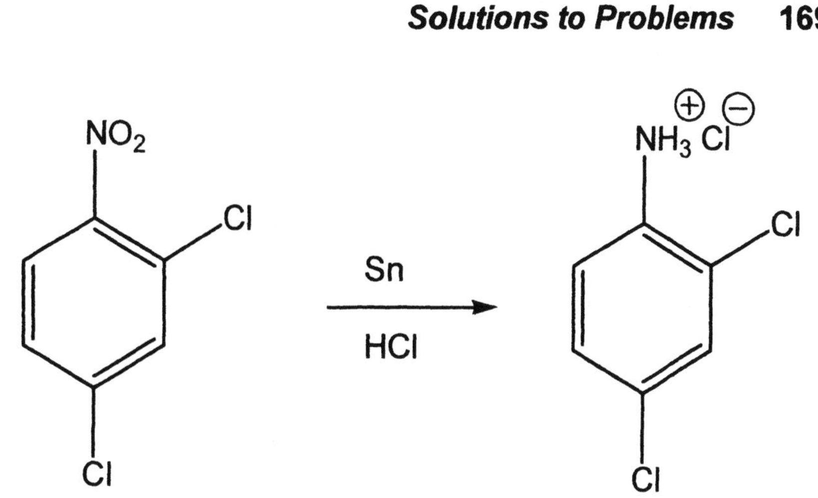

2,4-dichloro-1-
nitrobenzene

Sn
⟶
HCl

$\overset{\oplus}{NH_3}$ $\overset{\ominus}{Cl}$

2,4-dichloroanilium
chloride

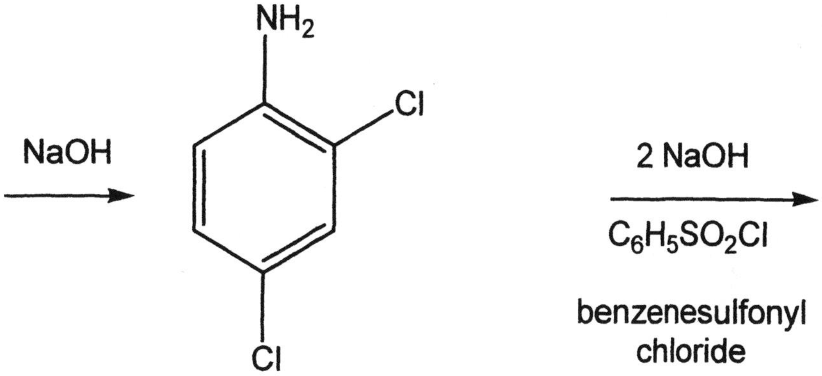

NaOH
⟶

NH_2

2,4-dichloroaniline

2 NaOH
⟶
$C_6H_5SO_2Cl$

benzenesulfonyl
chloride

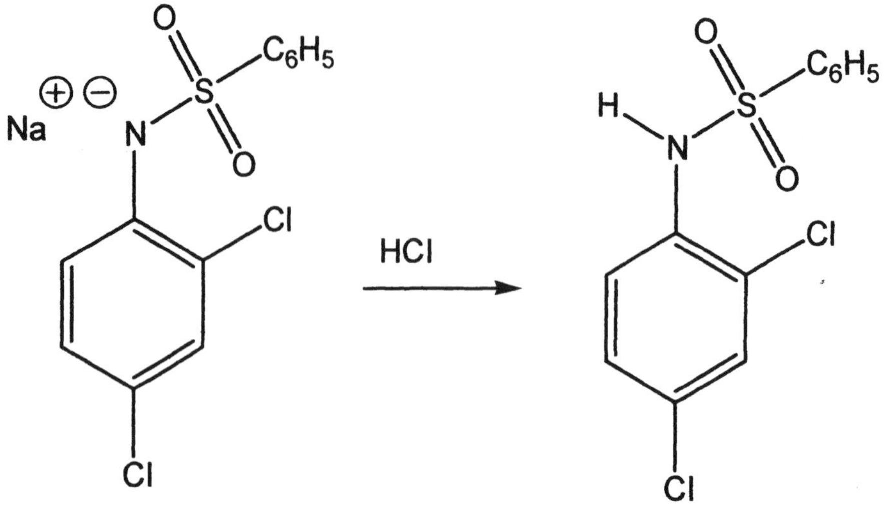

sodium salt of
N-(2,4-dichlorophenyl)-
benzenesulfonamide

N-(2,4-dichlorophenyl)-
benzenesulfonamide

32. **a.**

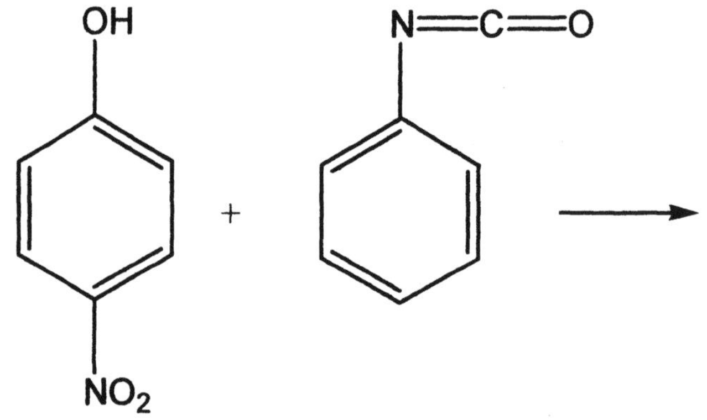

4-nitrophenol   phenyl isocyanate

phenylcarbamic acid 4-nitrophenyl ester

b.

4-nitrophenol   1-naphthyl isocyanate

naphthalen-2-yl-carbamic acid 4-nitrophenyl ester

c.

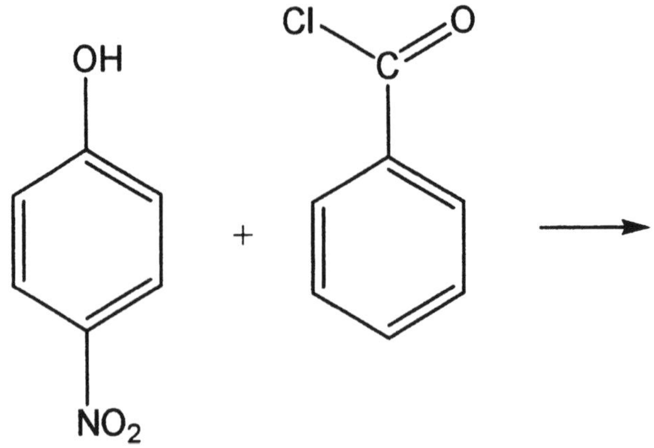

4-nitrophenol   benzoyl chloride

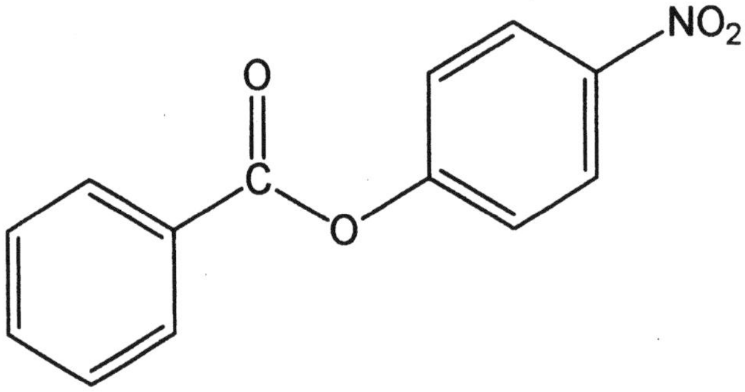

4-nitrophenyl benzoate

d.

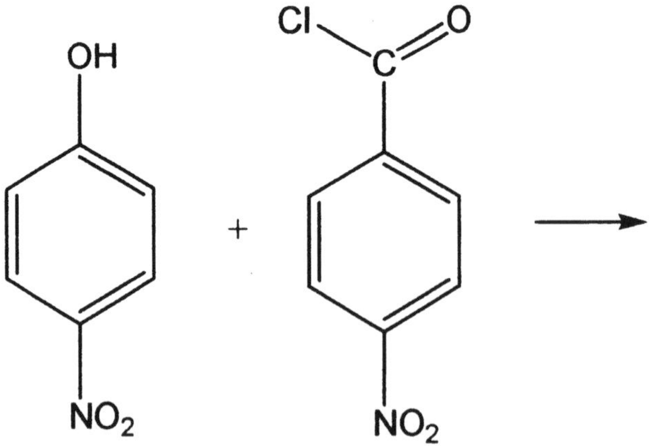

4-nitrophenol   4-nitrobenzoyl
                     chloride

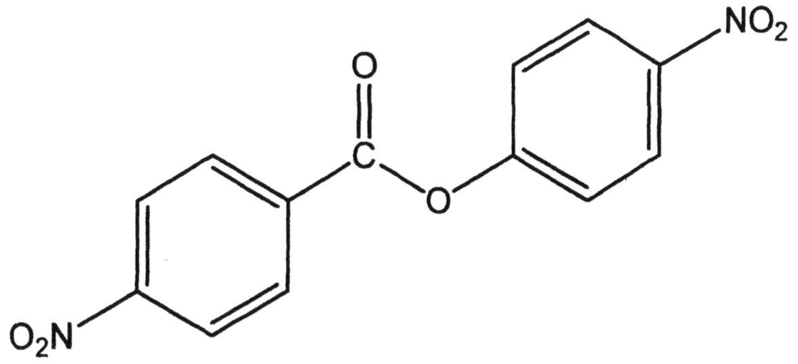

4-nitrophenyl 4-nitrobenzoate

e.

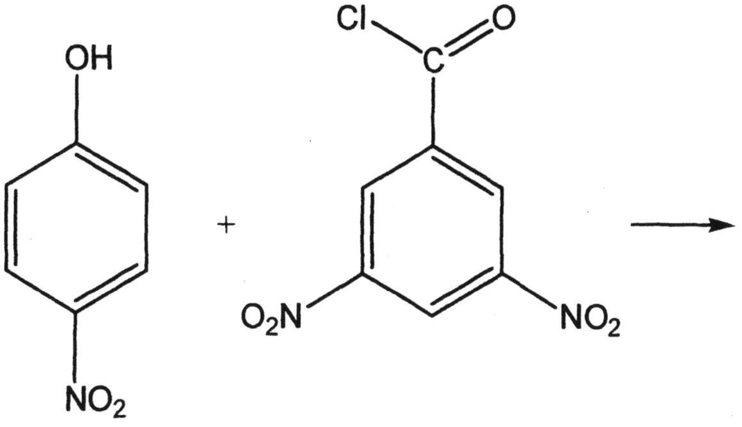

4-nitrophenol

3,5-dinitrobenzoyl
chloride

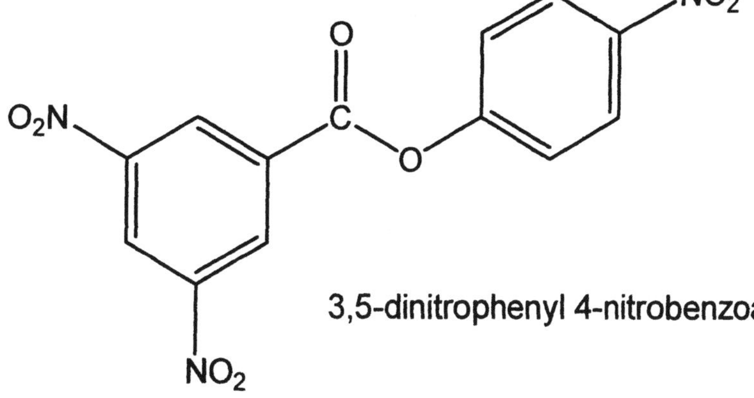

3,5-dinitrophenyl 4-nitrobenzoate

f.

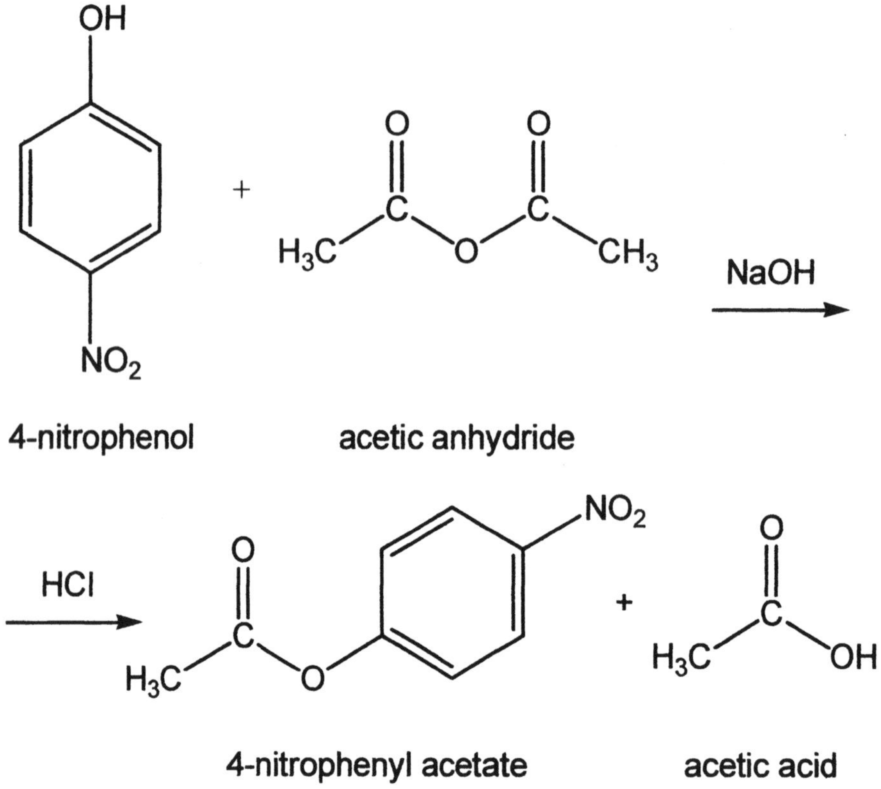
OH
NO2
+
O
O
H3C
C
O
C
CH3
NaOH
4-nitrophenol
acetic anhydride
HCl
O
C
H3C
O
NO2
+
O
C
H3C
OH
4-nitrophenyl acetate
acetic acid

g.

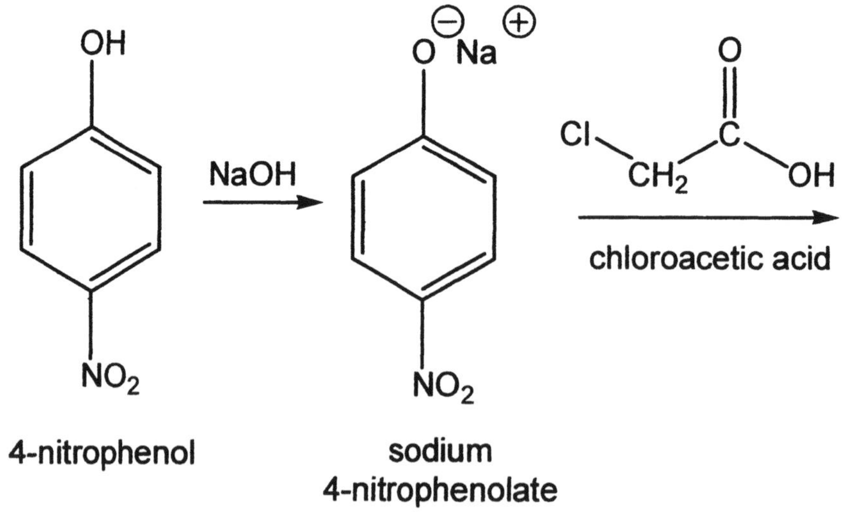
OH
NO2
NaOH
O Na
NO2
O
Cl
CH2
C
OH
chloroacetic acid
4-nitrophenol
sodium
4-nitrophenolate

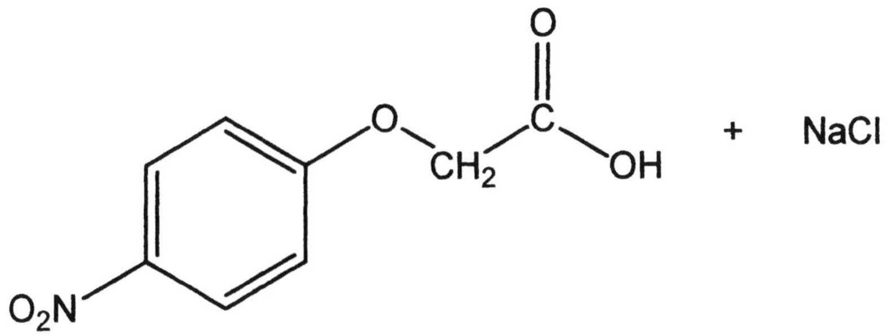

(4-nitrophenoxy)acetic acid   +   NaCl

h.

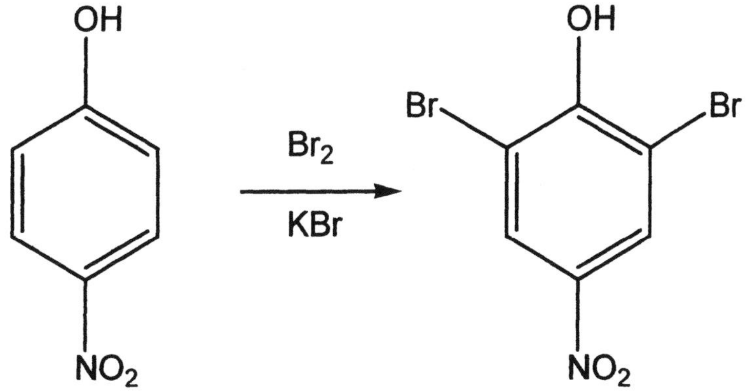

4-nitrophenol   2,6-dibromo-
                                       4-nitrophenol

33.   a.

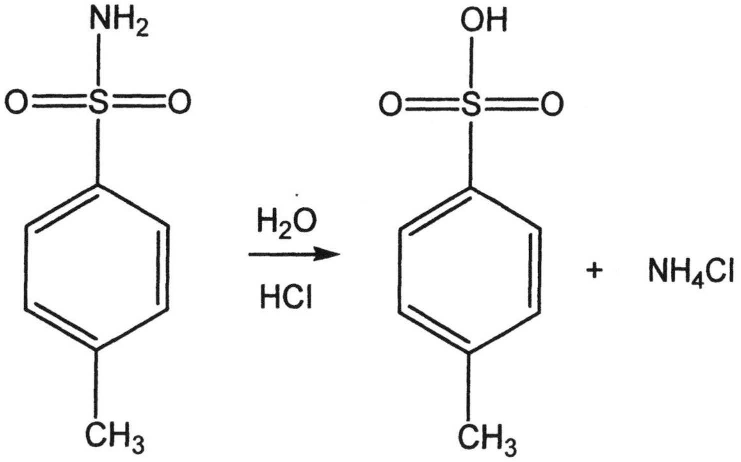

4-methylbenzene-   4-methylbenzene-
sulfonamide   sulfonic acid

b.

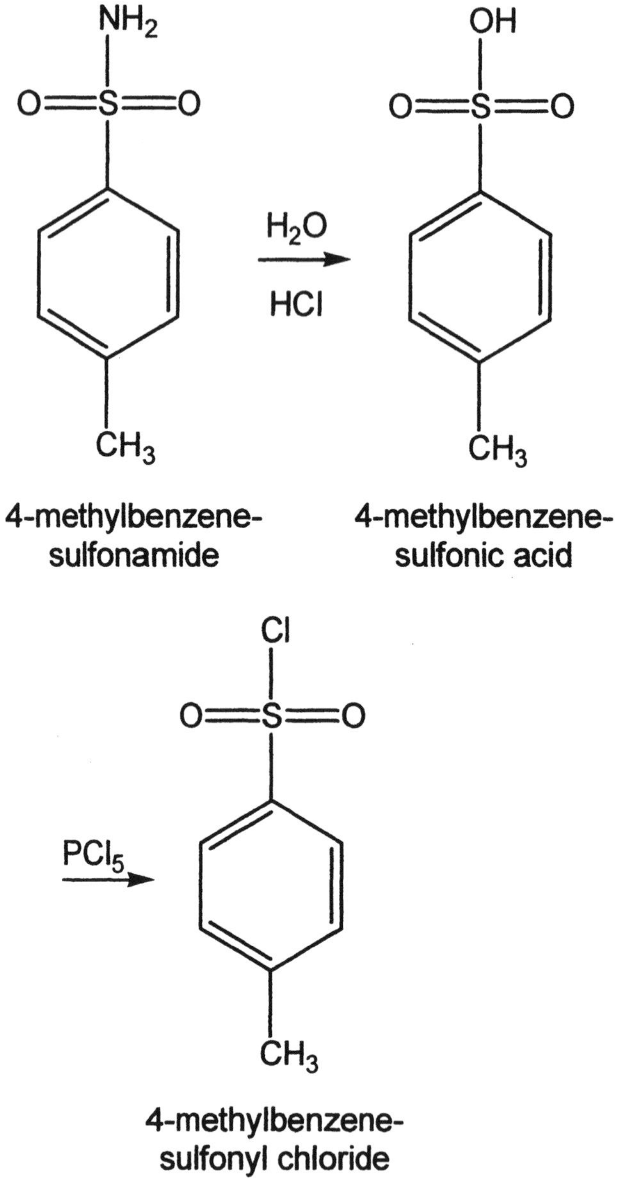
NH2
O=S=O
CH3
H2O
HCl
OH
O=S=O
CH3
4-methylbenzene-
sulfonamide
4-methylbenzene-
sulfonic acid
Cl
O=S=O
PCl5
CH3
4-methylbenzene-
sulfonyl chloride

c.

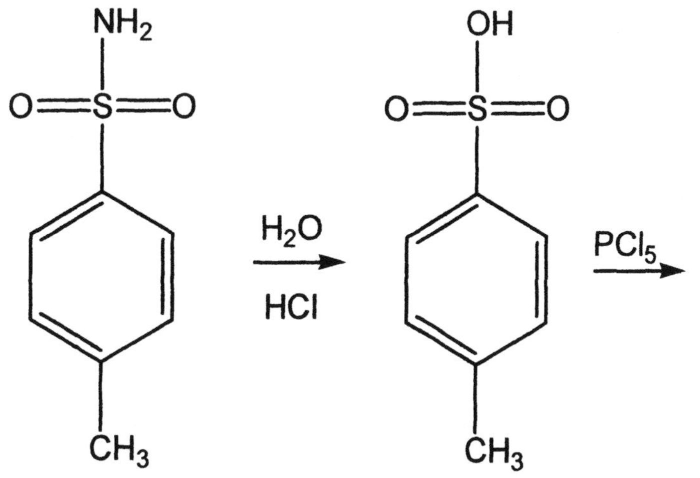

4-methylbenzene-
sulfonamide

H_2O / HCl →

4-methylbenzene-
sulfonic acid

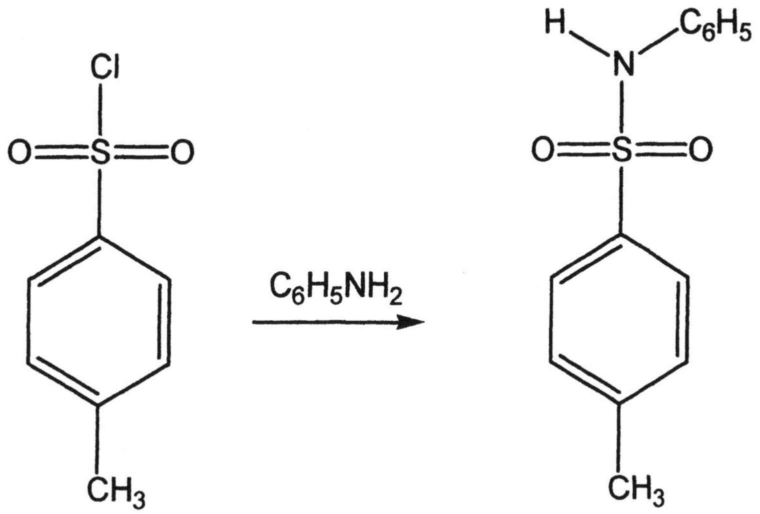

PCl_5 →

4-methylbenzene-
sulfonyl chloride

$C_6H_5NH_2$ →

4-methyl-*N*-phenyl-
benzenesulfonamide

d.

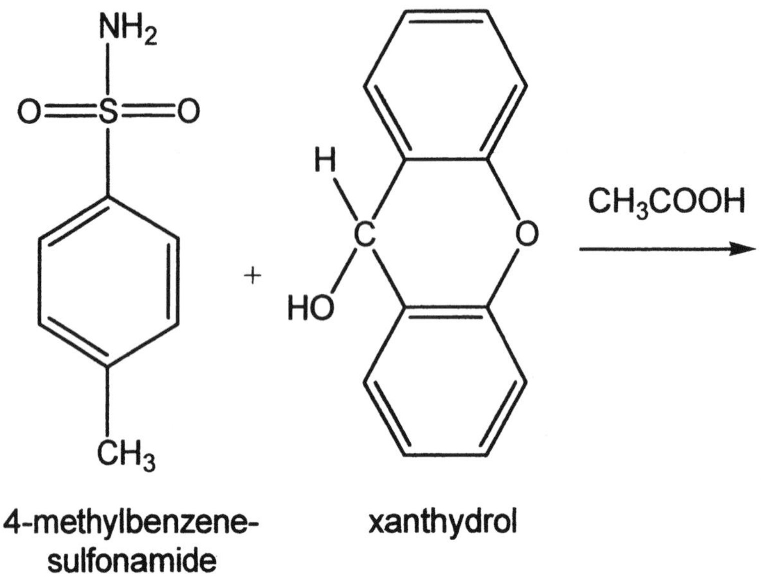

4-methylbenzene-
sulfonamide

xanthydrol

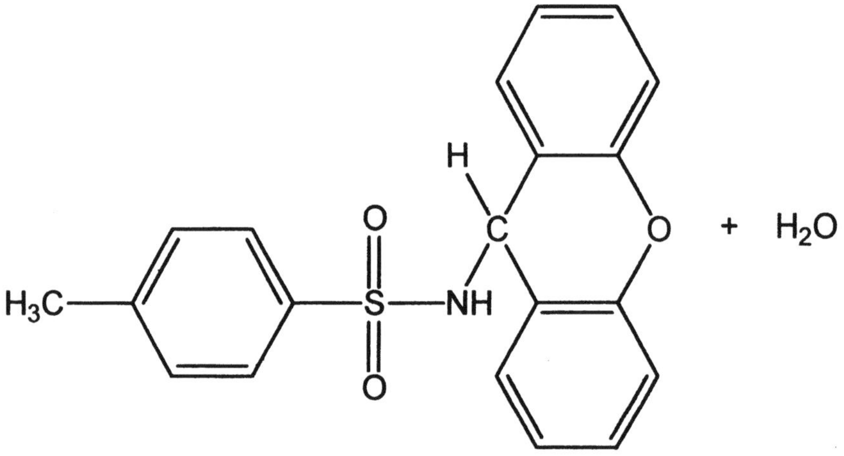

CH_3COOH

N-xanthyl-4-methylbenzenesulfonamide

34. **a.**

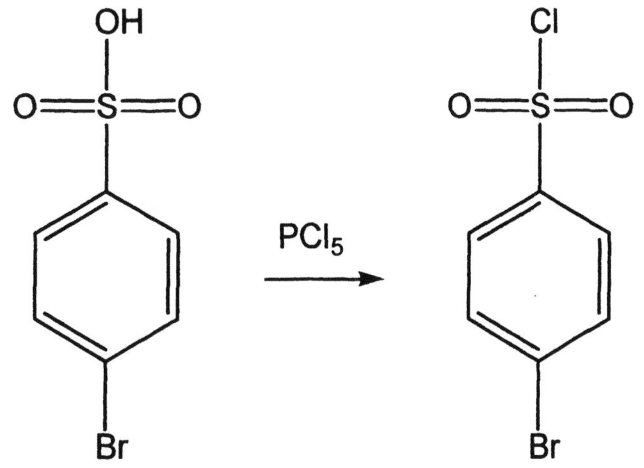

4-bromobenzene-
sulfonic acid

PCl₅

4-bromobenzene-
sulfonyl chloride

b.

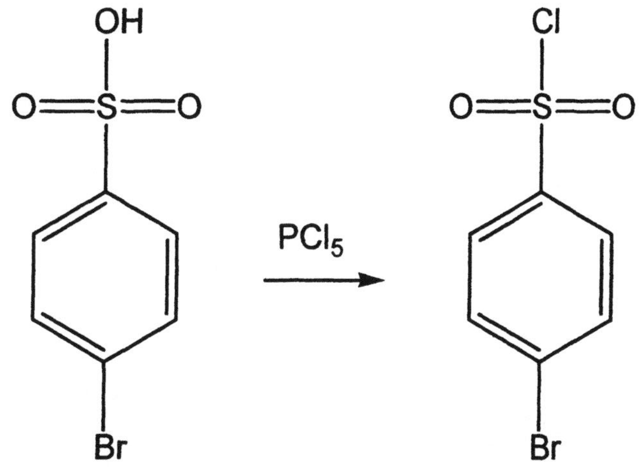

4-bromobenzene-
sulfonic acid

PCl₅

4-bromobenzene-
sulfonyl chloride

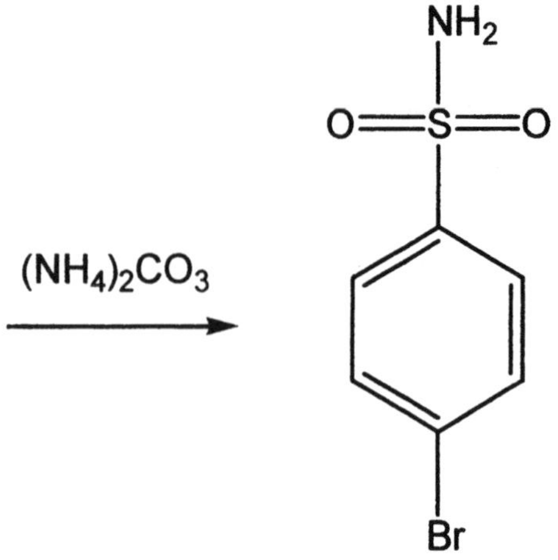

$(NH_4)_2CO_3$

4-bromobenzene-
sulfonamide

c.

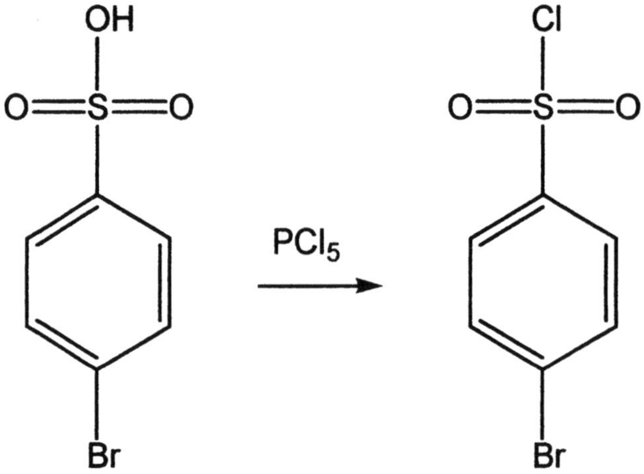

PCl_5

4-bromobenzene-
sulfonic acid

4-bromobenzene-
sulfonyl chloride

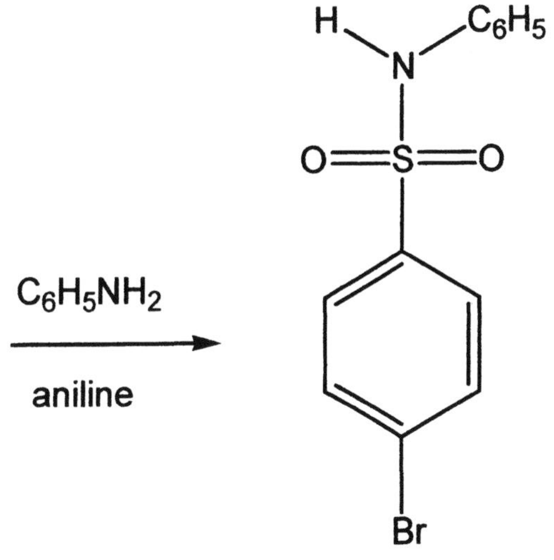

$$\xrightarrow[\text{aniline}]{C_6H_5NH_2}$$

4-bromo-*N*-phenyl-
benzenesulfonamide

d.

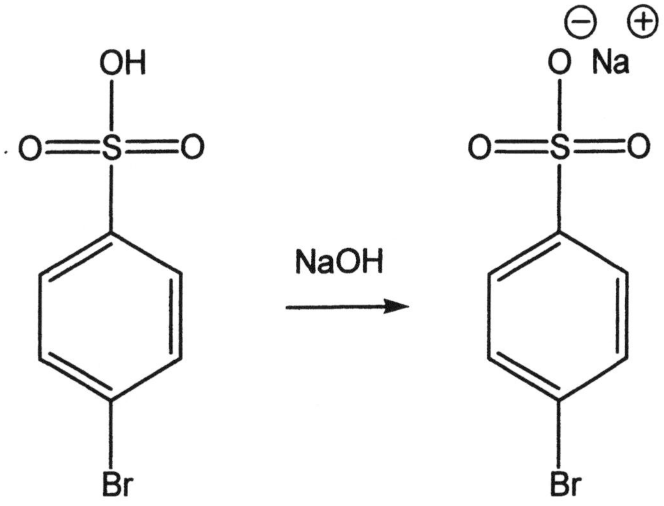

$$\xrightarrow{\text{NaOH}}$$

4-bromobenzene-
sulfonic acid

sodium salt of
4-bromobenzene-
sulfonic acid

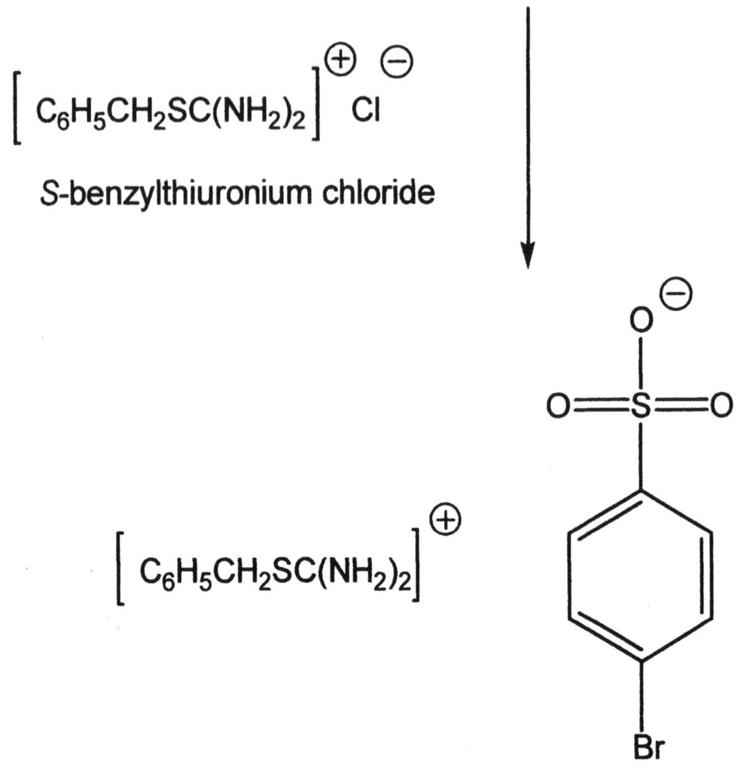

$$\left[C_6H_5CH_2SC(NH_2)_2 \right]^{\oplus} \overset{\ominus}{Cl}$$

S-benzylthiuronium chloride

$$\left[C_6H_5CH_2SC(NH_2)_2 \right]^{\oplus}$$

S-benzylthiuronium 4-bromobenzenesulfonate

e.

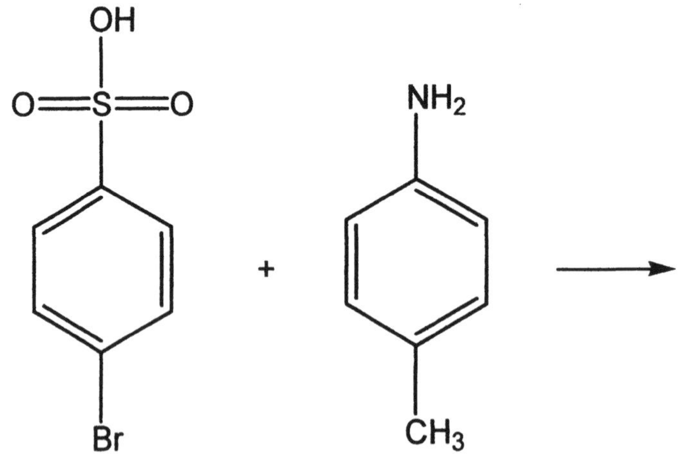

4-bromobenzene-
sulfonic acid

4-toluidine

4-toluidine salt of
4-bromobenzenesulfonic acid

CHAPTER ELEVEN
Structural Problems – Solution Methods and Exercises

Problem Set 1

1. Compound A is in solubility class I, which consists of saturated hydrocarbons, haloalkanes, aryl halides, other deactivated compounds, and diaryl ethers. Since a white precipitate was formed from the addition of silver nitrate to a sodium fusion solution of the unknown, a chlorine is indicated. Thus, compound A is chlorocyclohexane.

   Compound B is in solubility class I, which consists of saturated hydrocarbons, haloalkanes, aryl halides, other deactivated compounds, and diaryl ethers. A negative silver nitrate test indicates the absence of a halogen. Therefore, compound B is cyclohexane.

   Compound C is in solubility class N, which contains alcohols, aldehydes, ketones, esters with one functional group and more than five carbons but fewer than nine carbons, ethers, epoxides, alkenes, alkynes, and some aromatic compounds. A negative silver nitrate test indicates the absence of a halogen. Thus, compound C is diethyl ether.

2. Compound A is in solubility class A_1, A_2, B, MN, N, or I. It does not have a multiple bond (negative bromine test), or an active hydrogen (negative acetyl chloride test). It is not an aldehyde or a ketone (negative 2,4-dinitrophenylhydrazine test). Compound A is not a methyl ketone and it does not have a methyl next to a –CHOH (negative iodoform test). Compound A must be cyclohexane.

   Compound B is in solubility class A_1, A_2, B, MN, N, or I. It contains a multiple bond (positive bromine test), but does not have an active hydrogen (negative acetyl chloride test). It is not an aldehyde or a ketone (negative 2,4-dinitrophenylhydrazine test). Compound B is not a methyl ketone and it does not have a methyl next to a –CHOH (negative iodoform test). Compound B is cyclohexene.

   Compound C is in solubility class S_2, S_A, S_B, or S_1. It contains an active hydrogen (positive acetyl chloride test), but is not an aldehyde or a ketone (negative 2,4-dinitrophenylhydrazine test). It is either a methyl ketone or a methyl next to a –CHOH (positive iodoform test). Since compound C cannot be a ketone, compound C must be an alcohol that gives a positive iodoform test. Therefore, compound C is ethanol.

Compound D is in solubility class S_2, S_A, S_B, or S_1. It does not contain an active hydrogen (negative acetyl chloride test), but is an aldehyde or ketone (positive 2,4-dinitrophenylhydrazine test). It is either a methyl ketone or has a methyl next to a –CHOH (positive iodoform test). From the data, compound D must be acetone (propanone).

Compound E is in solubility class S_2, S_A, S_B, or S_1. It contains an active hydrogen (positive acetyl chloride test), but is not an aldehyde or a ketone (negative 2,4-dinitrophenylhydrazine test). Compound E is not a methyl ketone and it does not have a methyl next to a –CHOH (negative iodoform test). The only possibility for compound E is 2-methyl-2-propanol (*t*-butyl alcohol).

3. Compounds A, B, C are in solubility class S_1 which consists of monofunctional alcohols, aldehydes, ketones, esters, nitriles, and amides with five carbons or fewer. Compounds A, B, C do not contain aldehydes or ketones (negative 2,4-dinitrophenylhydrazine test), but contains an alcohol, phenol, 1^0 amine, or 2^0 amine (positive acetyl chloride test). Compound A is a 1^0 alcohol (positive chromium trioxide test, negative hydrochloric acid, zinc chloride test). Compound B is a 2^0 alcohol (positive chromium trioxide test, positive hydrochloric acid, zinc chloride test). Compound C is a 3^0 alcohol (negative chromium trioxide test, positive hydrochloric acid, zinc chloride test).

Compound A CH_3CH_2OH ethanol

Compound B $CH_3\!\!-\!\!\underset{\underset{\displaystyle CH_3}{|}}{C}HOH$ 2-propanol

$$CH_3$$
$$|$$
Compound C CH_3——C——CH_3 2-methyl-2-propanol
$$|$$
$$OH$$

4. Compounds A, B, and C are in solubility class I, which is composed of saturated hydrocarbons, haloalkanes, aryl halides, other deactivated compounds, and diaryl ethers. Compounds A, B, and C are not aromatic (negative chloroform and aluminum chloride test). Compound A contains chlorine (a white solid with silver nitrate). Compound B contains bromine (a pale-yellow solid with silver nitrate). Compound C contains iodine (a yellow solid with silver nitrate).

Compound A $ClCH_2CH_2CH_2CH_2CH_2CH_2CH_3$

1-chloroheptane

Compound B $BrCH_2CH_2CH_2CH_2CH_2CH_3$

1-bromohexane

Compound C $ICH_2CH_2CH_2CH_2CH_3$

1-iodopentane

5. Compounds A, B, and C are in solubility class S_1, which consists of monofunctional alcohols, aldehydes, ketones, esters, nitriles, and amides with five carbons or fewer. Compounds A, B, and C are not an alcohol, phenol, 1^0 amine, or 2^0 amine (negative acetyl chloride test), or an aldehyde or ketone (negative 2,4-dinitrophenylhydrazine

test and no IR peak between 1600 and 1800), but are nitriles (IR peak around 2250). Compound A contains a double bond or a triple bond (positive bromine test), but does not contain halogen. Compound B does not contain a double bond or a triple bond (negative bromine test), but contains halogen. Compound C does not contain a double bond or a triple bond (negative bromine test), but does not contain halogen.

Compound A $H_2C\!=\!\!\underset{H}{C}\!-\!C\!\equiv\!N$ propenenitrile

Compound B $FCH_2\!-\!C\!\equiv\!N$ fluoroethanenitrile

Compound C $CH_3\!-\!C\!\equiv\!N$ ethanenitrile

Problem Set 2

1. The unknown compound is in solubility class **N**, which consists of
    alcohols, aldehydes, ketones, esters with one functional group and
    more than five carbons but fewer than nine carbons, ethers, epoxides,
    alkenes, alkynes, and some aromatic compounds. The unknown did
    not have an active hydrogen (negative acetyl chloride test), was not
    an aldehyde or ketone (negative 2,4-dinitrophenylhydrazine test), and
    did not contain a double bond or triple bond (negative bromine test).
    The IR spectrum can be interpreted as follows.

frequency	bond	compound type
814	C-H out-of-plane bend	*p*-disubstituted aromatic
1361, 1385	C-H bend	isopropyl split
1455	C-H bend	methyl
2955	C-H stretch	alkyl

Interpretation of the ^{1}H NMR spectrum is listed below.

	chemical shift	splitting	integration	interpretation
(a)	1.165	d	6H	2 CH$_3$ adjacent to CH
(b)	2.180	s	3H	CH$_3$ isolated
(c)	2.737	sept	1H	CH adjacent to 2 CH$_3$
(d)	6.973	s	4H	disubstituted aromatic

4-Isopropyltoluene agrees with the IR and ^{1}H NMR spectra.

4-isopropyltoluene,
p-cymene

The C-13 NMR and DEPT spectra are interpreted below.

	chemical shift	DEPT	interpretation
(a)	20.036	CH$_3$	alkyl
(b)	23.331	CH$_3$	alkyl
(c)	33.069	CH	alkyl
(d)	125.423	CH	aromatic
(e)	128.305	CH	aromatic
(f)	133.940	C	aromatic
(g)	144.766	C	aromatic

4-isopropyltoluene,
p-cymene

2. The unknown compound is in solubility class **N**, which consists of alcohols, aldehydes, ketones, esters with one functional group and more than five carbons but fewer than nine carbons, ethers, epoxides, alkenes, alkynes, and some aromatic compounds. The unknown has an active hydrogen (positive acetyl chloride test), was not an aldehyde or ketone (negative 2,4-dinitrophenylhydrazine test), and is an ether (positive hydroiodic acid test).

The IR spectrum can be interpreted as follows.

frequency	bond	compound type
3319	O-H stretch	alcohol
1508, 1608	C=C stretch	aromatic
1032, 1243	C-O stretch	aryl alkyl ether
1455	C-H bend	methyl
814	C-H out-of-plane bend	p-disubstituted aromatic
2943	C-H stretch	aromatic

Interpretation of the ^{1}H NMR spectrum is listed below.

	chemical shift	splitting	integration	interpretation
(a)	3.602	s	3H	CH_3 isolated
(b)	4.511	s	2H	CH_2 isolated
(c)	5.206	s	1H	OH
(d)	6.813	d	2H	aromatic
(e)	7.267	d	2H	aromatic

4-Methoxybenzyl alcohol matches the spectral information and the classification tests.

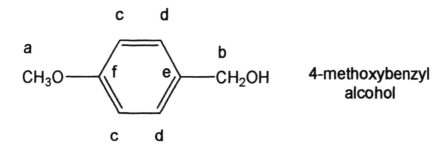

4-methoxybenzyl alcohol

The C-13 NMR and DEPT spectra are interpreted below.

	chemical shift	DEPT	interpretation
(a)	54.647	CH_3	ether
(b)	63.304	CH_2	alcohol
(c)	113.467	CH	aromatic
(d)	128.236	CH	aromatic
(e)	134.109	C	aromatic
(f)	158.608	C	aromatic

4-methoxybenzyl alcohol

4-Methoxybenzyl alcohol reacts with 4-nitrobenzoyl chloride to yield 4-methoxybenzyl 4-nitrobenzoate.

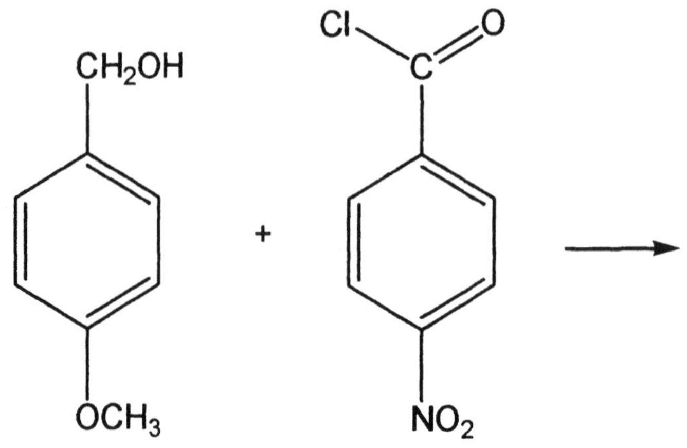

4-methoxybenzyl
alcohol

4-nitrobenzoyl
chloride

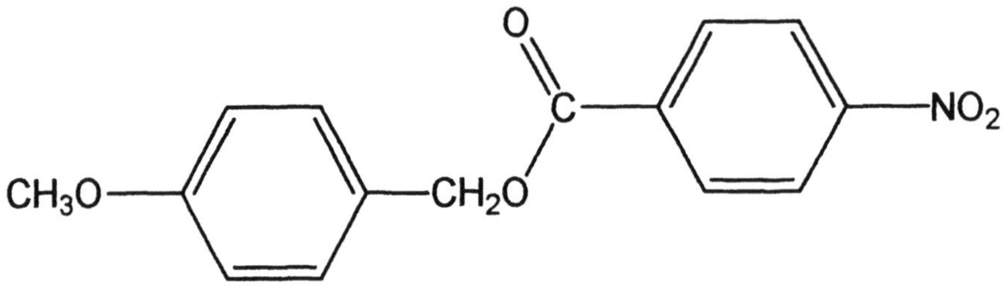

4-methoxybenzyl 4-nitrobenzoate

3.   The unknown is in solubility class A_1, which consists of strong organic acids, carboxylic acids with more than six carbons, phenols with electron-withdrawing groups in the *ortho* and/or *para* position(s); β-diketones, nitro compounds with α-hydrogens. The unknown has an active hydrogen (positive sodium test) and contains a 1° alcohol, 2° alcohol, or aldehyde.

The IR spectrum can be interpreted as follows.

frequency	bond	compound type
3201	O-H stretch	phenol
3060	C-H stretch	aromatic
2837	C-H stretch	aldehyde
2755	C-H stretch	aldehyde
1672	C=O stretch	aryl aldehyde
1390	C-H bend	aldehyde
1378	C-O stretch	phenol
1197	C-O stretch	phenol
761	C-H out-of-plane bend	*o*-disubstituted aromatic
761	O-H bend	phenol

Interpretation of the ^{1}H NMR spectrum is listed below.

	chemical shift	splitting	integration	interpretation
(a)	6.738-6.986	m	2H	aromatic
(b)	7.285-7.514	m	2H	aromatic
(c)	9.720	s	1H	CH isolated
(d)	11.116	bs	1H	-OH

Salicylaldehyde matches the spectra above.

salicylaldehyde

The C-13 NMR and DEPT spectra are interpreted below.

	chemical shift	DEPT	interpretation
(a)	117.464	CH	aromatic
(b)	119.939	CH	aromatic
(c)	120.858	C	aromatic
(d)	133.960	CH	aromatic
(e)	136.926	CH	aromatic
(f)	161.495	C	aromatic
(g)	196.967	CH	aldehyde

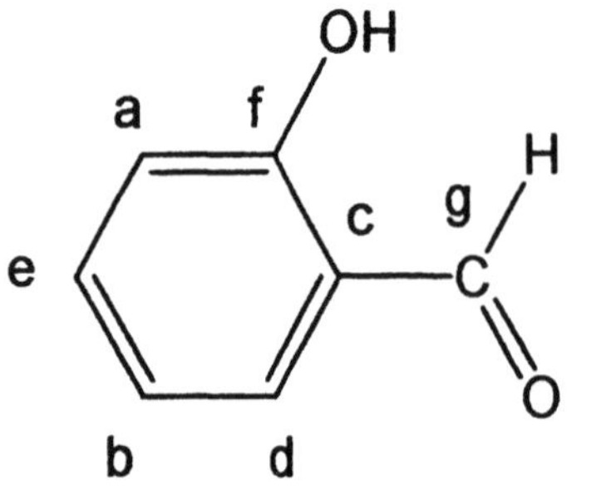

salicylaldehyde

Salicylaldehyde reacts with hydroxylamine hydro-chloride to produce salicylaldehyde oxime.

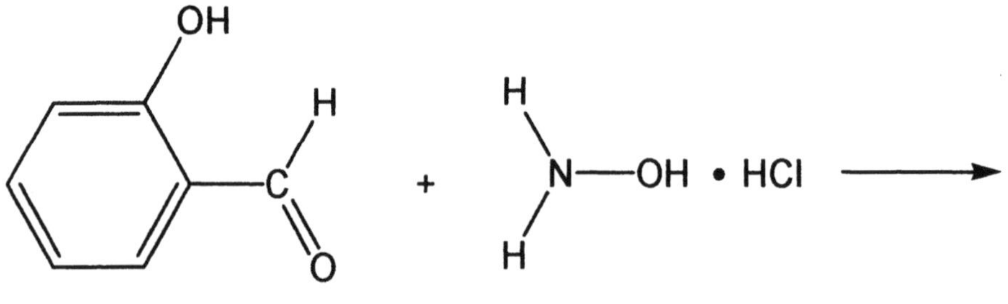

salicylaldehyde

hydroxylamine hydrochloride

salicylaldehyde oxime

Salicylaldehyde reacts with phenyl isocyanate to produce 2-formylphenyl phenyl carbamate.

salicylaldehyde   +   phenyl isocyanate   ⟶

2-formylphenyl phenylcarbamate

4. The unknown is in solubility class S_B, which consists of monofunctional amines with six carbons or fewer. The unknown does not have an active hydrogen (negative acetyl chloride and negative sodium test). The unknown compound is a 3^0 amine (benzene-sulfonyl chloride test).

A summary of the IR spectrum is listed below.

frequency	bond	compound type
1202	C-N stretch	3^0 amine
1378, 1448, 1467	C-H bend	methyl
2790	C-H stretch	alkyl

Interpretation of the ^{1}H NMR spectrum is listed below.

	chemical shift	splitting	integration	interpretation
(a)	0.966	t	3H	CH_3 adjacent to CH_2
(b)	2.442	q	2H	CH_2 adjacent to CH_3

Triethylamine matches the spectral information.

b a

CH_2CH_3

a b

CH_3CH_2——N   triethylamine

CH_2CH_3

b a

The C-13 NMR and DEPT spectra are interpreted below.

	chemical shift	DEPT	interpretation
(a)	11.680	CH_3	alkyl
(b)	46.154	CH_2	amine

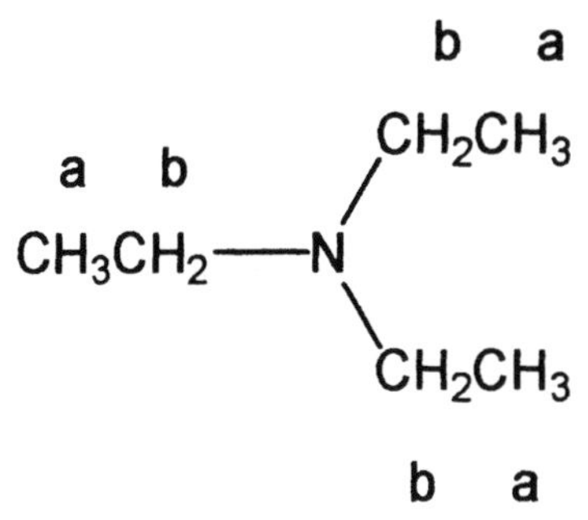

b   a
CH_2CH_3

a   b
CH_3CH_2—N   triethylamine

CH_2CH_3
b   a

Triethylamine reacts with picric acid to form a picrate.

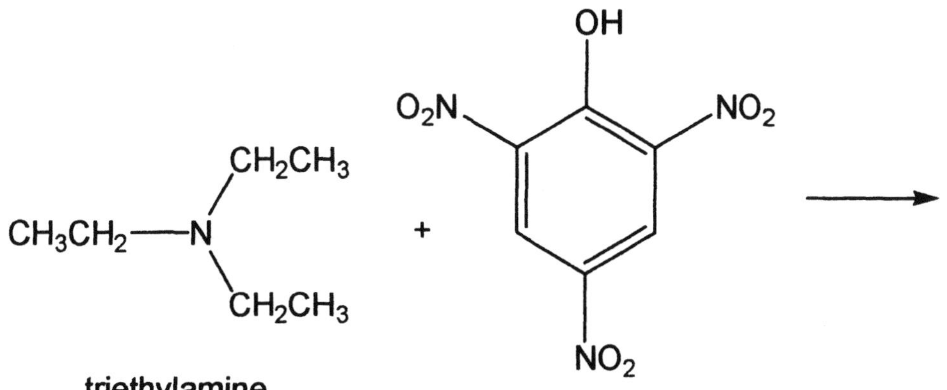

CH_2CH_3

CH_3CH_2—N   +

CH_2CH_3

triethylamine

OH

O_2N NO_2

NO_2

picric acid

→

triethylamine picrate

5. The unknown is in solubility class **N**, which consists of alcohols, aldehydes, ketones, esters with one functional group and more than five but fewer than nine carbons, ethers, epoxides, alkenes, alkynes, and some aromatic compounds. The unknown contains a 1^0 alcohol, 2^0 alcohol, or aldehyde (positive chromium trioxide test); an aldehyde or ketone (positive 2,4-dinitrophenylhydrazine test); an aldehyde (positive Tollens test); and an alkene or alkyne (positive potassium permanganate test). It does not contain an active hydrogen (negative sodium test).

A summary of the IR spectrum is listed below.

frequency	bond	compound type
3060	C-H stretch	aromatic
2813, 2743	C-H stretch	aldehyde
1678	C=O stretch	α,β-unsaturated aldehyde
1678	C=C stretch	trans RCH=CHR
1572, 1490	C=C stretch	aromatic
1390	C-H bend	aldehyde
973	C-H out-of-plane bend	trans RCH=CHR
743, 690	C-H out-of-plane bend	monosubstituted benzene

Interpretation of the ^{1}H NMR spectrum is listed below.

	chemical shift	splitting	integration	interpretation
(a)	6.530, 6.796	d of d	1H	CH split by 2 CHs
(b)	7.224-7.482	m	6H	aromatic plus alkene
(c)	9.688	d	1H	CH adjacent to CH

The C-13 NMR and DEPT spectra are interpreted below.

	chemical shift	DEPT	interpretation
(a)	128.688	2 CH	aromatic
(b)	128.826	CH	aromatic
(c)	129.262	2 CH	aromatic
(d)	131.340	CH	alkene
(e)	134.339	C	aromatic
(f)	152.537	CH	alkene
(g)	193.592	CH	aldehyde

The hydrogens in the ^{1}H NMR spectrum can be labeled as shown.

trans-cinnamaldehyde

The carbons in the ^{13}C NMR spectrum can be labeled as shown.

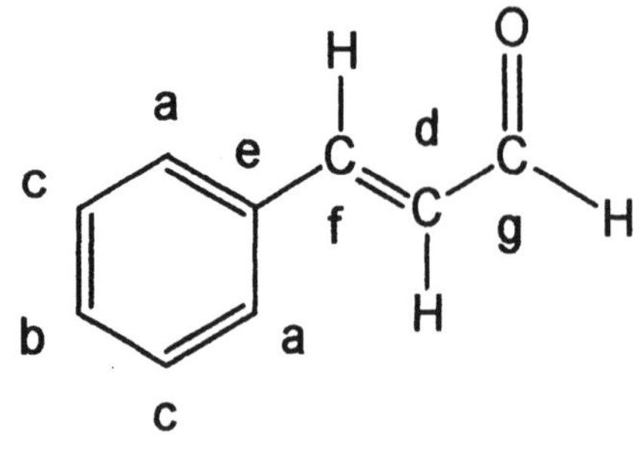

trans-cinnamaldehyde

Trans-cinnamaldehyde reacts with semicarbazide hydrochloride to yield the *trans*-cinnamaldehyde semicarbazone.

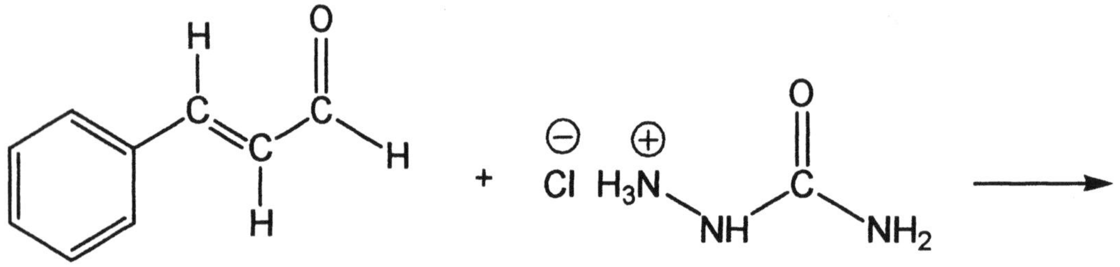

trans-cinnamaldehyde

semicarbazide
hydrochloride

trans-cinnamaldehyde semicarbazone

6.   The unknown is in solubility class **N**, which consists of alcohols, aldehydes, ketones, esters with one functional group and more than five but fewer than nine carbons, ethers, epoxides, alkenes, alkynes, and some aromatic compounds. The unknown does not contain nitrogen (positive ammonium polysulfide and ferric chloride test), sulfur (negative lead acetate test), or halogen (negative silver nitrate test).   It contains an aldehyde or a ketone (positive 2,4-dinitrophenylhydrazine test) and an anhydride, an acyl halide, an ester, an amide, a nitrile, a sulfonic acid, or a sulfonyl chloride (positive hydroxylamine and ferric chloride test).
    A summary of the IR spectrum is listed below.

frequency	bond	compound type
2978	C-H stretch	alkyl
1743	C=O ester	aliphatic ester
1713	C=O stretch	ketone
1372	C-H bend	methyl
1155	C-O stretch	ester
1149	C=O stretch and bend	ketone

Interpretation of the ^{1}H NMR spectrum is listed below.

	chemical shift	splitting	integration	interpretation
(a)	1.225	t	3H	CH_3 adjacent to CH_2
(b)	2.202	s	3H	CH_3 isolated
(c)	3.473	s	2H	CH_2 isolated
(d)	4.138	q	2H	CH_2 adjacent to CH_3

The C-13 NMR and DEPT spectra are interpreted below.

	chemical shift	DEPT	interpretation
(a)	13.285	CH_3	alkyl
(b)	28.910	CH_3	alkyl
(c)	49.175	CH_2	alkyl
(d)	60.248	CH_2	ester C-O
(e)	166.763	C	ester C=O
(f)	200.084	C	ketone

Ethyl acetoacetate is the only structure that matches the spectra above.

The hydrogens in the ^1H NMR spectrum can be labeled as shown.

ethyl acetoacetate

The carbons in the ^{13}C NMR spectrum can be labeled as shown.

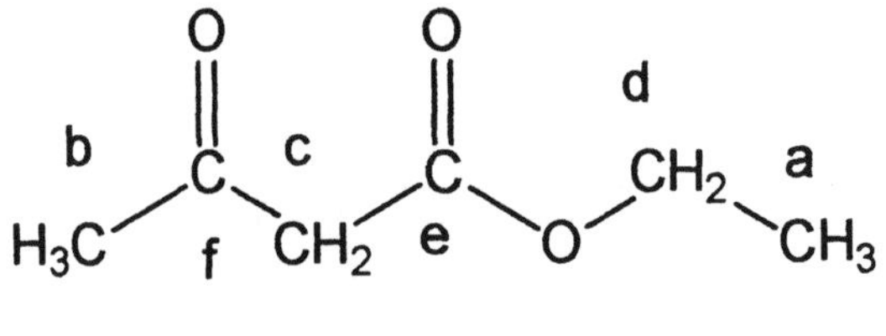

ethyl acetoacetate

Ethyl acetoacetate reacts with semicarbazide hydrochloride to form the semicarbazone.

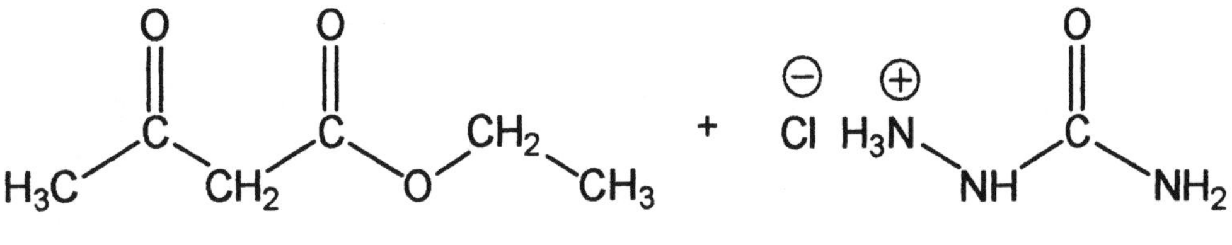

ethyl acetoacetate

semicarbazide
hydrochloride

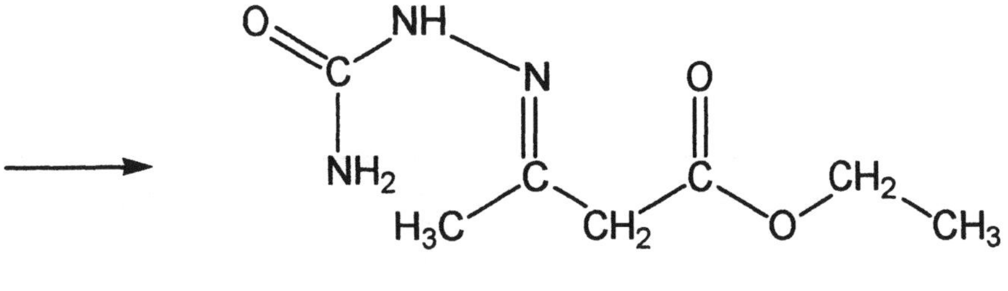

ethyl acetoacetate semicarbazone

Problem Set 3

1.   The unknown is in solubility class **N**, which consists of alcohols, aldehydes, ketones, esters with one functional group and more than five but fewer than nine carbons, ethers, epoxides, alkenes, alkynes, and some aromatic compounds. The unknown is a 1^0 alcohol, a 2^0 alcohol, or an aldehyde (positive Jones reagent), is not an aldehyde (negative Schiff's reagent), is not an aldehyde or a ketone (negative sodium bisulfite), but contains an active hydrogen (positive sodium test).

A summary of the IR spectrum is listed below.

frequency	bond	compound type
3319	O-H stretch	alcohol
3013	C-H stretch	aromatic
1014	C-O stretch	1^0 alcohol
732, 697	C-H out-of-plane bend	monosubstituted aromatic

Interpretation of the ^{1}H NMR spectrum is listed below.

	chemical shift	splitting	integration	interpretation
(a)	4.351	d	2H	CH_2 adjacent to H
(b)	5.388	t	1H	OH
(c)	7.133	s	5H	monosubstituted aromatic

From the spectra above, benzyl alcohol is the unknown.

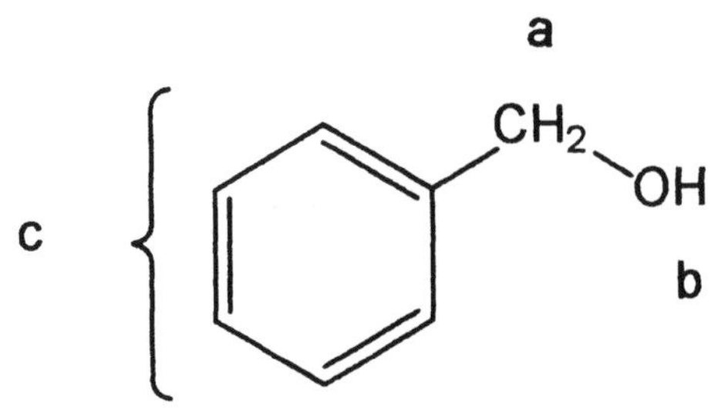

a

CH_2—OH   benzyl alcohol

c

b

The C-13 NMR and DEPT spectra are interpreted below.

	chemical shift	DEPT	interpretation
(a)	64.357	CH_2	alcohol
(b)	127.174	CH	aromatic
(c)	127.434	CH	aromatic
(d)	128.524	CH	aromatic
(e)	141.326	C	aromatic

The carbons in the ^{13}C NMR spectrum can be labeled as shown.

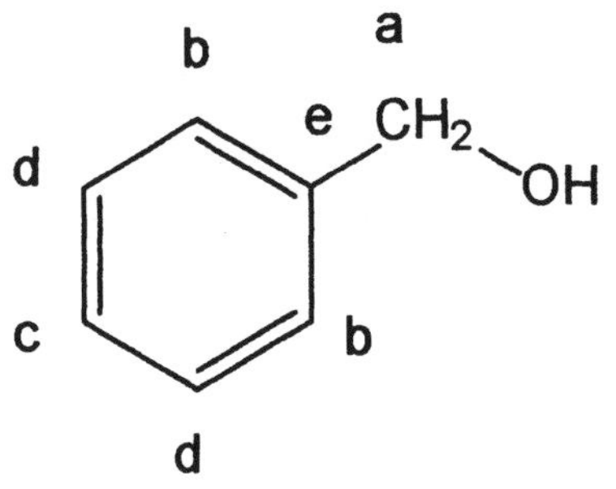

b   a

d   e CH_2—OH   benzyl alcohol

c   b

d

Benzyl alcohol reacts with 3,5-dinitrobenzoyl chloride to yield benzyl 3,5-dinitrobenzoate.

benzyl alcohol   +   3,5-dinitrobenzoyl
                         chloride

benzyl 3,5-dinitrobenzoate

2.   The unknown is in solubility class S_A, which consists of monofunctional carboxylic acids with five carbons or fewer. The unknown contains a carboxylic acid (positive sodium bicarbonate test).

A summary of the IR spectrum is listed below.

frequency	bond	compound type
2978	O-H stretch	carboxylic acid
1707	C=O stretch	aliphatic carboxylic acid
1414	O-H bend	carboxylic acid
1384, 1366	C-H bend	isopropyl split
1237	C-O stretch	carboxylic acid

Interpretation of the ^{1}H NMR spectrum is listed below.

	chemical shift	splitting	integration	interpretation
(a)	1.169	d	6H	2CH$_3$ adjacent to CH
(b)	2.558	sept	1H	CH adjacent to 2CH$_3$
(c)	11.857	s	1H	OH

2-Methylpropanoic acid matches the spectral information above.

2-methylpropanoic acid

The C-13 NMR and DEPT spectra are interpreted below.

	chemical shift	DEPT	interpretation
(a)	17.721	CH_3	alkyl
(b)	33.203	CH	alkyl
(c)	182.998	C	carboxylic acid

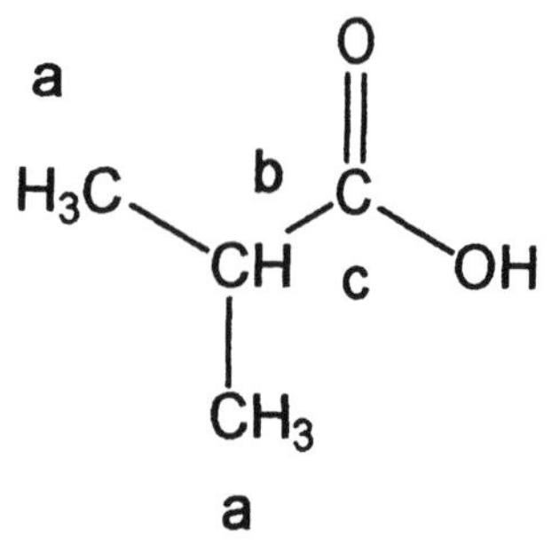

2-methylpropanoic acid

2-Methylpropanoic acid reacts with thionyl chloride to form the 2-methylpropanoyl chloride, which reacts with ammonia to yield the 2-methylpropanamide.

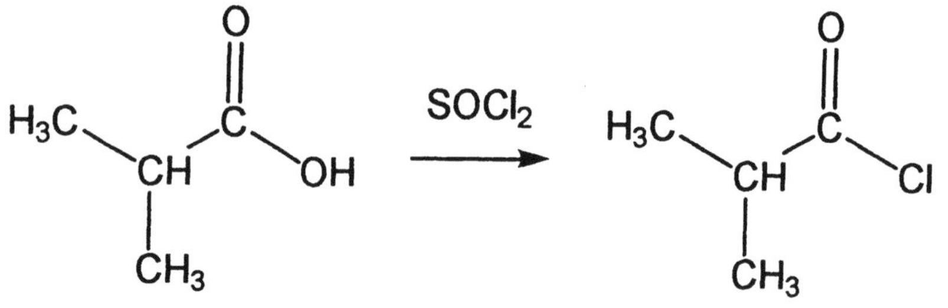

2-methylpropanoic acid

$SOCl_2$

2-methylpropanoyl chloride

NH_3

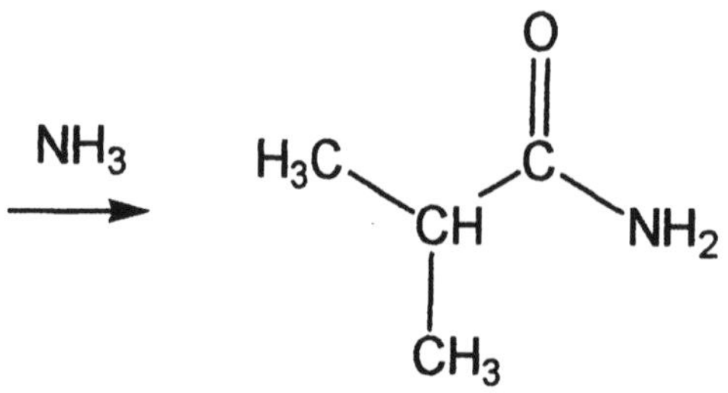

2-methylpropanamide

3. The unknown compound is in solubility class N, which consists of alcohols, aldehydes, ketones, esters with one functional group and more than five carbons but fewer than nine carbons, ethers, epoxides, alkenes, alkynes, and some aromatic compounds. The unknown does not contain an aldehyde or ketone (negative 2,4-dinitrophenylhydrazine test), an active hydrogen (negative acetyl chloride test), or an alkene or alkyne (negative potassium permanganate test and negative bromine test).

A summary of the IR spectrum is listed below.

frequency	bond	compound type
3060	C-H stretch	aromatic
2942, 2825	C-H stretch	alkyl
1596, 1490	C=C stretch	aromatic
1243, 1038	C-O stretch	aryl alkyl ether
756, 685	C-H out-of-plane bend	monosubstituted aromatic

Interpretation of the ^{1}H NMR spectrum is listed below.

	chemical shift	splitting	integration	interpretation
(a)	3.439	s	3H	CH_3 isolated
(b)	6.719-7.280	m	5H	monosubstituted aromatic

Anisole matches the spectral information above.

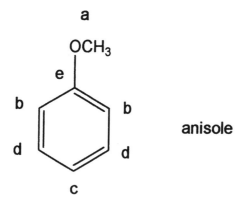

anisole

The C-13 NMR and DEPT spectra are interpreted below.

	chemical shift	DEPT	interpretation
(a)	53.969	CH_3	ether
(b)	113.518	CH	aromatic
(c)	120.098	CH	aromatic
(d)	129.014	CH	aromatic
(e)	159.486	C	aromatic

anisole

Anisole reacts with chlorosulfonic acid to give 4-methoxy-benzenesulfonyl chloride, which is treated with ammonium hydroxide to form 4-methoxybenzenesulfonamide.

4-methoxybenzenesulfonyl
chloride

4-methoxybenzene-
sulfonamide

4. The unknown compound is in solubility class N, which consists of alcohols, aldehydes, ketones, esters with one functional group and more than five carbons but fewer than nine carbons, ethers, epoxides, alkenes, alkynes, and some aromatic compounds. The unknown is not an aldehyde or a ketone (negative 2,4-dinitrophenylhydrazine test), does not have an active hydrogen (negative acetyl chloride test), contains oxygen (negative potassium thiocyanate). The unknown is an alkene or alkyne (positive bromine and potassium permanganate test).

A summary of the IR spectrum is listed below.

frequency	bond	compound type
3284	C-H stretch	alkyne
3013	C-H stretch	aromatic
2919	C-H stretch	alkyl
2120	C≡C stretch	alkyne
1460	C-H bend	methylene
744, 696	C-H out-of-plane bend	monosubstituted aromatic

Interpretation of the ^{1}H NMR spectrum is listed below.

	chemical shift	splitting	integration	interpretation
(a)	1.825-1.926	m	1H	CH isolated
(b)	2.086-2.400	m	2H	CH_2 adjacent to CH_2
(c)	2.567-2.821	m	2H	CH_2 adjacent to CH_2
(d)	7.101	s	5H	monosubstituted aromatic

The C-13 NMR and DEPT spectra are interpreted below.

	chemical shift	DEPT	interpretation
(a)	20.515	CH_2	alkyl
(b)	34.880	CH_2	alkyl
(c)	69.495	C	alkyne
(d)	83.787	CH	alkyne
(e)	126.367	CH	aromatic
(f)	128.440	2CH	aromatic
(g)	128.440	2CH	aromatic
(h)	140.486	C	aromatic

4-Phenyl-1-butyne matches the spectral information given above.

From the HETCOR spectrum, hydrogen *a* is attached to carbon *c*; hydrogen *b* is attached to carbon *a*; hydrogen *c* is attached to carbon *b*; and hydrogen *d* is attached to carbons *a*, *b*, and *c*.

The hydrogens can be labeled as follows.

4-phenyl-1-butyne

The carbons can be labeled as follows:

$$b \quad a \quad d \quad c$$

$$CH_2{-}CH_2{-}C{\equiv}CH$$

4-phenyl-1-butyne

5. The unknown is in solubility class B, which includes aliphatic amines with eight or more carbons; anilines; and some ethers. The unknown has an active hydrogen (positive acetyl chloride test) and is a 1^0 amine (benzenesulfonyl chloride test).

A summary of the IR spectrum is listed below.

frequency	bond	compound type
3354, 3284	N-H stretch	1° amine
3025	C-H stretch	aromatic
2966	C-H stretch	alkyl
1596	N-H bend	1° amine
1449, 1373	C-H bend	methyl
1020	C-N stretch	1° aliphatic amine
761, 703	C-H out-of-plane bend	monosubstituted aromatic

Interpretation of the ^{1}H NMR spectrum is listed below.

	chemical shift	splitting	integration	interpretation
(a)	1.211	d	3H	CH_3 adjacent to CH
(b)	1.832	s	2H	NH_2
(c)	3.883	q	1H	CH adjacent to CH_3
(d)	7.196	s	5H	monosubstituted aromatic

The C-13 NMR and DEPT spectra are interpreted below.

	chemical shift	DEPT	interpretation
(a)	25.880	CH_3	alkyl
(b)	51.143	CH	amino
(c)	125.853	CH	aromatic
(d)	126.479	2CH	aromatic
(e)	128.250	2CH	aromatic
(f)	148.350	CH	aromatic

α-Methylbenzylamine matches the spectral information above. The hydrogens are labeled on the structure below.

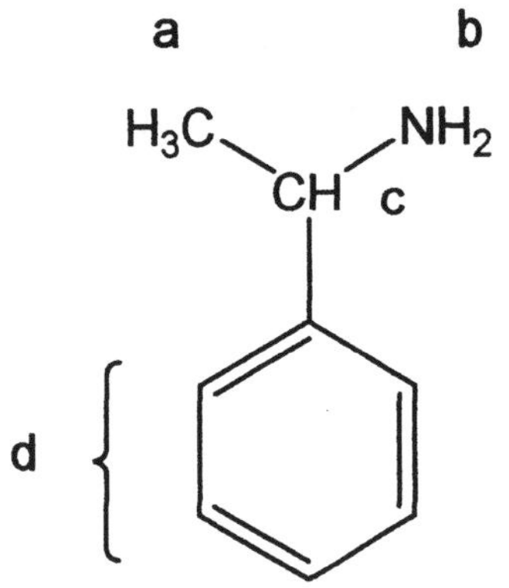

α-methylbenzylamine

The carbons are labeled on the structure below.

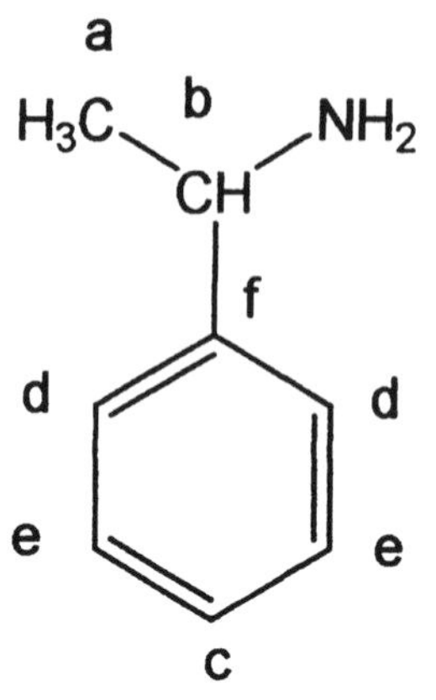

α-methylbenzylamine

6.   The unknown is in solubility class S_B, which consists of monofunctional amines with six or fewer carbons. It has an active hydrogen (positive sodium test) and is a 2^0 amine (benzenesulfonyl chloride test and positive nitrous acid)
A summary of the IR spectrum is listed below.

frequency	bond	compound type
3307	N-H stretch	2^0 amine
2954	C-H stretch	alkyl
1455	C-H bend	methyl
1384, 1367	C-H bend	isopropyl split
1173	C-N stretch	2^0 amine

Interpretation of the ^{1}H NMR spectrum is listed below.

	chemical shift	splitting	integration	interpretation
(a)	0.561	s	1H	NH
(b)	0.973	d	12H	2 ($2CH_3$ adjacent to CH)
(c)	2.867	sept	2H	2 (2CH adjacent to CH_3)

The C-13 NMR and DEPT spectra are interpreted below.

	chemical shift	DEPT	interpretation
(a)	23.184	CH_3	alkyl
(b)	44.789	CH	amine

Diisopropylamine matches the spectral information above. The hydrogens are labeled on the structure below.

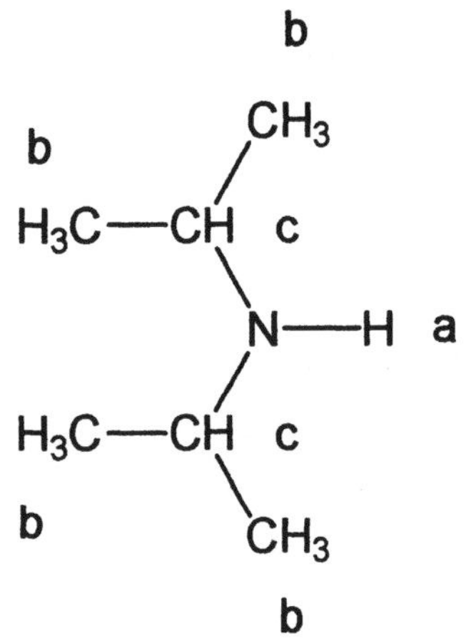

diisopropylamine

The carbons are labeled on the structure below.

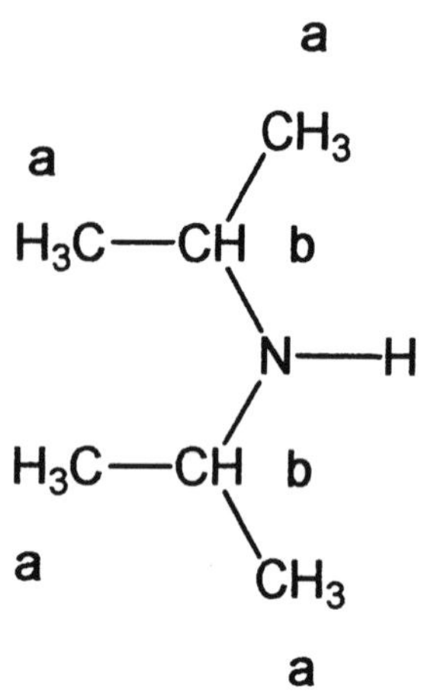

a

CH₃

a

H₃C—CH   b

N—H

H₃C—CH   b

a

CH₃

a

diisopropylamine

Problem Set 4

1. The unknown is in solubility class **N**, which consists of alcohols, aldehydes, ketones, esters with one functional group and more than five but fewer than nine carbons, ethers, epoxides, alkenes, alkynes, some aromatic compounds (especially those with activating groups). The unknown does not have an active hydrogen (negative sodium test) and is not an alkene or alkyne (negative potassium permanganate test).

    4-Heptanone matches the criteria above. 4-Heptanone reacts with 2,4-dinitrophenylhydrazine to form the 2,4-dinitrophenylhydrazone.

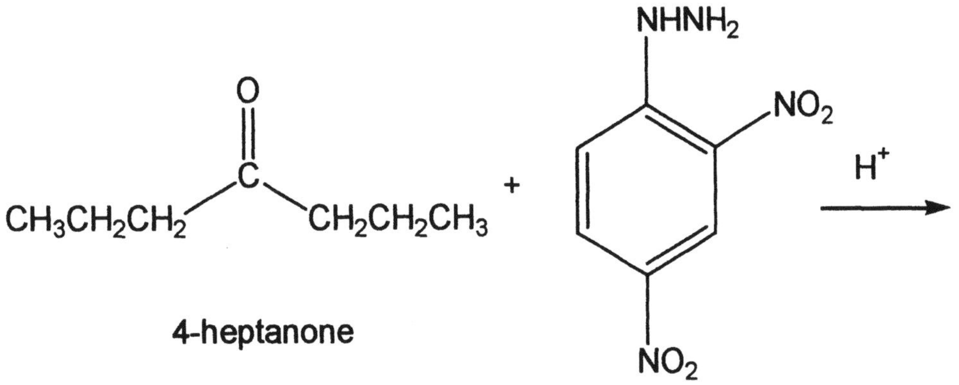

4-heptanone

2,4-dinitrophenylhydrazine

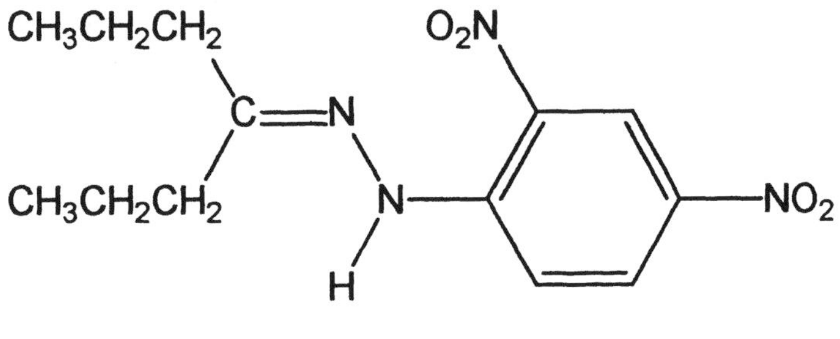

2,3-dinitrophenylhydrazine of 4-heptanone

2.   The unknown is in solubility class S_1, which consists of monofunctional alcohols, aldehydes, ketones, esters, nitriles, and amides with five carbons or fewer. The unknown contains nitrogen (positive sodium fusion filtrate with 10% ammonium polysulfide, 5% HCl, and 5% $FeCl_3$), does not have an active hydrogen (negative acetyl chloride test), is not an aldehyde or ketone (negative 2,4-dinitrophenylhydrazine test).

     N-Methylmethanamide matches the criteria above with a benzamide derivative.

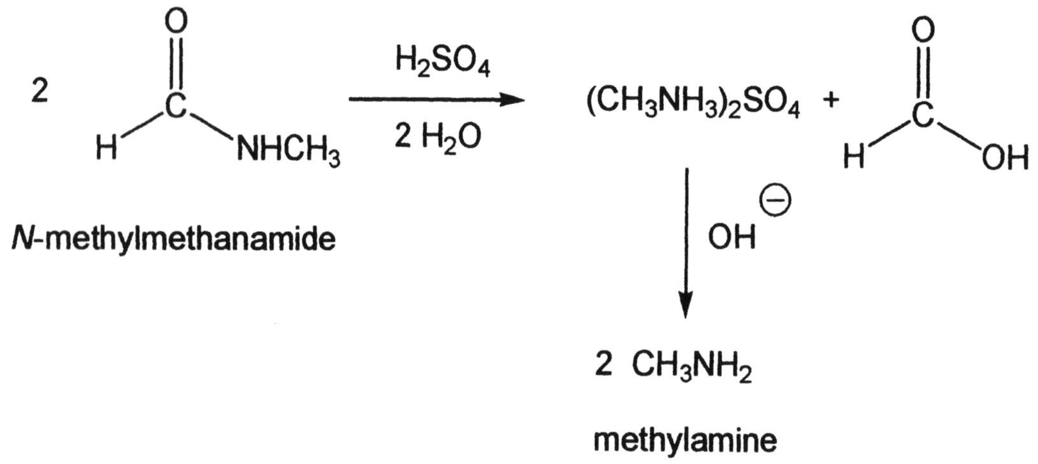

N-methylmethanamide

2 CH_3NH_2

methylamine

2 CH_3NH_2 +

methylamine   benzoyl chloride

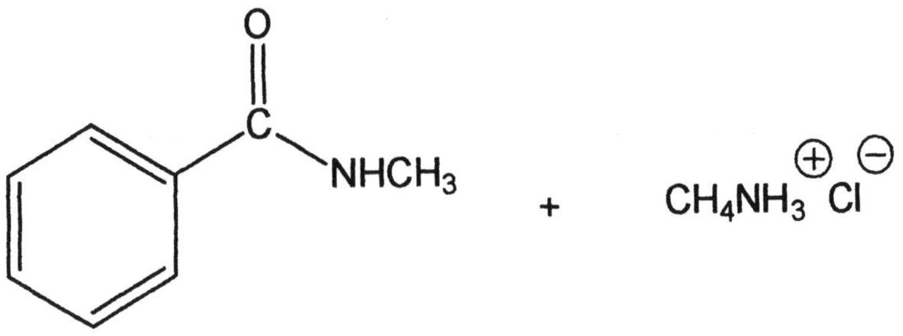

N-methylbenzamide

3.   The unknown is in solubility class S$_1$, which consists of monofunctional alcohols, aldehydes, ketones, esters, nitriles, and amides with five carbons or fewer. The unknown has a chlorine (positive silver nitrate test), is not an aldehyde or ketone (negative 2,4-dinitrophenylhydrazine test), and contains an active hydrogen (positive sodium test).

   2-Chloroethanol reacts with 4-nitrobenzoyl chloride to yield the 4-nitrobenzoate derivative.

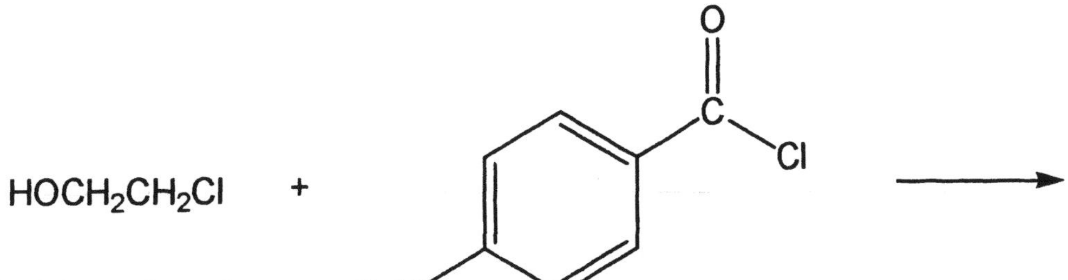

HOCH$_2$CH$_2$Cl   +

2-chloroethanol

4-nitrobenzoyl chloride

2-chloroethyl-4-nitrobenzoate

4. The unknown is in solubility class A_2, which consists of weak organic acids; phenols, enols, oximes, imides, sulfonamides, thiophenols, all with more than five carbons; β-diketones; nitro compounds with α-hydrogens. The unknown contains sulfur (positive sodium fusion filtrate with lead acetate test), nitrogen (positive sodium fusion filtrate with 10% ammonium polysulfide, 5% HCl, and 5% $FeCl_3$), and a nitro group (positive ferrous hydroxide test). The unknown does not have an active hydrogen (negative acetyl chloride test).

    The unknown must be 3-nitrobenzenesulfonamide. It reacts with phosphorus pentachloride, followed by aniline to form 3-nitro-*N*-phenylbenzenesulfonamide.

3-nitrobenzenesulfonamide   3-nitrobenzenesulfonic acid

3-nitrobenzenesulfonyl chloride

3-nitro-*N*-phenyl-
benzenesulfonamide

5. Compound A is in solubility class A$_2$, which consists of weak organic acids: phenols, enols, oximes, imides, sulfonamides, thiophenols, all with more than five carbons; β-diketones (1,3-diketones); and nitro compounds with α-hydrogens. Compound A has an active hydrogen (positive sodium test), contains a phenol or alcohol (positive ceric ammonium nitrate test), and a 1° amine (benzenesulfonyl chloride test). Compound A does not contain a halogen (negative silver nitrate test), a nitro group (negative ferrous hydroxide test), a carboxylic acid (negative sodium bicarbonate test).

Compound A is 3,4-diaminophenol. Compound A is treated with three equivalents of benzoyl chloride to produce the triacetyl derivative of 3,4-diaminophenol.

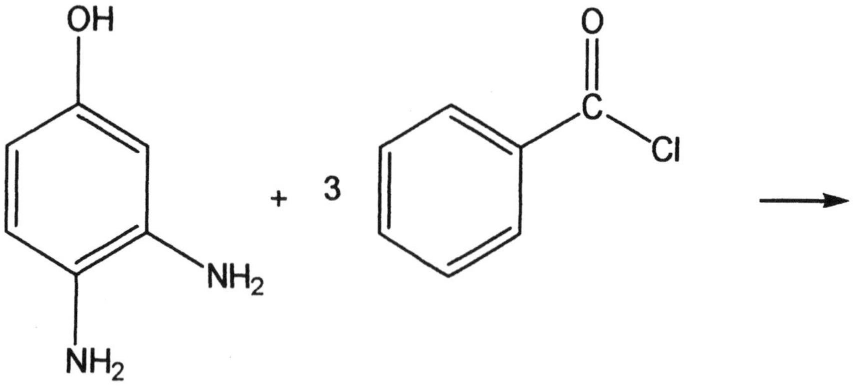

Compound A:
3,4-diaminophenol

benzoyl chloride

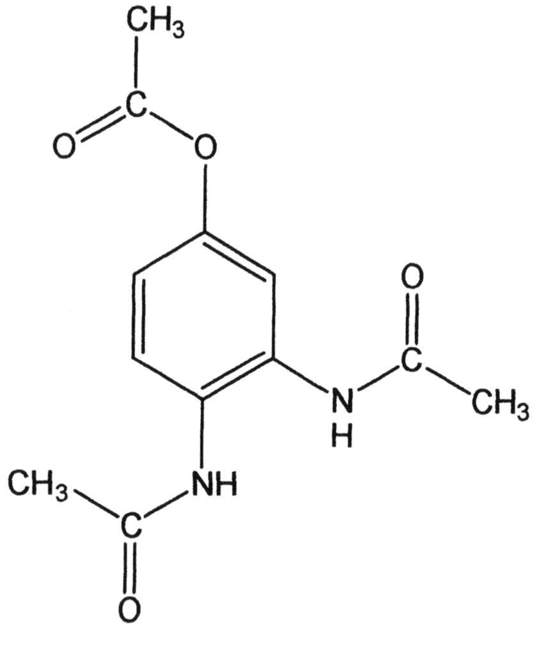

Compound B:
triacetyl derivative
of 3,4-diaminophenol

6. An unknown is in solubility class A$_2$, which consists of weak organic acids: phenols, enols, oximes, imides, sulfonamides, thiophenols, all with more than five carbons; β-diketones (1,3-diketones); nitro

compounds with α-hydrogens. The unknown does not contain sulfur, (negative sodium fusion test with lead acetate), or halogen (negative silver nitrate test). The unknown contains a phenol or alcohol (positive ceric ammonium nitrate test); a 1° alcohol, a 2° alcohol, or an aldehyde (positive chromium trioxide test); an active hydrogen (positive acetyl chloride test); and an aldehyde or a ketone (positive 2,4-dinitrophenylhydrazine test).

The unknown is vanillin.

4-hydroxy-3-methoxy-
benzaldehyde,
vanillin

7. Compound A is in solubility class **N**, which consists of alcohols, aldehydes, ketones, esters with one functional group and more than five but fewer than nine carbons, ethers, epoxides, alkenes, alkynes, some aromatic compound (especially those with deactivating groups). Compound A does not contain an alkene or alkyne (negative potassium permanganate test), an aldehyde or ketone (negative 2,4-dinitrophenylhydrazine test), or an active hydrogen (negative sodium test).

Compound B is in solubility class **A₁**, which consists of strong organic acids: carboxylic acids with more than six carbons; phenols with electron-withdrawing groups in the ortho and/or para position(s); β-diketones (1,3-diketones). Compound B contains a carboxylic acid (positive sodium bicarbonate test), but does not contain an active hydrogen (negative sodium test), or an aldehyde or ketone (negative 2,4-dinitrophenylhydrazine test).

Compound C is in solubility class A_2, which consists of weak organic acids: phenols, enols, oximes, imides, sulfonamides, thiophenols, all with more than five carbons; β-diketones (1,3-diketones); nitro compounds with α-hydrogens. Compound C does not contain a carboxylic acid (negative sodium bicarbonate test) or a nitro group (negative ferrous hydroxide test). Compound C contains an active hydrogen (positive sodium test).

Compound A is phenyl benzoate, Compound B is benzoic acid, and Compound C is phenol.

Compound A:
phenyl benzoate

Compound B:
benzoic acid

Compound C:
phenol

8.   The unknown is in solubility class N, which consists of alcohols, aldehydes, ketones, esters with one functional group and more than five but fewer than nine carbons, ethers, epoxides, alkenes, alkynes,

some aromatic compound (especially those with deactivating groups). The unknown does not contain a $1°$ alcohol, a $2°$ alcohol, or an aldehydes (negative chromium trioxide test); an active hydrogen (negative sodium test); an aldehyde (negative Schiff test); an aldehyde or a ketone (negative 2,4-dinitrophenylhydrazine test); an alkene or alkyne (negative bromine test); or an anhydride, an acyl halide, an ester, a nitrile, or an amide (negative hydroxamic acid and ferric chloride test). The unknown contains oxygen (positive ferric ammonium sulfate test).

The unknown is 1,4-dimethoxybenzene.

1,4-dimethoxybenzene

Problem Set 5

1.   The unknown is in solubility class B, which consists of aliphatic amines with eight or more compounds; anilines (only one phenyl group attached to nitrogen); some ethers.

      N-Ethylaniline (Compound A) reacts with benzoyl chloride to produce *N*-ethyl-*N*-phenylbenzamide (Compound B), which is saponified to make sodium benzoate and *N*-ethylaniline (Compound A).   Acidification of the sodium benzoate yielded benzoic acid (Compound C), which is treated with thionyl chloride to form benzoyl chloride (Compound D).   Benzoyl chloride (Compound D) was reacted with 4-toluidine to produce *N*-4-tolylbenzamide (Compound E).

Compound A
N-ethylaniline

benzoyl chloride

Compound B
N-ethyl-*N*-phenylbenzamide

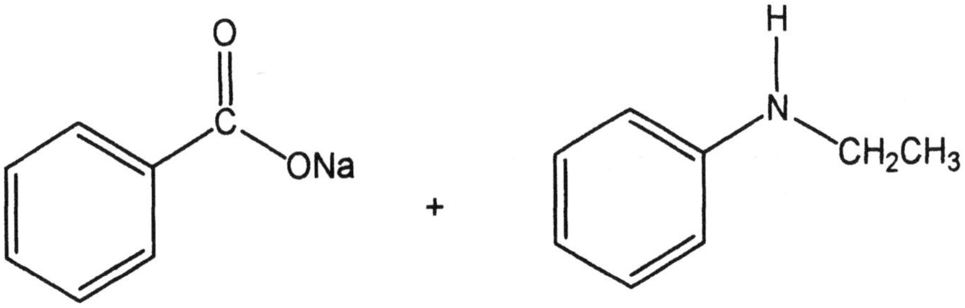

H

N
CH₂CH₃

sodium benzoate

+

Compound A
N-ethylaniline

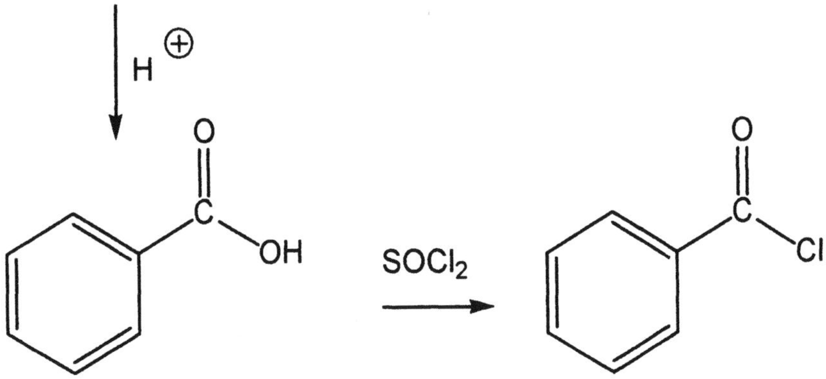

H⁺

SOCl₂

Compound C
benzoic acid

Compound D
benzoyl chloride

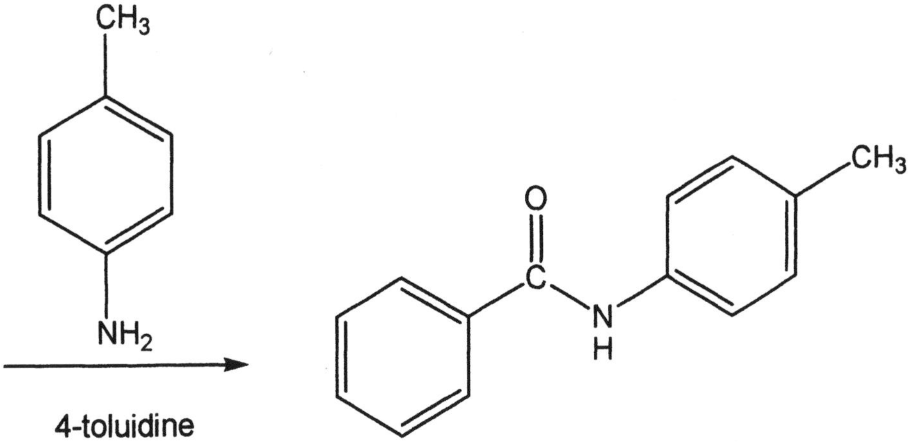

CH₃

NH₂

4-toluidine

CH₃

Compound E
N-4-tolylbenzamide

2. Compound A is in solubility class S$_1$, which consists of monofunctional alcohols, aldehydes, ketones, esters, nitriles, and amides with five carbons or fewer. Compound A does not contain nitrogen (negative ammonium polysulfide, hydrochloric acid, and ferric chloride test) but has an active hydrogen (positive acetyl chloride test). Compound A is not an aldehyde or ketone (negative 2,4-dinitrophenylhydrazine test).

   Compound B contains chlorine (positive silver nitrate test).

   Compound C is in solubility class S$_1$, which consists of monofunctional alcohols, aldehydes, ketones, esters, nitriles, and amides with five carbons or fewer. Compound C does not contain an active hydrogen (negative acetyl chloride test) or a labile halogen (negative silver nitrate test). Compound C is not an aldehyde or a ketone.

   1-Propanol (Compound A) reacts with propanoyl chloride (Compound B) to yield propyl propanoate (Compound C). Propyl propanoate (Compound C) reacts with hydrazine to produce propanoic acid hydrazide (Compound D).

$$CH_3CH_2CH_2OH \quad + \qquad \qquad \overset{\displaystyle O}{\underset{CH_3CH_2}{\overset{\|}{C}}}\diagdown Cl \qquad \longrightarrow$$

Compound A	Compound B
1-propanol	propanoyl chloride

$$\underset{CH_3CH_2}{\overset{\displaystyle \overset{O}{\|}}{C}}\diagdown OCH_2CH_2CH_3 \qquad \xrightarrow{\;NH_2NH_2\;}$$

Compound C
propyl propanoate

Compound D
propanoic acid hydrazide

3.   Compound A is in solubility class MN, which consists of miscellaneous neutral compounds containing nitrogen or sulfur and having more than five carbon atoms. Compound A contains sulfur (positive lead acetate test), nitrogen (positive ammonium polysulfide, hydrochloric acid, and ferric chloride), and chlorine (positive for silver nitrate).

   4-Methyl-3-nitrobenzenesulfonyl chloride (Compound A) is hydrolyzed to 4-methyl-3-nitrobenzenesulfonic acid (Compound B). 4-Methyl-3-nitrobenzenesulfonic acid (Compound B) is treated with phosphorus pentachloride to give 4-methyl-3-nitrobenzenesulfonyl chloride (Compound A). 4-Methyl-3-nitrobenzenesulfonyl chloride (Compound A) is reacted with ammonium hydroxide to give 4-methyl-3-nitrobenzenesulfonamide (Compound C).

Compound A
4-methyl-3-nitrobenzenesulfonyl
chloride

Compound B
4-methyl-3-nitrobenzene-
sulfonic acid

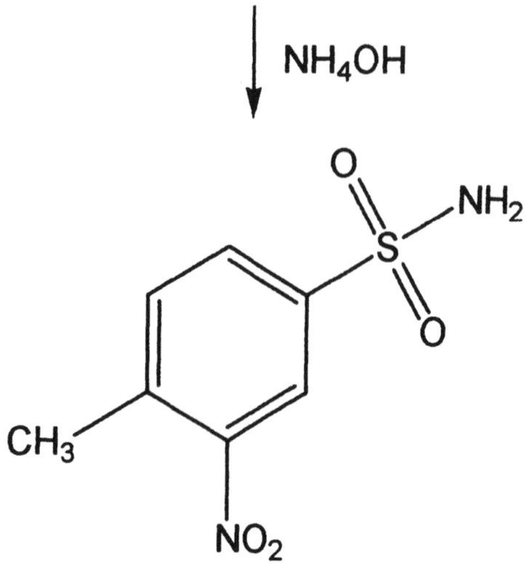

$$\downarrow \text{NH}_4\text{OH}$$

Compound C
4-methyl-3-nitro-
benzenesulfonamide

4.   The unknown is in solubility class S_2, which consists of salts of
organic acids, amine hydrochlorides, amino acids, polyfunctional
compounds with hydrophilic functional groups, carbohydrates
(sugars), polyhydroxy compounds, and polybasic acids.   The
unknown contains an active hydrogen (positive acetyl chloride test)
but does not contain nitrogen (negative ammonium polysulfide,
hydrochloric acid, and ferric chloride test).
     The unknown is sucrose.

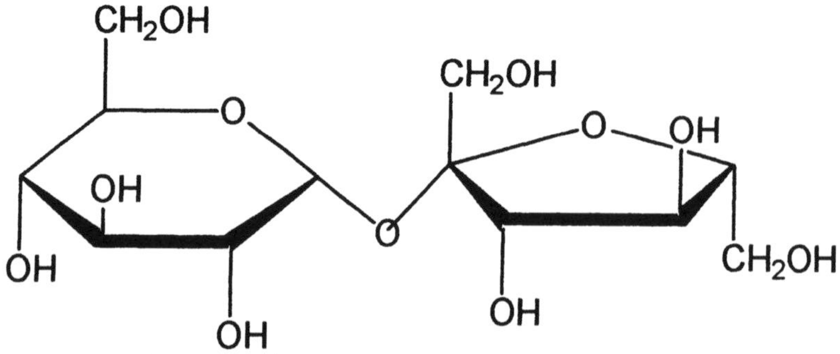

sucrose, α-D-glucopyranosyl-β-D-fructofuranose

5. The unknown contains an active hydrogen (positive acetyl chloride), a phenol (positive ferric chloride and pyridine test), a nitro group (positive ferrous hydroxide test), and 1^0 amine (positive benzenesulfonyl chloride test). The unknown does not contain a 1^0 alcohol, 2^0 alcohol, and aldehyde (negative chromium trioxide test).
The unknown is 2-amino-4,6-dinitrophenol.

2-amino-4,6-dinitrophenol

6. The unknown contains an alkene or alkyne (positive bromine test), a carboxylic acid (positive sodium bicarbonate test), and an aldehyde or ketone (2,4-dinitrophenylhydrazine test). The unknown does not contain an aldehyde (negative Tollens test) or 1^0 alcohol, 2^0 alcohol, aldehyde (negative chromium trioxide test).
The unknown is 3-benzoylbenzoic acid.

3-benzoylpropenoic acid

7.   The unknown contains an active hydrogen (positive acetyl chloride test); a 1^0 alcohol, a 2^0 alcohol, or an aldehyde (positive chromium trioxide test); and chlorine (positive silver nitrate test). The unknown does not contain an aldehyde or a a ketone (negative 2,4-dinitrophenylhydrazine test); or an alkene or an alkyne (negative bromine test).

The unknown is 2,3-dichloro-1-propanol.

$$\underset{\text{Cl}}{\overset{\displaystyle \text{Cl}}{|}}$$

CICH$_2$——CH—CH$_2$-OH   2,3-dichloro-1-propanol

8.   The unknown contains an alkene or alkyne (positive bromine test and positive potassium permanganate test); an aldehyde or a ketone (positive 2,4-dinitrophenylhydrazine test); an aldehyde (positive Tollens test); or a 1^0 alcohol, a 2^0 alcohol, or an aldehyde (positive chromium trioxide test).

CH$_3$CH$_2$CH$_2$CH$=$CH

C$=$O   2-hexenal

H

Problem Set 6

1. Aldehydes are oxidized to an acid by hydrogen peroxide. The peak at δ 9.0 in the ^{1}H NMR spectrum is indicative of an aldehyde. A yellow color indicates conjugation, thereby indicating that the original compound is a α-ketoaldehyde.

α-ketoaldehyde   α-ketoacid

2. The carbonyl groups in aldehydes and ketones react with phenylhydrazine. Water is eliminated when a phenylhydrazine undergoes reaction with a carbonyl group. Ethanol can be eliminated if the phenylhydrazone of a β-ketoester cyclizes. Thus the unknown is a β-ketoester.

β-ketoester   phenylhydrazine

+ H_2O

phenylhydrazone of β-ketoester

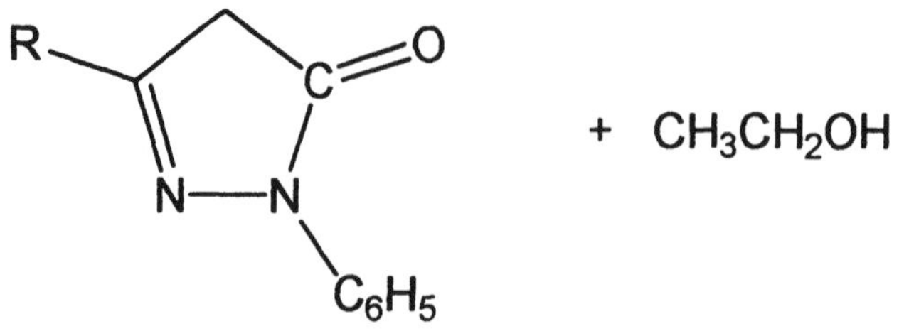

5-alkyl-2-phenyl-2,4-
dihydropyrazol-3-one

$$+ \ CH_3CH_2OH$$

3. Reaction with acetyl chloride indicated the presence of an active hydrogen in such compounds as alcohols and amines. The unknown cannot be an aldehyde or ketone (negative phenylhydrazine test). Periodic acid oxidizes compounds such as

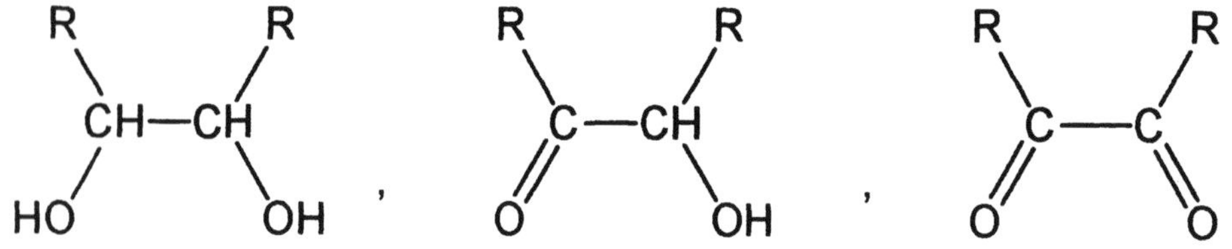

1,2-diol   α -hydroxyketone   1,2-diketone

The original compound must be a 1,2-diol, since it must contain an active hydrogen and is not an aldehyde or ketone.

Treatment of a 1,2-diol with periodic acid will produce aldehydes and/or ketones that will give a positive test with phenylhydrazine. An example is shown below.

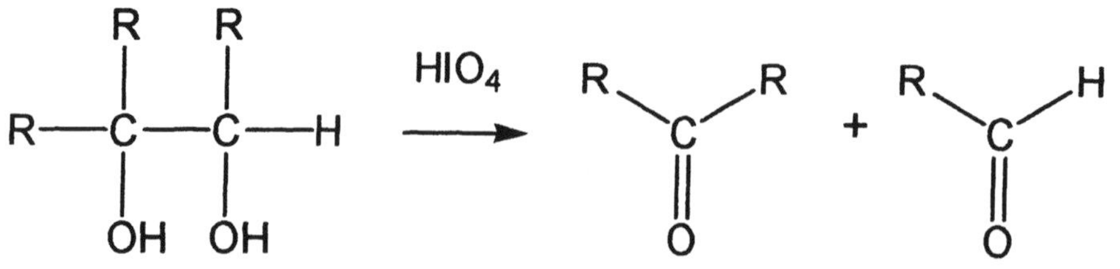

1,2-diol   ketone   aldehyde

aldehyde   phenylhydrazine   phenylhydrazone

ketone   phenylhydrazine   phenylhydrazone

4.   An IR band at 1710 cm^{-1} is near the frequency for a C=O stretching for a ketone. A yellow color indicates conjugation. A α-diketone is a possibility.

α-diketone   *o*-phenylenediamine   quinoxaline

5.   An alcohol that gives a positive test with iodoform indicates a structure of the type

A negative Lucas test indicates a primary alcohol. The only possibility is ethanol.

   ethanol   sodium iodate   iodoform   sodium formate

6.   The original compound reacted with acetyl chloride, so it must have an active hydrogen. The original compound yielded a negative test with phenylhydrazine, thereby indicating that it is not an aldehyde or a ketone. Treatment of the unknown with mineral acid yielded a compound which is not an alcohol (negative acetyl chloride test), but is an aldehyde or ketone (positive phenylhydrazine test). Compounds containing carbon-carbon double bonds, methyl ketones, and similarly sized ketones give a positive test with bromine.
     The ^{1}H NMR spectrum of the original compound contains two singlets, indicating that there are no hydrogens on the adjacent carbons. In the presence of mineral acid, 2,3-dimethylbutane-2,3-diol (pinacol) rearranges to 3,3-dimethyl-2-butanone.

Thus, the original compound (2,3-dimethylbutane-2,3-diol) is an alcohol, but not an aldehyde or ketone and the product (3,3-dimethyl-2-butanone) after treatment with mineral acid gave a positive test for a ketone and negative test for an alcohol.

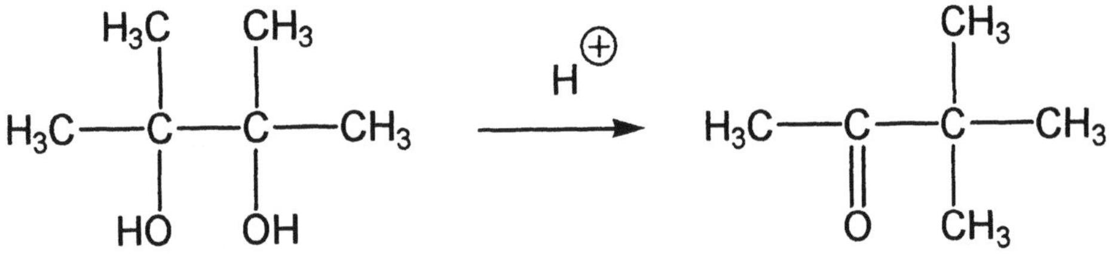

2,3-dimethylbutane-2,3-diol
(pinacol)

3,3-dimethyl-2-butanone

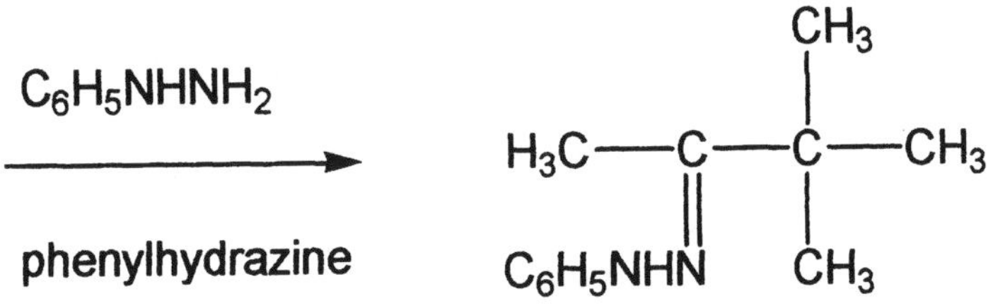

3,3-dimethyl-2-butanone
phenylhydrazone

7. The secondary amine must also contain another functional group that makes it soluble in base, such as an alcohol group or a carboxyl group. Possibilities would be an amino acid or an amino alcohol.

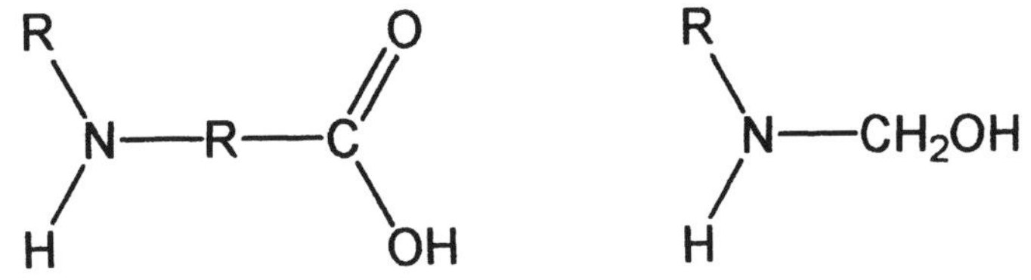

amino acid

amino alcohol

8. Aldehydes and ketones react with ethanol to yield acetals and ketals.

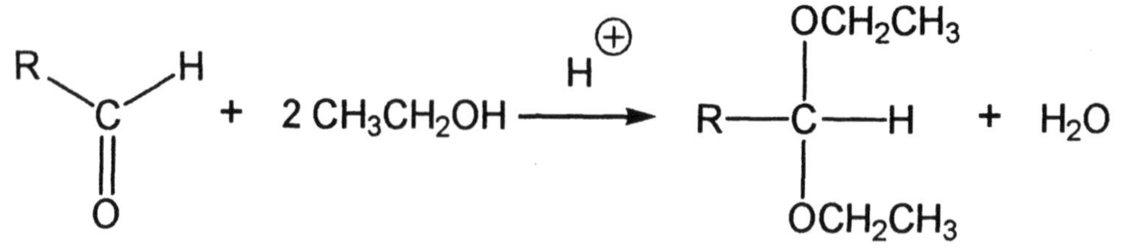

aldehyde ethanol acetal

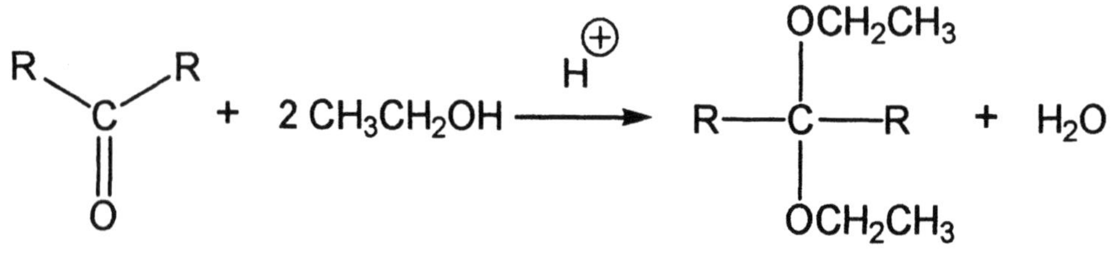

ketone ethanol ketal

A signal of $\delta\,9$–$\delta\,10$ in the ^{1}H NMR spectrum is indicative of an aldehyde group. After the treatment with ethanol and acid, the C–H is moved to $\delta\,5.0$.

9. The unknown is benzaldehyde (or a substituted benzaldehyde), which, in the presence of sodium cyanide, undergoes a self-condensation reaction to yield benzoin (or a substituted benzoin).

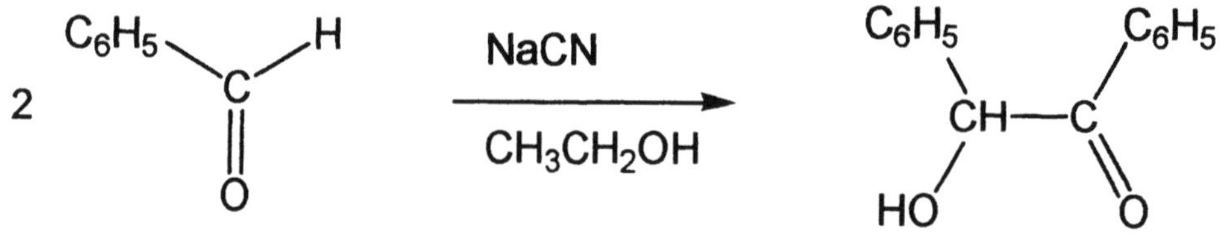

benzaldehyde benzoin

2 Ar—C(=O)—H →[NaCN / CH₃CH₂OH]→ Ar—CH(OH)—C(=O)—Ar

substituted benzaldehyde   substituted benzoin

The IR bands at 2700 and 2800 cm^{-1} correspond to the C–H stretching of an aldehyde group.

10. Tertiary amines do not react with benzenesulfonyl chloride. Nitrous acid undergoes reaction with *N,N*-dialkylanilines to form nitroso compounds.

N,N-dialkylaniline   4-nitroso-*N,N*-dialkylaniline

Another indication that the unknown compound is an *N,N*-dialkylaniline, is the lack of an N–H band near 3333 cm^{-1}. Aromatic nitroso compounds yield N=O stretching bands near 1550 cm^{-1} in an IR spectrum.

11. An ester with a SE of 59 ± 1 would have a formula of $C_2H_4O_2$. The only possibility that would match the integration ratio of 3:1 in the ^{1}H NMR spectrum would be methyl methanoate.

methyl methanoate
MW = 60, SE = 60

12. The original acid must contain more than one –COOH group in order to have an increase in neutralization equivalent. Since CO_2 is lost with heat, the two –COOH groups must have originally been attached to the same carbon. The following equation can be set up to calculate the number of –COOH groups, with n = the number of original –COOH groups. Carbon dioxide has a molecular weight of 44.

$$(n \times 54) \ — \ (n-1)\,59 \ = \ 44$$

$$54n \ — \ 59n \ + \ 59 \ = \ 44$$

$$-5n \ = \ -15$$

$$n \ = \ 3$$

Thus, there must be 3 –COOH groups present in the original carboxylic acid. The initial molecular weight must be approximately 162 (3 x 54). To calculate the remainder of the molecule, the 3 –COOH groups can be subtracted from the molecular weight, leaving a value of 27, which can be a CH_2CH.

$$162 \ — \ (3 \times 45) \ = \ 27$$

The only possibility is 2-carboxybutanedioic acid.

2-carboxybutanedioic acid
MW = 162, NE = 54

butanedioic acid
(succinic acid)
MW = 118, NE = 59

13. An insoluble barium salt is barium sulfate. Amines form hydrochloride salts. Thus, the original compound must contain both a sulfate group and an amine group. The only compound that meets these criteria with a neutralization equivalent of 142 ± 1 is anilinium sulfate, $(C_6H_5NH_3^+)_2SO_4^{-2}$. The N-H stretch of the amine salt is in the range of 2500-3333 cm^{-1}.

$$(C_6H_5\overset{\oplus}{NH_3})_2SO_4^{-2} \quad + \quad BaCl_2 \quad \longrightarrow$$

anilinium sulfate   barium chloride
MW = 284,
NE = 142

$$2 \; C_6H_5\overset{\oplus}{NH_3} \; \overset{\ominus}{Cl} \quad + \quad BaSO_4 \, (s)$$

anilinium chloride   barium sulfate

$$C_6H_5\overset{\oplus}{NH_3} \; \overset{\ominus}{Cl} \quad \xrightarrow{\text{NaOH}} \quad C_6H_5NH_2 \quad \xrightarrow{\text{HCl}} \quad C_6H_5\overset{\oplus}{NH_3} \; \overset{\ominus}{Cl}$$

anilinium chloride   aniline   anilinium chloride
MW = 129.5,
NE = 129.5

14. Only secondary amines give a precipitate with benzenesulfonyl chloride, followed by sodium hydroxide solution. In the ^{1}H NMR spectrum, a quartet indicates that there are three hydrogens on the adjacent carbons and a triplet indicates that there are two hydrogens on the adjacent carbons, thus concluding that two ethyl groups are present. A broadened singlet indicates an N–H group. The only compound that fits this data is diethylamine.

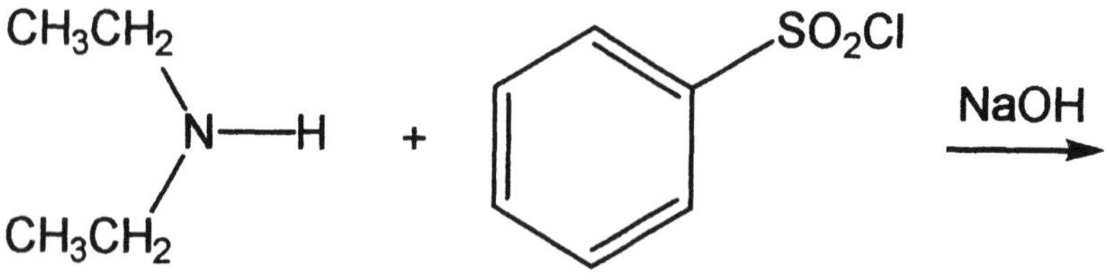

diethylamine
MW = 73, NE = 73

benzenesulfonyl
chloride

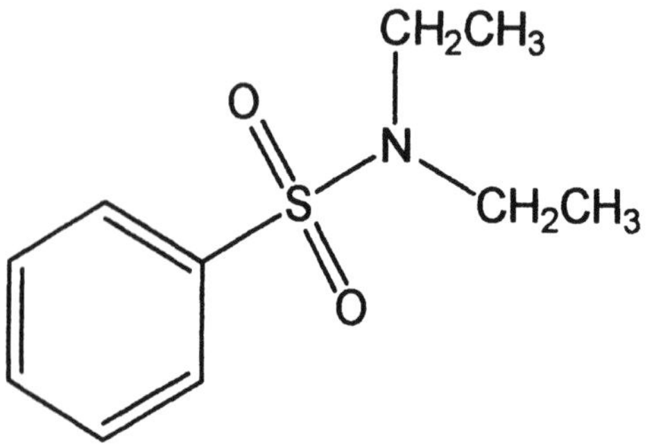

N,N-diethylbenzene-
sulfonamide

15. The following calculations need to be performed first.

$$\frac{12\ g}{124\ g/mole} = 0.096 \text{ moles of compound } C_8H_8O$$

$$\frac{60\ g}{160\ g/mole} = 0.375 \text{ moles of } Br_2 \text{ initially}$$

$$\frac{16.2\ g}{81\ g/mole} = 0.200 \text{ moles of HBr evolved}$$

$$\frac{12\ g}{160\ g/mole} = 0.075 \text{ moles of } Br_2 \text{ unused}$$

$0.375 - 0.075 = 0.300$ moles of Br_2 used

Therefore, 0.200 moles of Br_2 reacted with the compound to yield 0.200 moles of HBr. Another 0.100 moles of Br_2 were added across a multiple bond in the compound. Bromine substitutes on phenols with the evolution of hydrogen bromide. Additionally, bromine adds across multiple bonds. Thus, the unknown is a phenol containing an aliphatic double bond. With a formula of C_8H_8O, the only possibilities would be 2-hydroxystyrene, 3-hydroxystyrene, or 4-hydroxystyrene. The equation below is given for 4-hydroxystyrene.

4-hydroxystyrene

3,5-dibromo-4-
(1,2-dibromoethyl)phenol

The ^1H NMR spectrum of the original compound would include aromatic signals and the protons would be in an integration ratio of 2(CH_2):1(CH): 4(aromatic). In the IR spectrum, C=C stretching and bending, O–H stretching and bending, and aromatic frequencies would be present. The product would have an integration ratio of 2(CH_2):1(CH): 2(aromatic) in the ^1H NMR spectrum and would contain the C=C aliphatic stretching and bending in the IR spectrum.

Problem Set 7

I. A negative test with phenylhydrazine indicates that the unknown is not an aldehyde or ketone. A positive test with potassium permanganate and a negative test with bromine in carbon tetrachloride signifies an aldehyde or an alcohol. An active hydrogen is indicated by the positive test for acetyl chloride. A positive periodic acid test indicates that a carbonyl group is adjacent to a carbonyl group, a carbonyl group is adjacent to an hydroxy group, or a hydroxy group is adjacent to a hydroxy group.

Carboxylic acids have neutralization equivalent and 4-nitrobenzyl ester derivatives. Thus, the unknown must contain a carboxylic acid group adjacent to a hydroxy group. In the carboxylic acid derivative tables, (±)-mandelic acid matches the melting point of 117-118°C and the melting point of 123°C for the 4-nitrobenzyl ester derivative.

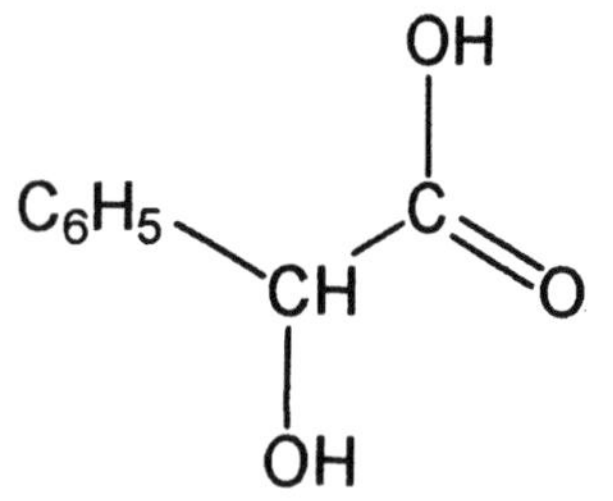

mandelic acid
MW = 152, NE = 152

The equation for the formation of the 4-nitrobenzyl ester is shown below.

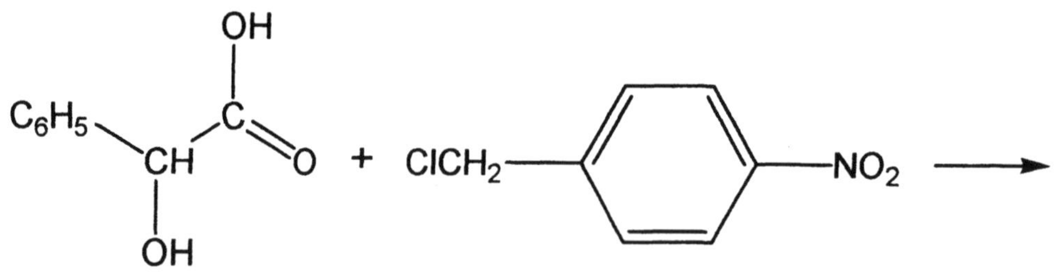

mandelic acid   4-nitrobenzyl chloride

4-nitrobenzyl 1-hydroxy-1-phenyl acetate

The IR spectrum of mandelic acid contains signals at 2400–3333 cm^{-1} for the O–H stretching of the acid and the alcohol; 1706 cm^{-1} for the C=O stretching of the acid; and 1190 and 1200 cm^{-1} for the C–O stretching of the 2° alcohol. In the ^{1}H NMR spectrum, the singlet at δ5.22 signifies the C–H and the multiplet at δ6.93–δ7.88 indicates five aromatic protons and both –OH protons.

II.   The unknown is in the solubility class of N, indicating an alcohol, aldehyde, ketone, ester, ether, alkene, alkyne, or aromatic.

A negative hydroxylamine hydrochloride test indicates that the unknown is not an aldehyde or a ketone. A negative acetyl chloride test indicates the absence of an active hydrogen. Negative results from the silver nitrate and the sodium iodide tests show that the bromine is not labile. Both the potassium permanganate and bromine tests give negative results, indicating the absence of a multiple bond.

The elemental analysis yielded the presence of bromine, but the tests above indicate that there is no labile halogen present. Therefore, the halogen must be attached to an aromatic ring. From the results of the classification tests, alcohol, aldehyde, ketone, alkene, and alkyne are eliminated as possibilities. Ester, ether, and aromatic compounds are left as possibilities.

Saponification equivalents indicate the presence of an ester. Hydrazine and 3,5-dinitrobenzoic acid react with esters to yield derivatives. From the derivative tables, by comparison of the boiling point of 259-261°C of the original compound and of the melting points of the hydrazine derivative (164°C) and the 3,5-dinitrobenzoate derivative (92°C), the unknown must be ethyl 4-bromobenzoate. The product from the sodium hydroxide hydrolysis is 4-bromobenzoic acid (mp 250°C).

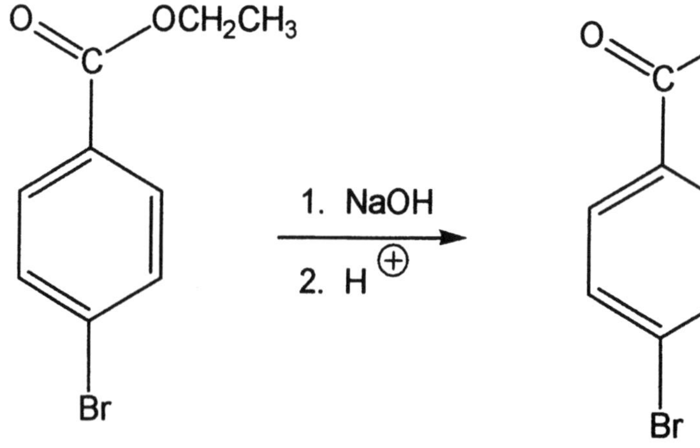

ethyl 4-bromobenzoate
MW = 229, SE = 229

4-bromobenzoic acid

The equations for the preparation of the derivatives are shown below.

ethyl
4-bromobenzoate

hydrazine

4-bromobenzoic acid
hydrazide

ethyl 4-bromobenzoate   3,5-dinitrobenzoic acid

4-bromobenzoic acid   ethyl 3,5-dinitrobenzoate

III.   The solubility tests give the unknown compound an A_2 solubility class indicating a weak organic acid such as a phenol, an enol, an oxime, an imide, a sulfonamide, a thiophenol, a β-diketone, or a nitro compound.

     The unknown has an active hydrogen from a positive acetyl chloride test, but cannot be an aldehyde or a ketone from negative results in the phenylhydrazine test. A precipitate in the bromine in water test and a violet color in the iron (III) chloride test indicate a phenol. A primary, secondary, or tertiary alcohol is not present due to the negative cerium (IV) test.

The compound must be a phenol from the classification and elemental tests. A compound that matches the boiling point of 198-200°C and the melting point of 102-103°C for the chloroacetic acid derivative is 3-hydroxytoluene.

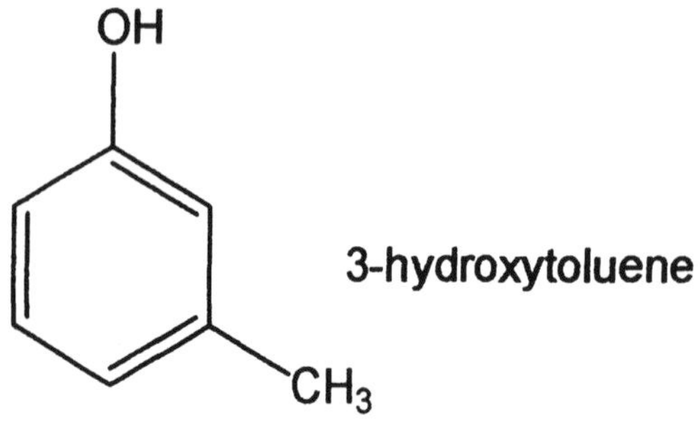

OH

3-hydroxytoluene

CH₃

The equation for the preparation of the aryloxyacetic acid is illustrated below.

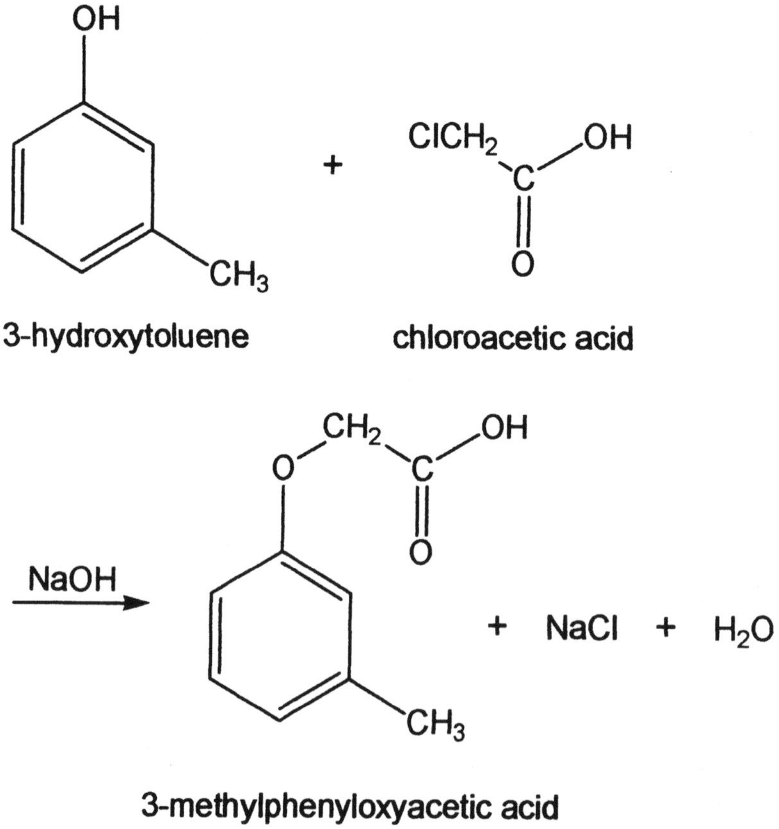

OH

CH₃

3-hydroxytoluene

+

ClCH₂—C—OH

‖

O

chloroacetic acid

NaOH

CH₂—C—OH

O—

‖

O

CH₃

+ NaCl + H₂O

3-methylphenyloxyacetic acid

The IR spectrum of this compound indicates an O–H stretching of an alcohol at 3390 cm^{-1}, a C–O stretching of a phenol at 1163 cm^{-1}, and C–H out-of-plane bending for *m*-disubstituted aromatic compounds at 690 and 780 cm^{-1}. The ^{1}H NMR spectrum indicates an isolated CH$_3$ at δ 2.25, the –OH group at δ 5.67, and four aromatic peaks in the range of δ 6.5–δ 7.3.

IV. The solubility class would be I, which consists of saturated hydrocarbons, haloalkanes, aryl halides, diaryl ethers, and deactivated aromatic compounds. The negative classification tests of sulfuric acid with sulfur trioxide and aluminum chloride in chloroform show that the unknown is not an aromatic compound. The negative test with bromine indicates no multiple bond. The boiling point of 194-195°C, the specific gravity of 0.8963, and the refractive index of 1.4811 all indicate that the sample is *cis*-decalin.

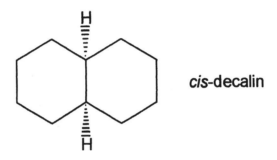

cis-decalin

The ^{13}C NMR spectrum indicates three sets of equivalent protons that are consistent with the structure of *cis*-decalin: δ 24.6 (CH$_2$), δ 29.6 (CH$_2$), and δ 36.9 (CH). *Cis*-decalin agrees with the mass spectrum formula of C$_{10}$H$_{18}$.

V. The solubility and elemental tests indicate that the unknown is a carboxylic acid. The negative results in the potassium permanganate and the bromine tests show that the unknown does not have a multiple bond. The unknown is not an aldehyde or a ketone due to the negative phenylhydrazine test. A negative acetyl chloride test indicates that an active hydrogen of an alcohol is not present. The neutralization equivalent of 59 confirms the presence of a carboxylic acid.

The melting points of the original compound (187-188°C) and the 4-bromophenacyl ester derivative (209-210°C), and the neutralization equivalent of 59 indicate that the compound is butanedioic acid or succinic acid.

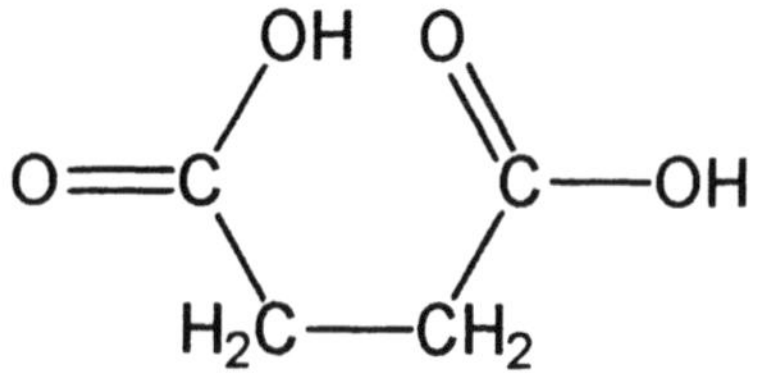

butanedioic acid
(succinic acid)
$Mw = 118$, $NE = 59$

The IR spectrum indicates the O–H stretching of a carboxylic acid (3222–$3333 cm^{-1}$), C=O stretching of an aliphatic carboxylic acid ($1695 cm^{-1}$), O–H bending of a carboxylic acid ($1418 cm^{-1}$), and the C–H bending of a CH_2 ($1307 cm^{-1}$). In the ^1H NMR spectrum, the signals at δ 2.43 (s) and δ 11.80 (bs) in the integration ratio of 2:1, indicate the CH_2 to COOH comparison.

The equations for the preparation of the derivative are shown below.

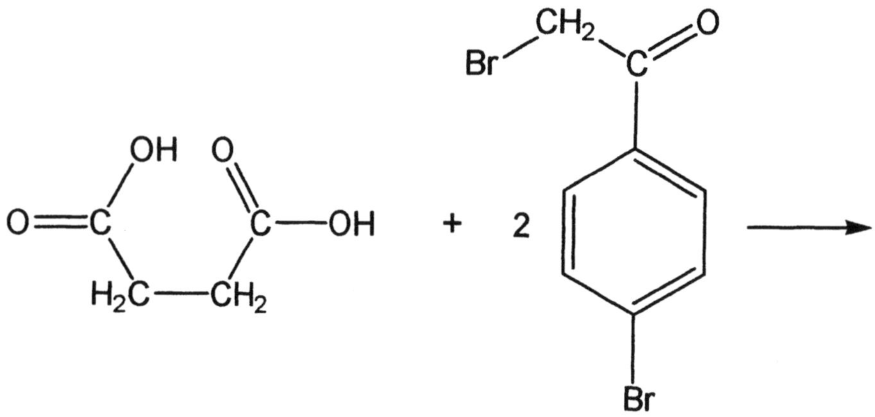

butanedioic acid   4-bromophenacyl bromide

+ 2 HBr

di-4-bromophenacyl ester of butanedioic acid

VI. The unknown is in solubility class B, which consists of amines of eight or more carbons, anilines, or ethers. A positive potassium permanganate test shows that the unknown contains a multiple bond, an alcohol group, or is a phenol. A precipitate in the bromine in carbon tetrachloride and the bromine water tests indicates an aromatic amine. The unknown contains an aldehyde or ketone group from a positive phenylhydrazine test. A tertiary amine is indicated from the benzenesulfonyl chloride test, followed by hydrochloric acid. The negative iron (II) hydroxide test shows the absence of a nitro group. A positive result with the Tollens reagent means that an aldehyde is present.

From the above information, a tertiary amine is present which contains an aldehyde group. Phenylhydrazone and semicarbazone derivatives are prepared from aldehydes. From the melting points of the unknown (72°C) and the phenylhydrazone (148°C) and semicarbazone (224°C) derivatives, the unknown is 4-(*N*,*N*-dimethylamino)benzaldehyde.

4-(N,N-Dimethylamino)benzaldehyde (Compound A) reacts with base in a Cannizzaro reaction to yield 4-(*N*,*N*-dimethylamino)-benzoic acid (Compound B), (mp 236-240°C), and 4-(*N*,*N*-dimethylamino)benzyl alcohol (Compound C).

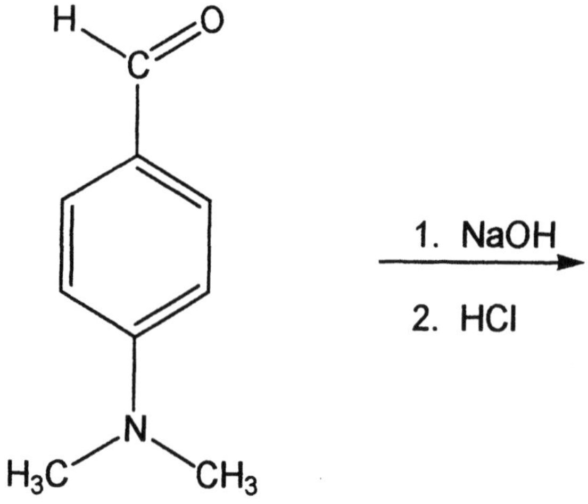

Compound A:
4- (*N*,*N*-dimethyl-
amino)benzaldehyde

1. NaOH
2. HCl

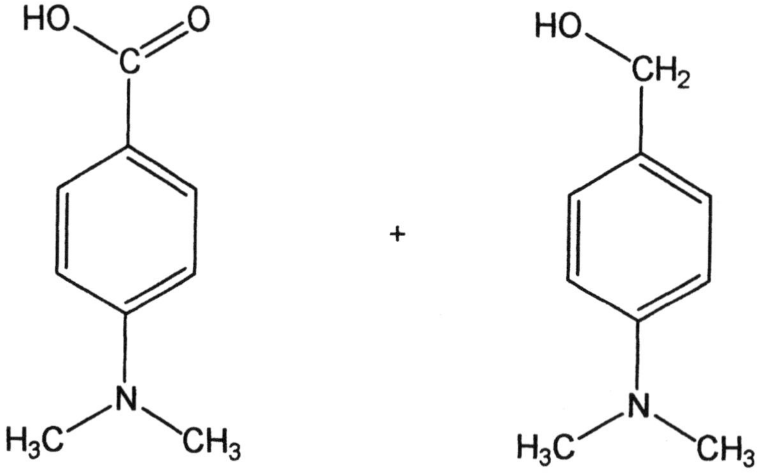

Compound B:
4-(*N*,*N*-dimethyl-
amino)benzoic acid

+

Compound C:
4-*N*,*N*-dimethyl-
aminobenzyl alcohol

4-(*N*,*N*-Dimethylamino)benzaldehyde (Compound A) reacts
with acetone and sodium hydroxide in a crossed aldol condensation

to yield 4-[4-(*N,N*-dimethylamino)phenyl]-3-buten-2-one (Compound D) (mp 134-135°C).

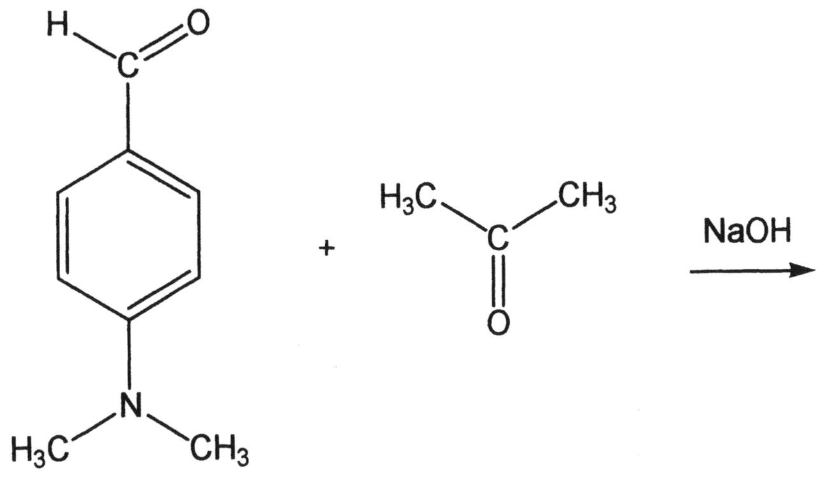

Compound A:
4-(*N,N*-dimethylamino)-
benzaldehyde

acetone

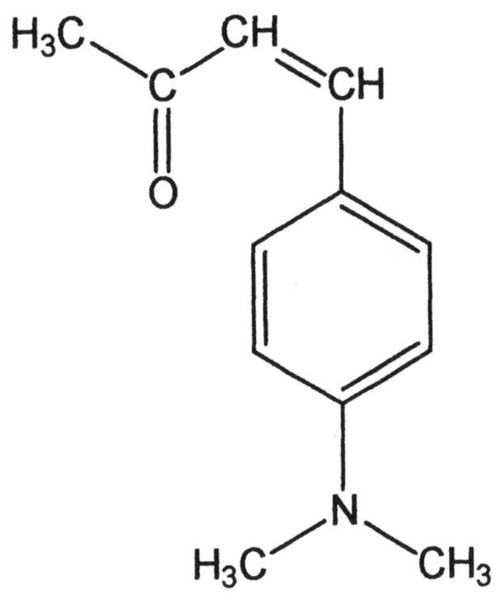

Compound D:
4-[4-*N,N*-dimethylamino)-
phenyl-3-buten-2-one

The equations for the formation of the derivatives are shown below.

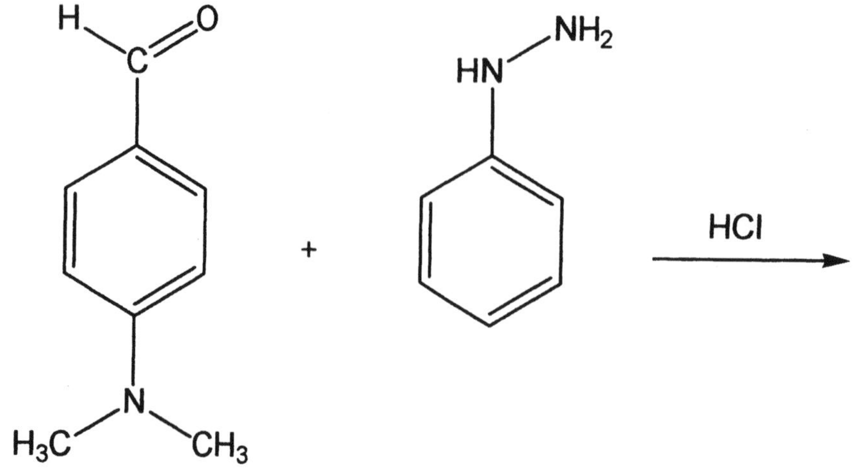

4-(*N,N*-dimethylamimo)-
benzaldehyde

phenylhydrazine

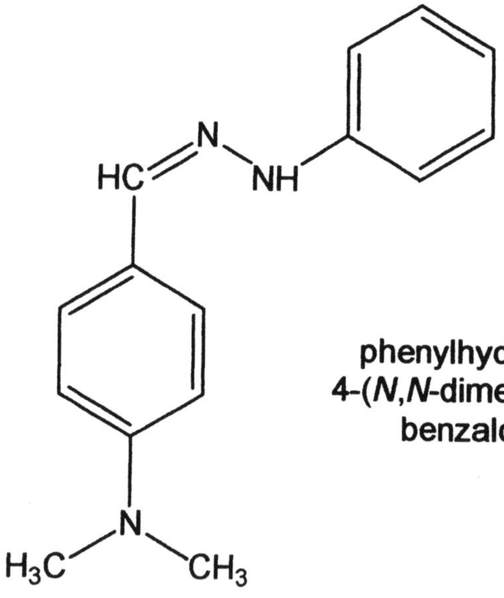

phenylhydrazone of
4-(*N,N*-dimethylamino)-
benzaldehyde

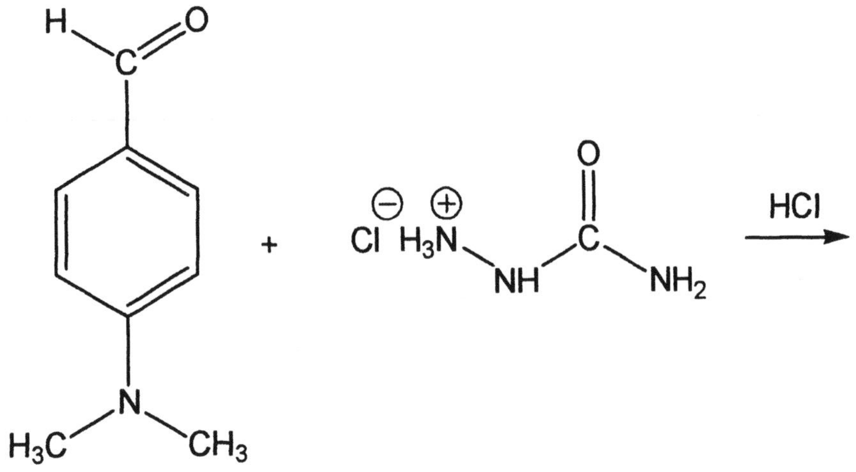

4-(*N*,*N*-dimethylamino)-
benzaldehyde

semicarbazide
hydrochloride

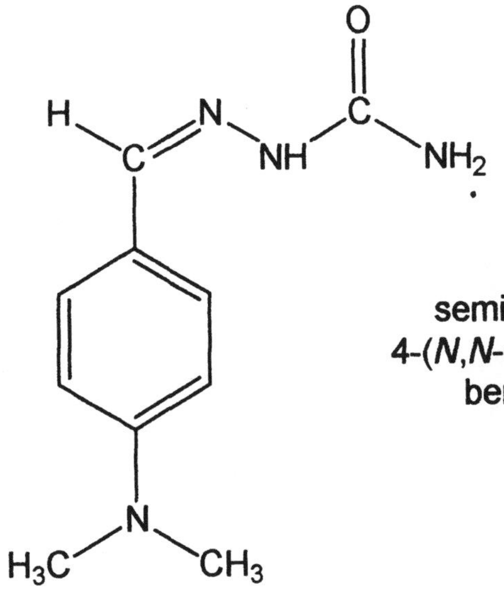

semicarbazone of
4-(*N*,*N*-dimethylamino)-
benzaldehyde

The IR spectrum has no absorption near 3333 cm⁻¹, showing that it is not a primary or secondary amine. A frequency at 1653 cm⁻¹ indicates the C=O stretching of an aromatic aldehyde. The ¹H NMR spectrum indicates two CH₃ groups at δ 3.05, *p*-disubstituted aromatic signals at δ6.69 and δ7.71, and the aldehyde signal at δ 9.70.

Problem Set 8

1. The solubility class for this unknown is B, which means it is an amine, an aniline, or an ether. A negative acetyl chloride test means that there is not an active hydrogen present. The compound is not a primary or secondary amine, since it did not react with benzenesulfonyl chloride. A tertiary amine is present (positive nitrous acid test).

   A tertiary amine that fits the boiling point of 193-195°C is *N,N*-dimethylaniline (Compound I). This compound is treated with nitrous acid to produce 4-nitroso-*N,N*-dimethylaniline (Compound II) (mp 164°C), which undergoes a reaction with hot sodium hydroxide solution to yield *N,N*-dimethylamine (Compound III) and a product which is acidified to produce 4-nitrosophenol (Compound V) (mp 125-126°C). Dimethylamine (Compound III) reacts with phenyl isocyanate to form 1,1-dimethyl-3-phenyl-2-thiourea (Compound IV) (mp 134-135°C).

Compound I: *N,N*-dimethylaniline		Compound II: 4-nitroso-*N,N*- dimethylaniline

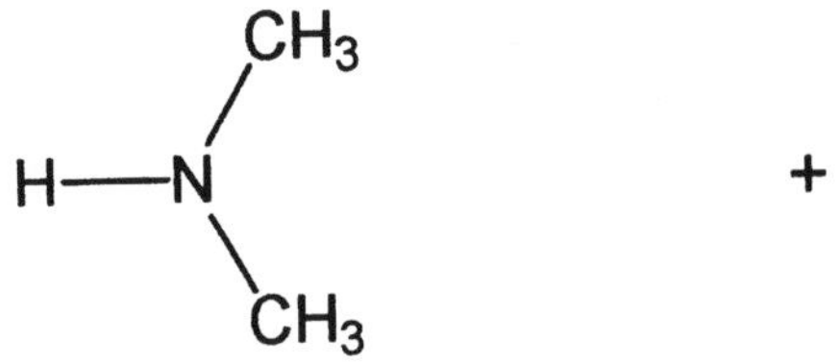

Compound III:
N,N-dimethylamine

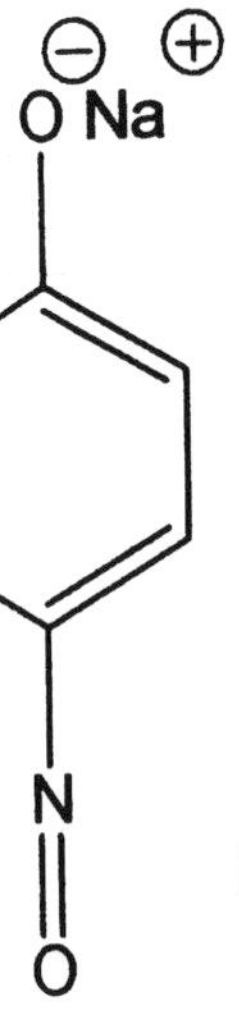

+

sodium
4-nitro-
phenolate

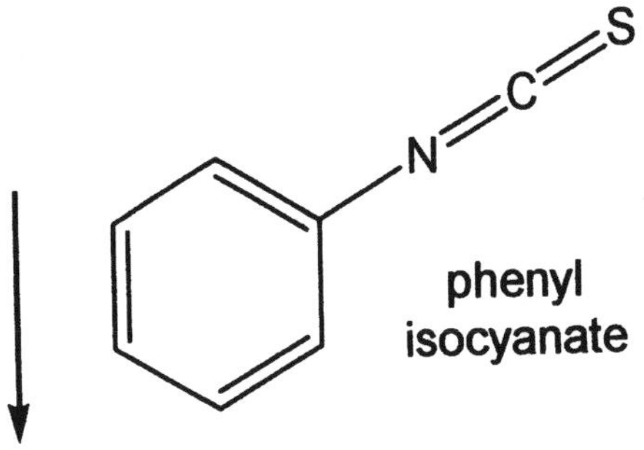

phenyl
isocyanate

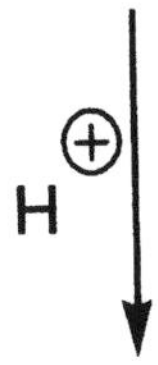

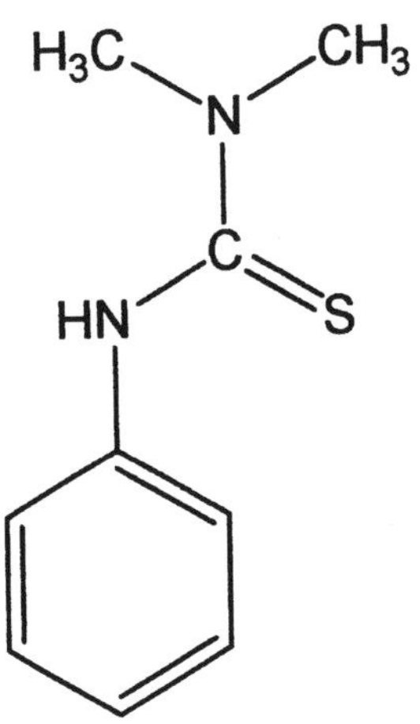

Compound IV:
1,1-dimethyl-3-
phenyl-2-thiourea

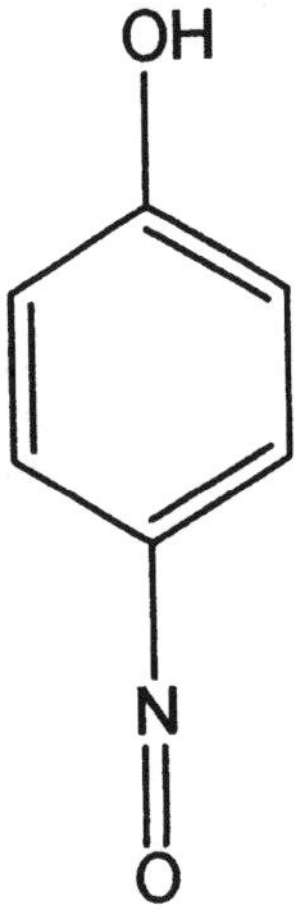

Compound V:
4-nitrosophenol

The ¹H NMR spectrum for compound V shows a pair of doublets (4H) in the aromatic region at δ6.63 and δ7.67, indicating a p-disubstituted aromatic ring. The broad singlet at δ8.7 is the –OH group.

2. The compound must be in solubility class S_A, S_B, or S_1, which includes carboxylic acids, arylsulfonic acids, amines, alcohols, aldehydes, ketones, esters, nitriles, and amides. Arylsulfonic acids, amines, nitriles, and amides are eliminated as possibilities because of the elemental tests. The compound must contain a multiple bond, since it yielded positive potassium permanganate and bromine tests. The presence of an active hydrogen is indicated by the reaction with acetyl chloride and sodium. From a negative iodoform test, the unknown is neither a methyl ketone nor a secondary alcohol with a methyl group on the carbon adjacent to the carbon bearing the –OH group. A negative phenylhydrazine test indicates that the compound is not an aldehyde or a ketone.

From the boiling point of 94-96°C of the compound and the melting point of 47–48°C for the 3,5-dinitrobenzoate derivative, the compound must be 2-propen-1-ol or allyl alcohol.

The ¹H NMR spectrum of the original compound can be analyzed as follows: δ3.58, 1H, s (**a**); δ4.13, 2H, m (**b**); δ5.13, 1H, m (**c**); δ5.25, 1H, m (**d**); and δ6.0, 1H, 10 lines (**e**). The protons **c** and **d** are in slightly different environments and are observed at different chemical shifts.

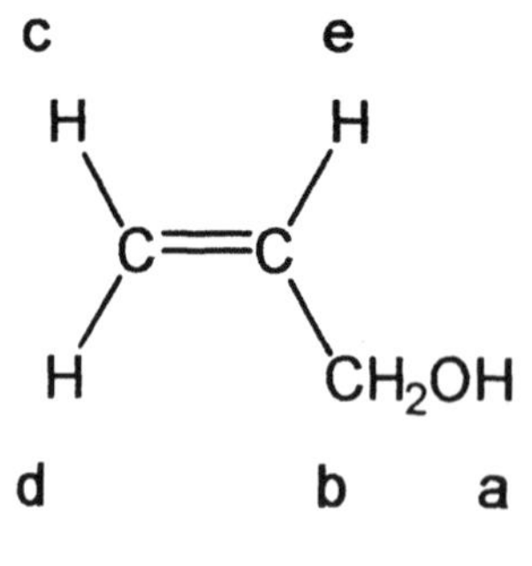

2-propen-1-ol
(allyl alcohol)

The equation for the synthesis of the derivative is shown below.

allyl alcohol   3,5-dinitrobenzoyl chloride

3,5-dinitrobenzoate
of allyl alcohol

3.   Compound I is in solubility class of B, which contains amines of eight or more carbons, anilines, or ethers. Compound II is in solubility class of A_2, which includes phenols, enols, oximes, imides, sulfonamides, thiophenols, and nitro compounds. Compounds I and II must have a nitro group since the products obtained by treatment of I and II with zinc and boiling solution of ammonium chloride readily reduced Tollens reagent.

      A compound that is an amine, has a melting point of 113-114°C, and has a nitro group is 3-nitroaniline. Treatment of 3-nitroaniline (Compound I) with nitrous acid produces 3-nitrophenol (Compound II) (mp 95-96°C).

Compound I: 3-nitroaniline   Compound II: 3-nitrophenol

The reaction of 3-nitroaniline (Compound I) with benzenesulfonyl chloride in the presence of sodium hydroxide produced *N*-(3-nitrophenyl)benzenesulfonamide (Compound III) (mp 135-136°C).

Compound I: 3-nitroaniline   benzenesulfonyl chloride

Compound III: *N*-(3-nitrophenyl)benzenesulfonamide

The ^{1}H NMR spectrum of compound II indicates a disubstituted aromatic compound (4H) in the range of δ7.0–δ7.75 and a singlet at δ9.8 which could be attributed to the –OH.

4. A solubility class of A_1 is indicated from the data for compound I, suggesting a carboxylic acid, a phenol, or a β-diketone.

    Compound I does not contain a multiple bond (negative bromine and potassium permanganate tests), an active hydrogen (negative acetyl chloride test), or an aldehyde or ketone (negative phenylhydrazine test).

    A carboxylic acid that matches data given for compound I, has a melting point of 186-187°C, a neutralization equivalent of 179, and contains nitrogen is N-benzoylaminoethanoic acid.

    Compound II, with its melting point of 120-121°C and neutralization equivalent of 122, corresponds to benzoic acid.

    Compound III is in solubility class S_2, which consists of salts of organic acids, amine hydrochlorides, amino acids, and polyfunctional compounds. Compound III contains a labile halogen (positive silver nitrate test) and is a 1° amine (nitrous acid test). Compound III is glycine hydrochloride because of the data given and the reaction of N-benzoylaminoethanoic acid (Compound I) with boiling hydrochloric acid. Treatment of glycine hydrochloride (Compound III) with benzenesulfonyl chloride produced the benzenesulfonamide of glycine hydrochloride (Compound IV) (mp 164-165°C).

Compound I:
N-benzoylamino-
ethanoic acid
MW = 179, NE = 179

Compound II:
benzoic acid
MW = 12, NE = 122

Compound III:
glycine
hydrochloride

Compound III:
glycine hydrochloride

+

benzenesulfonyl
chloride

1. NaOH
2. HCl

Compound IV:
benzenesulfonamide
of glycine

The ^{1}H NMR spectrum for *N*-benzoylaminoethanoic acid (Compound I) shows an isolated CH_2 at δ 4.06 and a monosubstituted aromatic ring (5H) at δ 7.5–8.0. Benzoic acid (Compound II) has a monosubstituted aromatic ring (5H) at δ 7.3–δ 7.7 and δ 8.0–δ 8.25, with a singlet for the –OH group at δ 12.8. The ^{1}H NMR spectrum for glycine hydrochloride (Compound III) shows only the CH_2 signal at δ 3.58.

5. Compound I is in solubility class of S_2, which includes salts of organic acids, amino hydrochlorides, amino acids, and polyfunctional compounds. Compound I contains an active hydrogen (positive acetyl chloride test), has a multiple bond or is an alcohol (positive potassium permanganate test), and is an aldehyde (positive Fehling's solution and Tollens reagent tests). Also, many carbohydrates are optically active. A carbohydrate that has the melting point of 168°C is *D*-galactose. Its osazone derivative (mp 199-201°C) is Compound II.

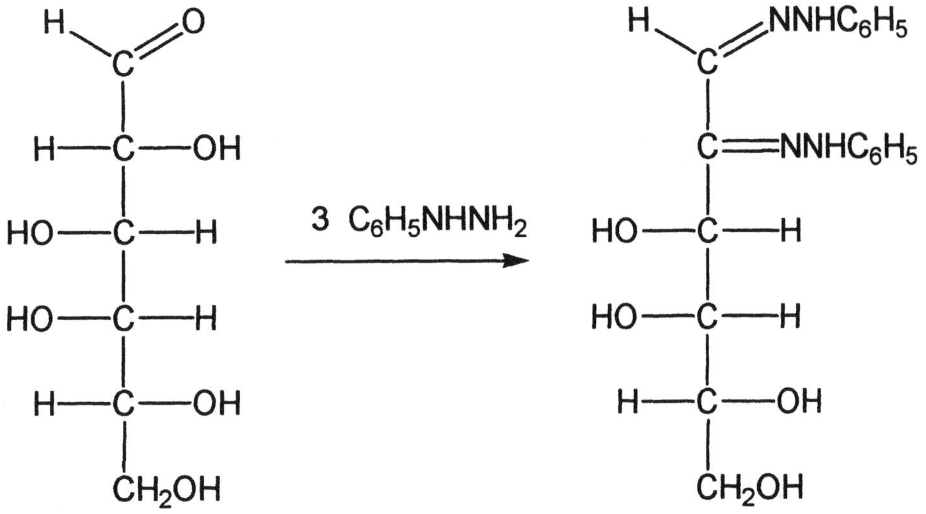

Compound I:
D-galactose

Compound II:
osazone of *D*-galactose

Nitric acid oxidizes both terminal carbons of *D*-galactose to –COOH groups to yield mucic or galactaric acid (Compound III) (mp 212–213°C).

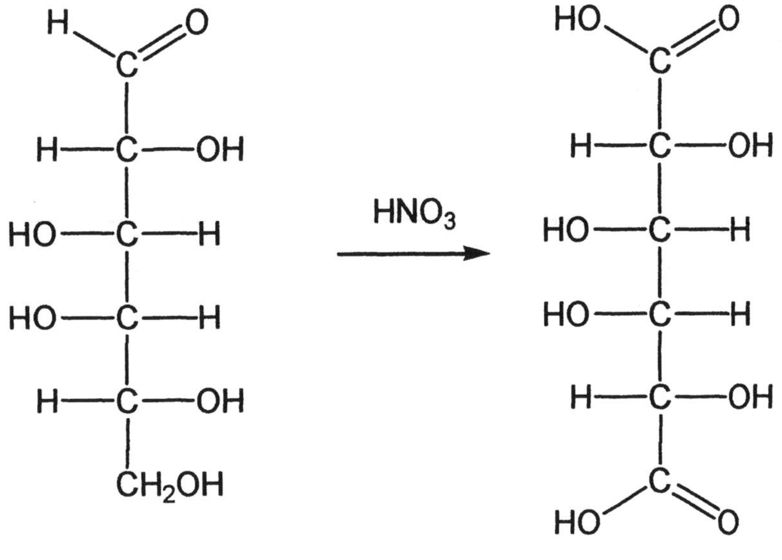

Compound I:
D-galactose

Compound III:
mucic acid
MW = 210, NE = 105

Mucic acid gives a negative phenylhydrazine test for aldehydes and ketones, but a positive test for active hydrogen with acetyl chloride.

After prolonged heating mucic acid (III) cyclizes to furoic acid (Compound IV) (mp 132-133°C), which reacts with 4-bromophenacyl bromide to yield the 4-bromophenacyl ester (Compound V) (mp 137-138°C).

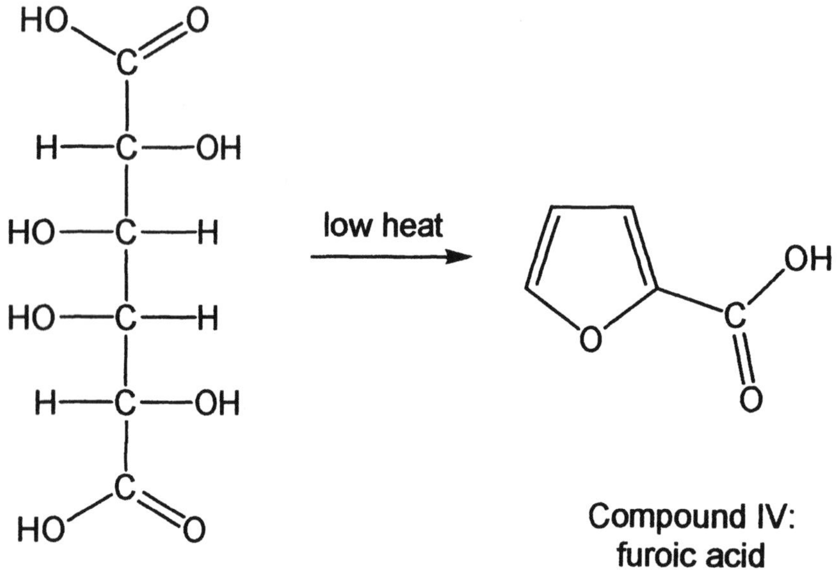

Compound III: mucic acid

Compound IV:
furoic acid
MW = 112, NE = 112

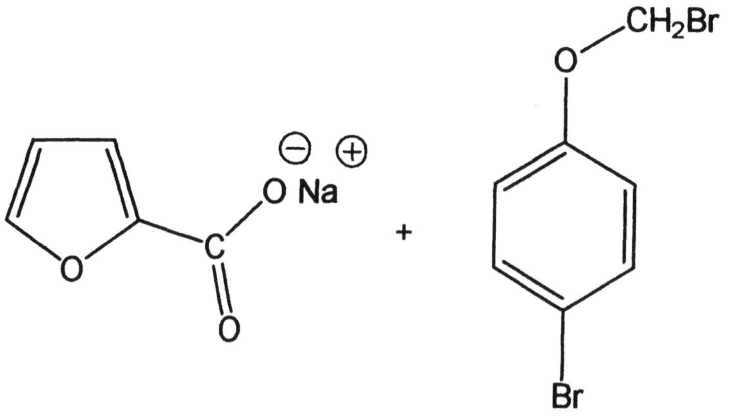

sodium furoate   4-bromophenacyl bromide

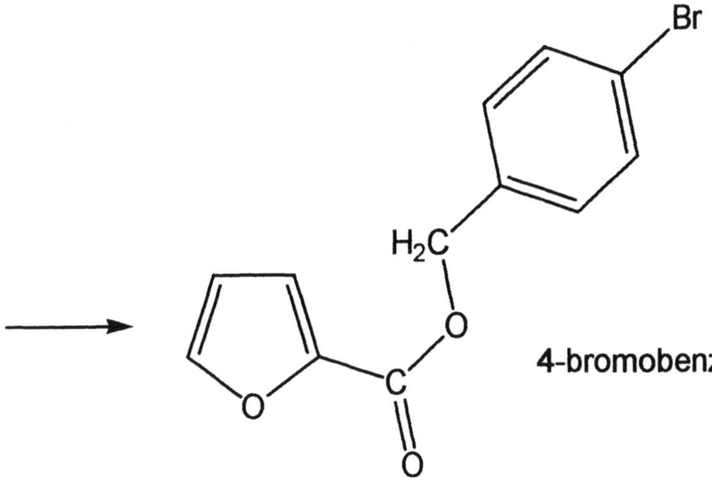

4-bromobenzyl furoate

The ^{1}H NMR spectrum of the sodium salt of compound IV indicates a conjugated system. The signals in the range of $\delta 6.59$–δ 7.64 are the hydrogens attached to the carbons in the ring. The protons are identified as follows: $\delta 6.59$, 1H, m (**a**); $\delta 7.05$, 1H, m (**b**); and $\delta 7.64$, 1H, m (**c**).

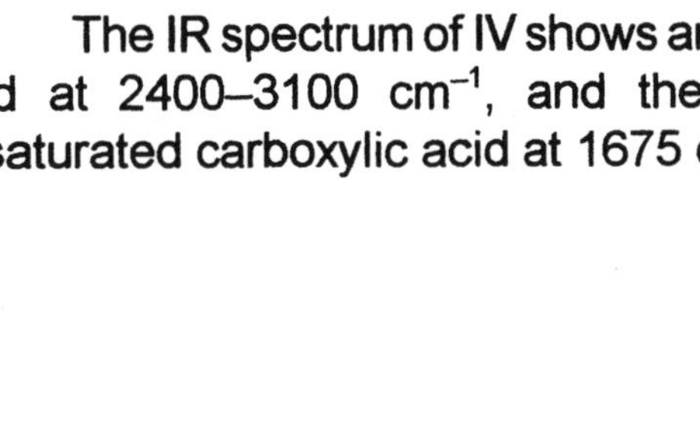

sodium furoate

The IR spectrum of IV shows an –OH stretching of a carboxylic acid at 2400–3100 cm^{-1}, and the C=O stretching of an α,β-unsaturated carboxylic acid at 1675 cm^{-1}.

Problem Set 9

1. The compound is in the solubility class of N, which means that it is an alcohol, aldehyde, ketone, ester, ether, epoxide, alkene, alkyne, or aromatic. A benzenesulfonamide derivative that is soluble in base is prepared from a primary amine. From the ^{1}H NMR spectrum, the product of the reaction of the compound with tin and hydrochloric acid is aniline, with the $-NH_2$ appearing as a singlet at δ 3.32, and the 5 aromatic protons at δ 6.44–δ 7.0. Either nitrobenzene or nitroso-benzene can be treated with tin and hydrochloric acid to yield aniline. However, the melting point of 68°C agrees with nitrosobenzene.

nitrosobenzene

1. Sn, HCl

2. neutralized

aniline

benzenesulfonyl chloride

N-phenylbenzenesulfonamide

The reaction of nitrosobenzene with zinc and sodium hydroxide yields hydrazobenzene (mp 130°C).

nitrosobenzene   hydrazobenzene

2. This compound is in solubility class A$_1$ or A$_2$. The only class of compounds that contain sulfur in these categories is thiophenols. Thiophenol has a boiling point of 166-169°C and reacts with 2,4-dinitrochlorobenzene to form 2,4-dinitrophenyl phenyl sulfide with a melting point of 118-119°C.

thiophenol   2,4-dinitrochlorobenzene

2,4-dinitrophenyl phenyl sulfide

Thiophenol will oxidize to a disulfide (mp 60-61°C).

thiophenol   diphenyl disulfide

The ^{1}H NMR spectrum of thiophenol indicated a monosubstituted benzene (5H) at δ 7.12 and a proton on the sulfur at δ 3.39. The disulfide only contains aromatic protons.

3. This compound is in solubility class MN, N, or I, which includes nitrogen or sulfur compounds, alcohols, aldehydes, ketones, esters, ethers, epoxides, alkenes, alkynes, aromatic compounds, saturated hydrocarbons, haloalkanes, aryl halides, or diaryl ethers. The compound does not have a nitroso or nitro group (negative tin and hydrochloric acid test). However, since the compound contains a nitrogen, it is probably an amide or an amine. Amides hydrolyze after prolonged treatment with sodium hydroxide to yield amines and acids. The amine is soluble in dilute hydrochloric acid and reacts with acetyl chloride to produce an acetamide.

Comparing the melting points of the compound (145-146°C), the acetamide (111-112°C), and the carboxylic acid (120-121°C), the compound must be *N*-benzoyl-2-methylaniline. *N*-Benzoyl-2-methylaniline hydrolyses to 2-methylaniline (Compound I) (bp 200°C) and benzoic acid (mp 120-121°C). 2-Methylaniline (Compound I) reacts with acetyl chloride to form 2-methylacetanilide (Compound II) (mp 111-112°C).

N-benzoyl-2-
methylaniline

NaOH

Compound I: 2-mehylaniline,
2-aminotoluene, *o*-toluidine

sodium benzoate

Compound II:
2-methylacetanilide

benzoic acid
MW = 122, NE = 122

The ^{1}H NMR spectrum of benzoic acid indicates a monosubstituted benzene (5H) in the aromatic range of δ7.4–δ8.3 and the –OH (1H) of the –COOH at δ12.8. The ^{1}H NMR spectrum of 2-methylaniline (I) shows a singlet indicating a –CH$_3$ at δ2.15, a broad singlet indicating a –NH$_2$ at δ3.48, and a disubstituted benzene (4H) at δ6.45–δ6.8 and δ6.8–δ7.15.

4.   This compound is in solubility class I, consisting of hydrocarbons, haloalkanes, aryl halides, diaryl ethers, or deactivated aromatic compounds. The unknown contains an aromatic ring, since it

dissolved in fuming sulfuric acid. The chlorine is attached to an aromatic ring or an alkene carbon (negative silver nitrate test). Hot potassium permanganate solution causes oxidation of a –R group attached to the benzene ring, resulting in a benzoic acid derivative. 2-Chlorobenzoic acid agrees with the melting point of 138-139°C and a neutralization equivalent of 156.5.

According to the ^{1}H NMR spectrum the –R group that is oxidized must be a methyl group, since a singlet of three protons appears at δ2.37. The aromatic region with four protons at δ7.0–δ 7.35 indicates a disubstituted benzene ring. The IR spectrum of the acid shows an O–H stretching of a carboxylic acid at 2381–3333 cm^{-1}, a C=O stretching of a carboxylic acid at 1678 cm^{-1} and C–H out-of-plane bending for an aromatic o-disubstituted benzene ring at 742 cm^{-1}. 2-Chlorotoluene (bp 159-161°C) is the original compound.

2-chlorotoluene

2-chlorobenzoic acid
MW = 156.5, NE = 156.5

5.   The compound is in solubility class N, which consists of alcohols, aldehydes, ketones, esters, ethers, epoxides, alkenes, alkynes, or aromatic compounds. A negative test with phenylhydrazine eliminates aldehydes and ketones as possibilities. The compound does not have an active hydrogen (negative acetyl chloride test) or a multiple bond (negative bromine test). Boiling alkalies cause the compound to break apart. Of the choices listed above, only esters, upon treatment with boiling alkali, break into two organic products, alcohols and acids.

The distillate of the alkali reaction is an alcohol. An alcohol that matches the boiling point of 129-130°C with a 1-naphthylurethan derivative of melting point 65-66°C is 3-methyl-1-butanol. The residue is a salt of the carboxylic acid, which upon acidification yields the acid. The acid is treated with thionyl chloride, followed by aniline to form an anilide. 3-Methylbutanoic acid will react with thionyl chloride and aniline to form an anilide (mp 108-109°C). The original compound is 3-methylbutyl 3-methylbutanoate or isopentyl isovalerate (bp 188-192°C).

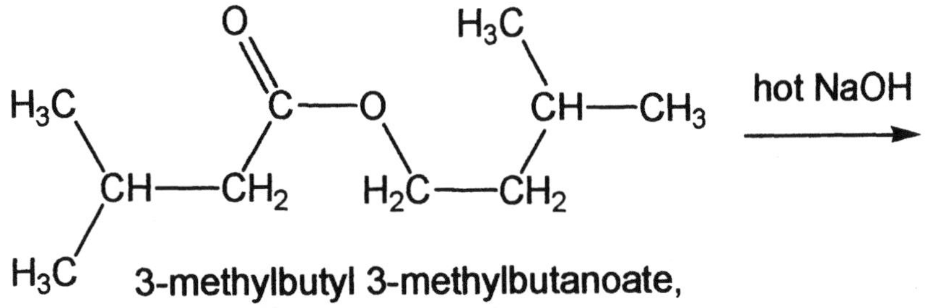

3-methylbutyl 3-methylbutanoate,
isopentyl isovalerate

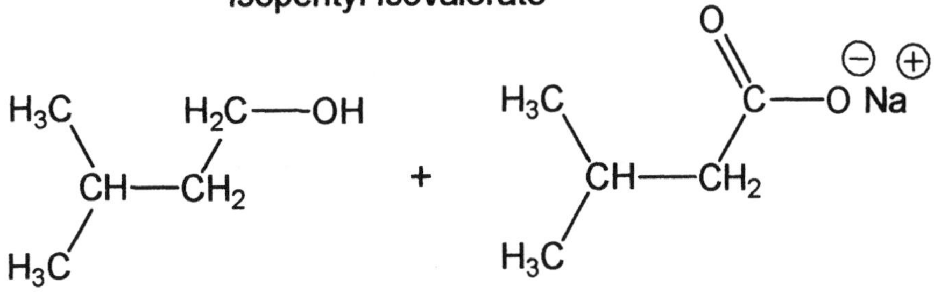

3-methyl-1-butanol, isopentyl alcohol   sodum 3-methylbutanoate

The equations for the derivatives are shown below.

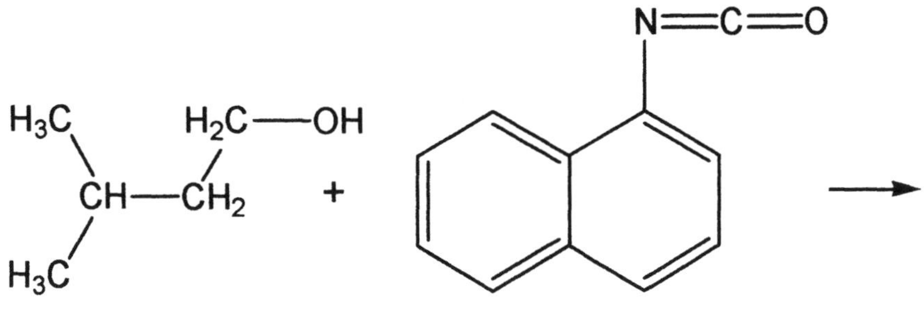

3-methyl-1-butanol   1-naphthyl isocyanate

1-naphthylurethan of 3-methyl-1-butanol

sodium 3-methylbutanoate   3-methylbutanoic acid

3-methylbutanoyl chloride

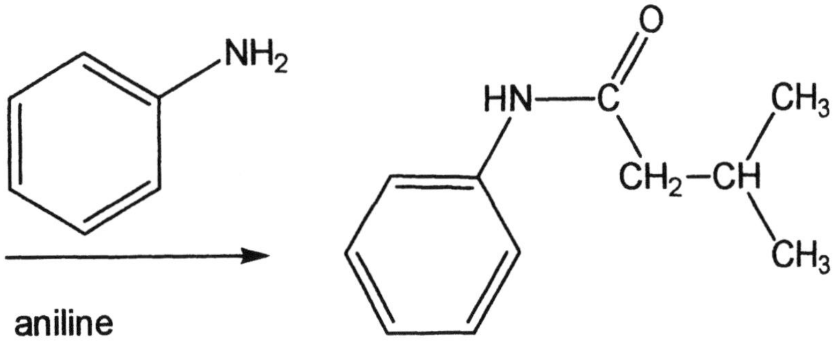

aniline

N-phenyl-3-methylbutanamide

Treatment of the original compound with lithium aluminum hydride produced the same alcohol as from the saponification reaction. The conclusion is reached that the acyl portion and the alkyl portion of the ester each must have five carbons arranged as an isopentyl group.

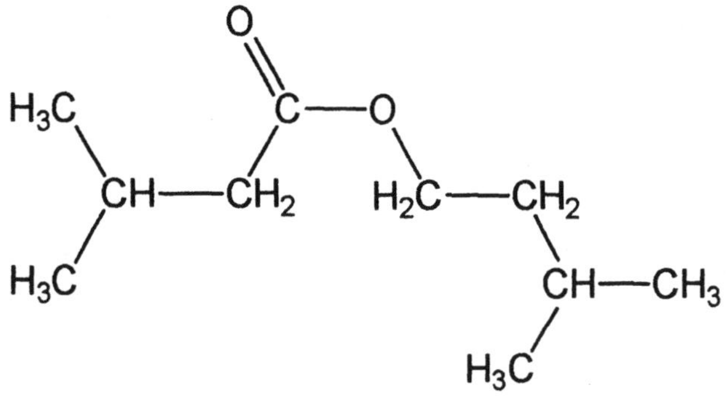

3-methylbutyl 3-methylbutanoate

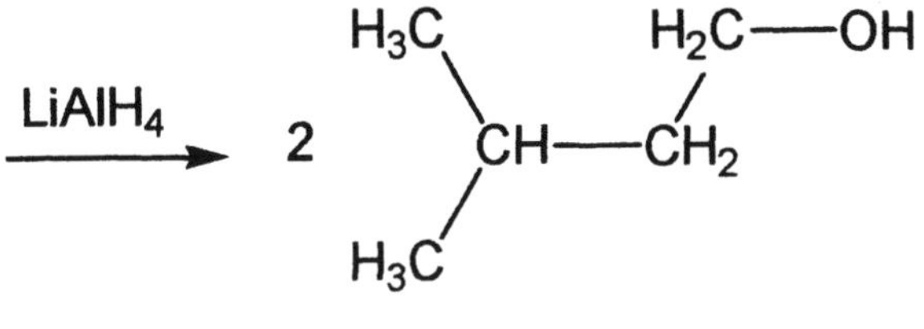

3-methyl-1-butanol

The ^{1}H NMR spectrum agrees with the structure of 3-methyl-1-butanol. The six protons at δ 0.92 that are split into a doublet are the two methyls adjacent to a methine proton. The three protons at δ 1.2–δ 1.7 are the protons attached to carbons two and three. The two hydrogens split into a triplet at δ 3.63 corresponds to the methylene group at carbon one. The –OH is seen as a broad singlet at δ 2.13.

6. A smoky flame indicates that the compound is aromatic. The unknown is in solubility class B, which consists of amines, anilines, and ethers. The compound reacted with benzenesulfonyl chloride to yield a benzenesulfonamide which was soluble in alkali, indicating that the original compound was a primary amine. Comparing the melting point (112-114°C) of the original compound and the melting points of the benzenesulfonamide (101-102°C) and acetamide (132°C) derivatives, the original compound must be 2-amino-naphthalene.

The equations for the preparation of the derivatives are shown below.

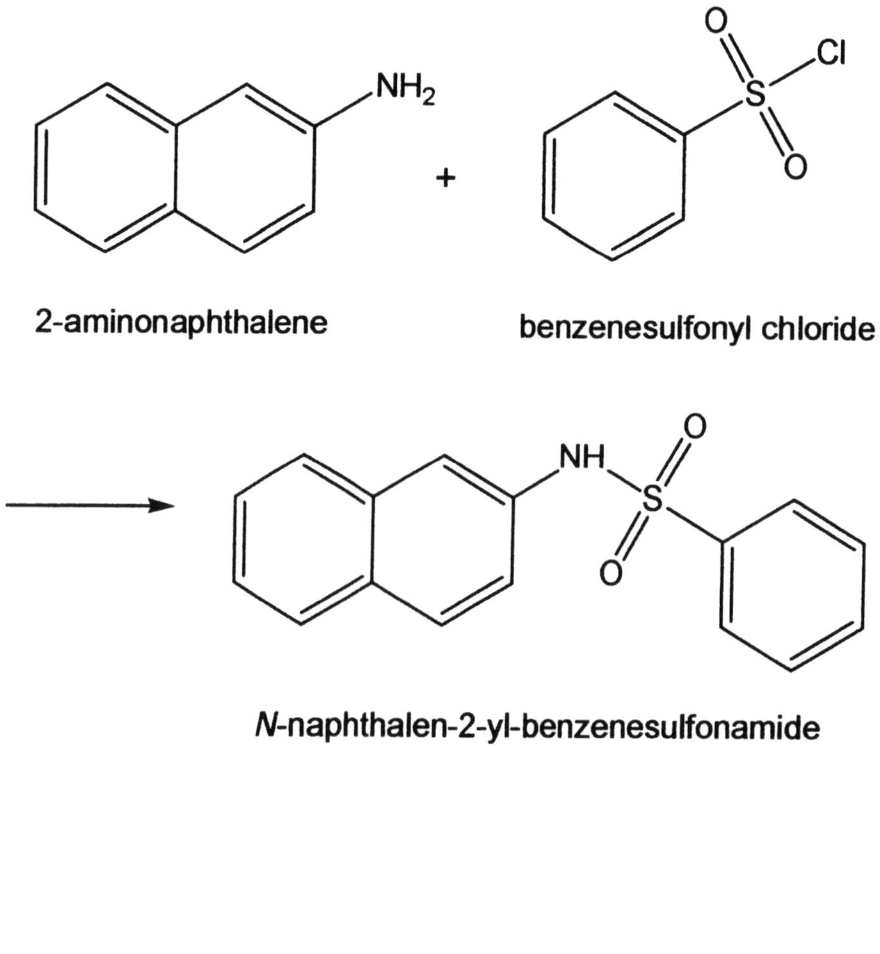

2-aminonaphthalene   benzenesulfonyl chloride

N-naphthalen-2-yl-benzenesulfonamide

2-aminonaphthalene   acetyl chloride

N-naphthalen-2-yl-acetamide

The IR spectrum shows the N–H stretching of a primary amine at 3350 and 3400 cm^{-1}, C–N stretching of a primary aromatic amine at 1280 and 1290 cm^{-1}, and N–H bending of a primary amine at 1640 cm^{-1}.

Problem Set 10

1.   Compound A is not an aldehyde or ketone (negative phenylhydrazine test). Since compound A gave a positive iodoform test, one of the following fragments must be present in the structure.

However, it cannot be a ketone according to the negative phenyl-hydrazine test.   If 45 (CH_3CHOH) and 45 (COOH) are subtracted from the neutralization equivalent of 103 ± 1, then 13 ± 1 is left. This is probably a CH_2. Therefore, compound A must be 3-hydroxy-butanoic acid.

3-Hydroxybutanoic acid (Compound A) loses water with sulfuric acid to yield 2-butenoic acid (Compound B). Compound B is an alkene or alkyne (positive bromine and potassium permanganate tests).   Treatment of 3-hydroxybutanoic acid (Compound A) with iodoform produces propanedioic acid (Compound C).

Compound A:
3-hydroxybutanoic acid
MW = 104, NE = 104

Compound B:
2-butenoic acid,
crotonic acid
MW = 86, NE = 86

Compound A:
3-hydroxybutanoic acid
MW = 104, NE = 104

Compound C:
propanedioic acid
MW = 104, NE = 52

+ CHI₃

iodoform

The ¹H NMR spectrum of compound B shows the –CH₃ at δ 1.90 (d of d), the –CH on carbon 3 at δ 5.83 (d of q), the –CH on carbon 2 at δ 7.10 (m), and the –OH of the acid at δ 12.18 (s).

2. The ¹H NMR spectrum of the second acid indicates a symmetrical aromatic dicarboxylic acid. The lack of reactivity for bromination indicates the presence of deactivating groups. Since a 2:1 ratio is present between the aromatic region (δ 8.08) and the –OH group (δ 11.0), the second carboxylic acid must be 1,4-benzenedicarboxylic acid (terephthalic acid). The first carboxylic acid cannot contain a labile hydrogen since substitution with bromine does not occur thus must be 2-(4-carboxyphenyl)-2-oxoethanoic acid.

vigorous
oxidation

4-oxalylbenzoic acid
MW = 194, NE = 97

1,4-benzenedicarboxylic acid,
terephthalic acid
MW = 166, NE = 83

3. This compound is in the solubility class of **N**, which consists of alcohols, aldehydes, ketones, esters, ethers, epoxides, alkenes, alkynes, or aromatic compounds. Since the compound is a hydrocarbon, all oxygen-containing compounds are eliminated as possibilities. The compound must contain a multiple bond since it decolorized bromine and potassium permanganate solutions.

    Since the molecular ion in the mass spectrum was at *m/z* of 68, thus the formula must be C_5H_8. Therefore, the structure contains two rings, two carbon-carbon double bonds, a carbon-carbon triple bond, or a carbon-carbon double bond and a ring. For oxidation to yield a carboxylic acid, a carbon-carbon double bond must be broken. Both the original compound and the carboxylic acid product contain the same number of carbons. The original compound must be cyclopentene or 3-methylcyclobutene. The product after oxidation must be a dicarboxylic acid.

cyclopentene → oxidation → pentanedioic acid
MW = 132, NE = 66

3-methylcyclobutene → oxidation → 2-methylbutanedioic acid
MW = 132, NE = 66

4. The original compound is not an alkene or alkyne (negative bromine test). For the neutralization equivalent to increase in value from 66 to 88, a carbon dioxide molecule must be lost when the original compound was heated.

The ^{1}H NMR spectrum of the second carboxylic acid agrees with the structure of 2-methylpropanoic acid. The methyl groups appear as a doublet at δ 1.20, the –CH on carbon two appears as a septet at δ 2.57, and the carboxylic acid hydrogen is at δ 12.4.

For carbon dioxide to be lost upon heating of the first carboxylic acid, two –COOH groups must be on carbon two. The original acid must be 2,2-dimethylpropanedioic acid.

2,2-dimethylpropanedioic acid, dimethylmalonic acid
MW = 132, NE = 66

2-methylpropanoic acid
MW = 88, NE = 88

5. From the ^{1}H NMR spectrum, both compounds have five protons in the aromatic region (acid--δ 7.52 and δ 8.14, 5H; base--δ 7.23, 5H) and thus contain monosubstituted benzene rings. The spectrum of the base indicate a –NH$_2$ group at δ 0.91, and two –CH$_2$ groups centered at δ 2.78. The spectrum of the acid shows only one other proton at δ 12.82, which is the carboxylic acid proton. The base must be phenylethylamine and the acid is benzoic acid.

phenylethylamine
MW = 121, NE = 121

benzoic acid
MW = 122, NE = 122

6. This compound does not contain a multiple bond, since it is unaffected by bromine. A positive iodoform test indicates the presence of one of the following groups.

Compounds that fit this data are 2, 3, or 4 (1-hydroxyethyl)benzoic acid.

2-(1-hydroxyethyl)-benzoic acid	2-(1-hydroxyethyl)-benzoic acid	2-(1-hydroxyethyl)-benzoic acid
MW = 166, NE = 166	MW = 166, NE = 166	MW = 166, NE = 166

7. Compound A belongs to the solubility class of A₁ or A₂, but does not contain sulfur, nitrogen, or halogen. Therefore the possibilities are carboxylic acids, phenols, enols, or β-diketones. Compound A is not a phenol (negative ferric chloride test) and does not have a multiple bond (negative potassium permanganate test).

The first product from the treatment of compound A with hydrobromic acid is in the solubility class of I, which means that this compound could be a saturated hydrocarbon, haloalkane, aryl halide, diaryl ether, or aromatic compound. This compound contained a labile halogen (positive sodium iodide in acetone test).

Compound B contains a multiple bond as indicated from the positive bromine test and is a phenol as indicated from the positive ferric chloride test. The difference in the neutralization equivalent between compounds A and B is 43 ($180 - 137 = 43$), which could be a propyl group. The ^{1}H NMR spectrum of compound B shows a *p*-disubstituted benzene with doublets of two hydrogens each at $\delta 6.84$ and $\delta 7.86$, and two broad –OH signals at $\delta 8.2$. From the ^{1}H NMR spectrum and the neutralization equivalent, compound B must be 4-hydroxybenzoic acid and compound A must be 4-propoxybenzoic acid.

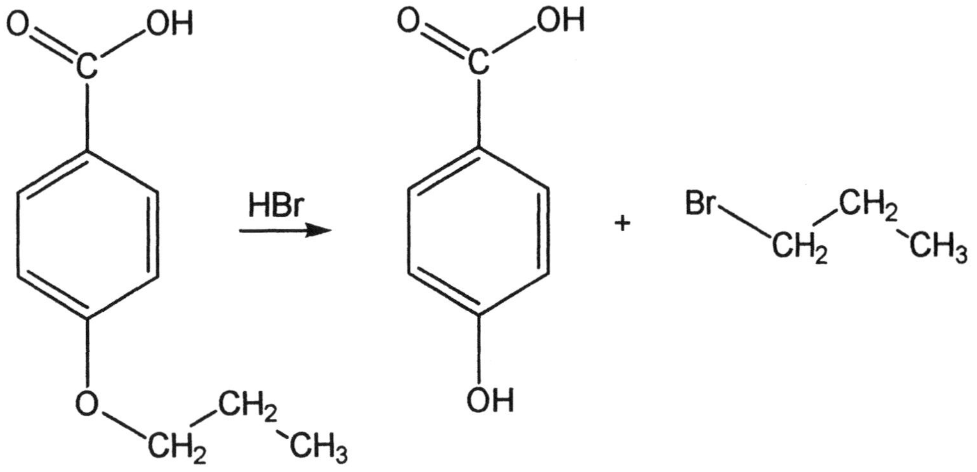

Compound A:	Compound B:	1-bromopropane
4-propoxybenzoic acid	4-hydroxybenzoic acid	
MW = 180, NE = 180	MW = 138, NE = 138	

8. The acid cannot contain an aromatic ring, due to the value of the molecular weight. The carboxylic acid group has a molecular weight of 45, which leaves 53 ($98 - 45 = 53$). The molecular weight of 53 could be C_4H_5. Possibilities from the formula C_4H_5COOH include the following structures.

cyclobut-2-enecarboxylic acid

2-methylbut-3-ynoic acid

2-methylcycloprop-2-
enecarboxylic acid

3-methylcycloprop-1-
enecarboxylic acid

9.   The original compound (I) is an aldehyde or a ketone, since it yielded a positive 2,4-dinitrophenylhydrazine test. The distillate from the sodium hydroxide treatment contains an active hydrogen (positive sodium test), is a methyl ketone or a methyl secondary alcohol (positive iodoform test), and is not a tertiary or secondary alcohol (negative Lucas test).

The residue from the steam-distillation is a carboxylic acid. The 4-bromophenacyl ester had a saponification equivalent of 257. If the 4-bromophenacyl group

is subtracted from this saponification equivalent, a value of 59 (257 – 198 = 59) is left. This fragment must be

$$—O\diagdown C(\diagup CH_3)(=O)$$

From the mass spectrum, the following peaks can be interpreted as follows:

m/z	fragment
15	$CH_3{}^+$
29	$CH_3CH_2{}^+$
43	$H_3C—C{\equiv}O^+$
45	$CH_3CH_2O^+$
85	$H_3C\diagdown C(=O)\diagup CH_2{-}C{\equiv}O^+$
130	$H_3C\diagdown C(=O)\diagup CH_2{\diagdown}C(=O)\diagup O\diagdown CH_2{\diagup}CH_3 \overset{+}{\cdot}$

With this information, the original compound (I) must be ethyl acetoacetate. This compound undergoes a reverse aldol condensation to yield sodium acetate and ethanol.

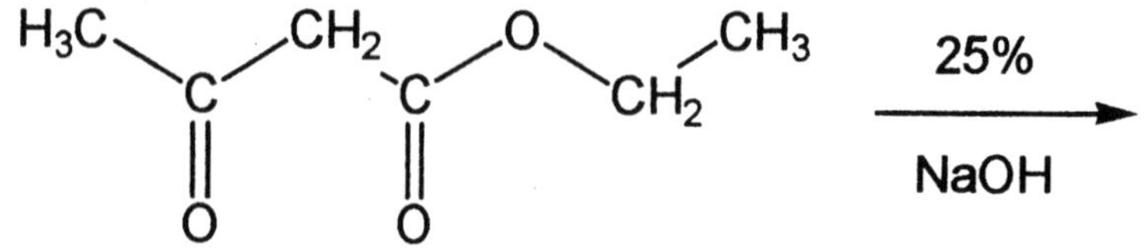

ethyl acetoacetate

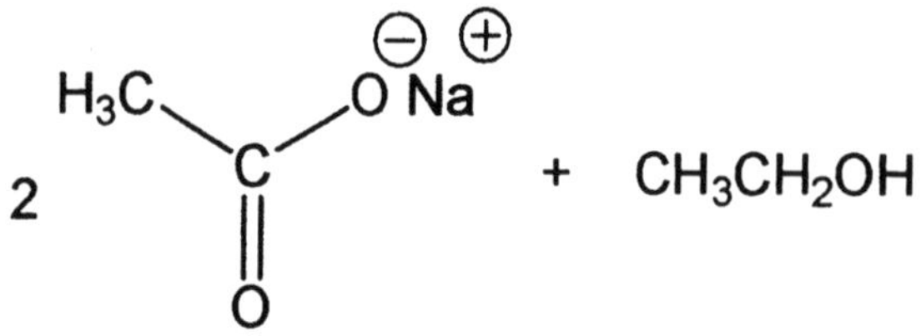

2   sodium acetate   + CH_3CH_2OH

sodium acetate   ethanol

The equation for the preparation of the 4-bromophenacyl ester is given below.

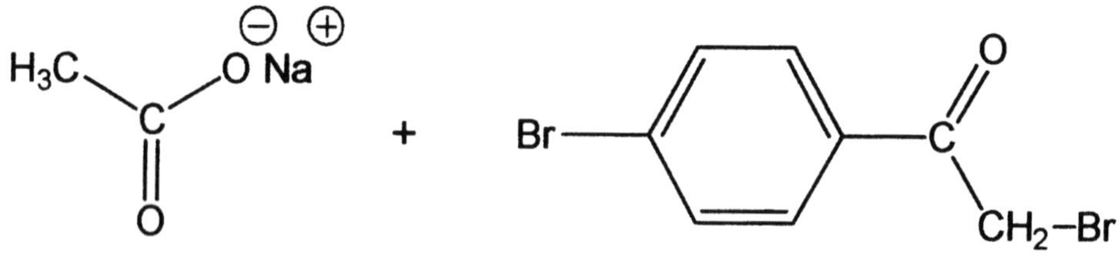

sodium acetate   4-bromophenacyl bromide

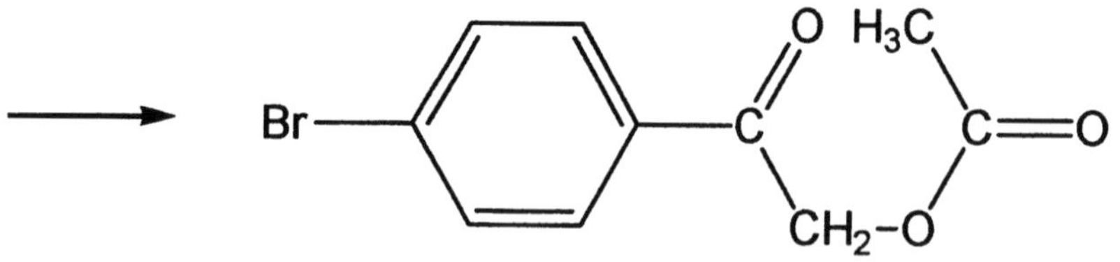

2-(4-bromophenyl)-2-oxoethyl acetate

The IR spectrum indicated at 1715 cm^{-1} the C=O stretch of an aliphatic ketone, at 1634 cm^{-1} the C=O stretch of a β-keto ester, at 1250 cm^{-1} the C–O stretch of an acetate, and at 2933 cm^{-1} the C–H stretch of an alkyl group.

Problem Set 11

1. The reactions can be summarized as follows:

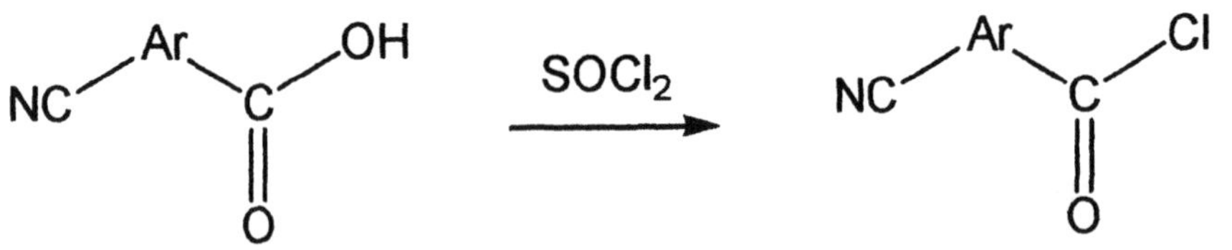

Compound I:
a cyanoacid

SOCl₂ →

a cyano acyl chloride

NH₄OH →

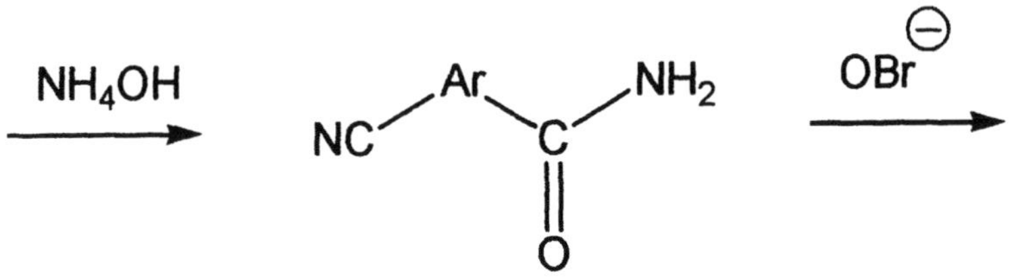

Compound II:
a cyanoamide

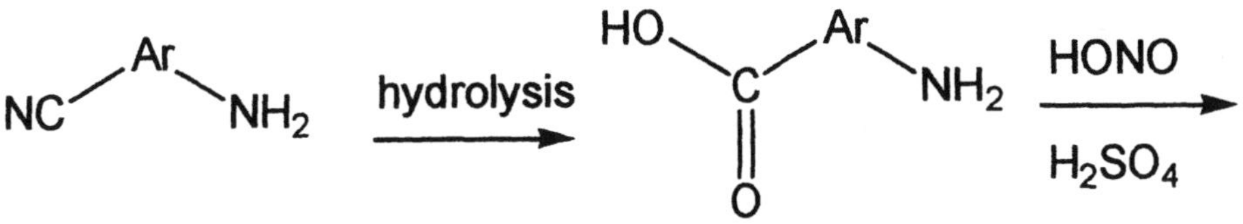

OBr⁻ →

Compound III:
a cyanoamine

hydrolysis →

Compound IV:
an amino acid

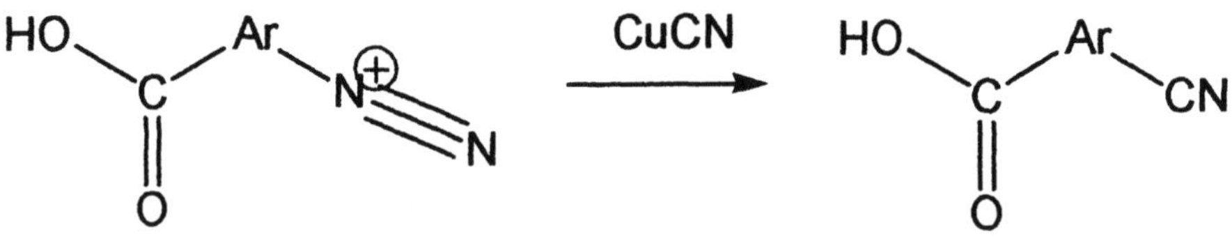

HONO / H₂SO₄ →

a diazonium acid

CuCN →

Compound I:
a cyano acid

An aromatic ring must be present for the diazotization to occur. Since the final product contains a –CN and a –COOH group, the initial compound also contains these groups. The aryl group can be determined from the neutralization equivalent.

$$197 - (26 + 45) = 126$$
$$-CN, -COOH$$

A value of 126 corresponds to $C_{10}H_6$, which is a disubstituted naphthalene. The original acid (Compound I) must be 8-cyano-1-naphthoic acid.

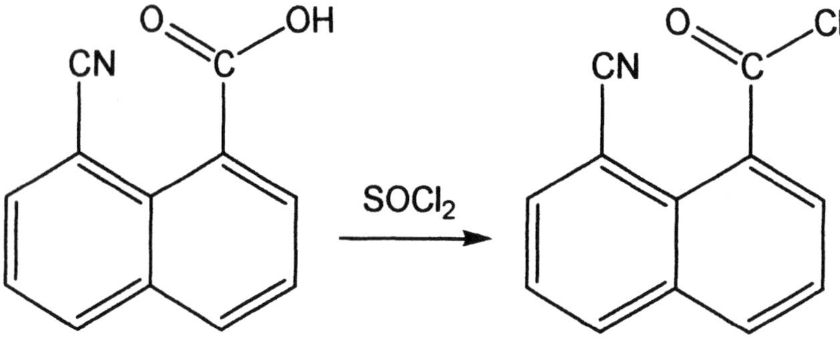

Compound I:
8-cyano-1-naphthoic acid
MW = 197, NE = 197

8-cyano-1-naphthoyl
chloride

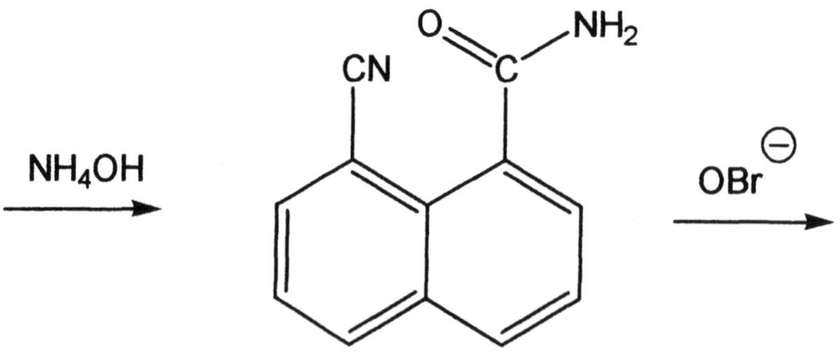

Compound II:
8-cyano-1-naphthamide

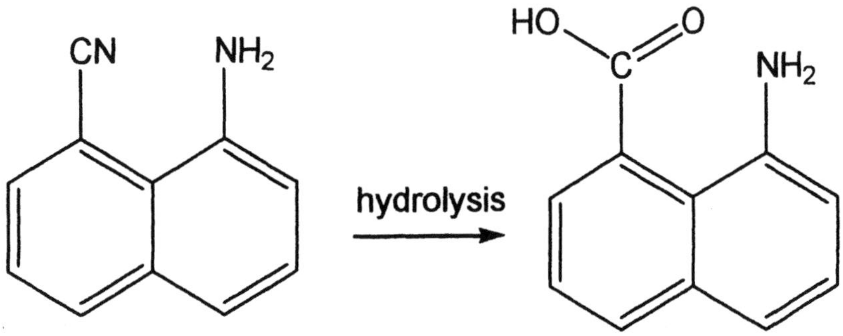

Compound III:
1-amino-8-cyanonaphthalene

hydrolysis

Compound IV:
8-amino-1-naphthoic acid
MW = 187, NE = 187

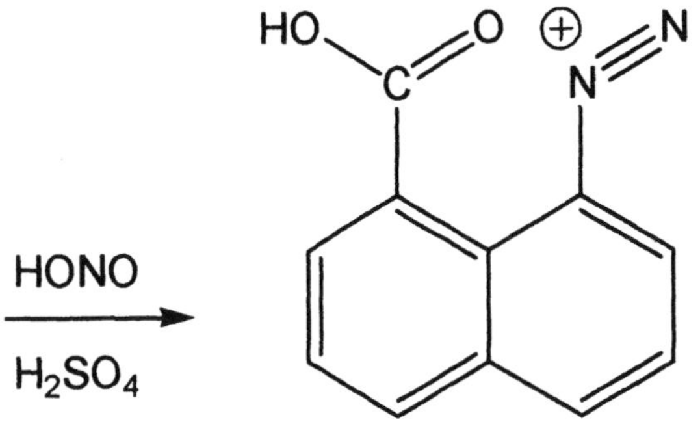

HONO

H$_2$SO$_4$

8-carboxy-2-naphthalene
diazonium salt

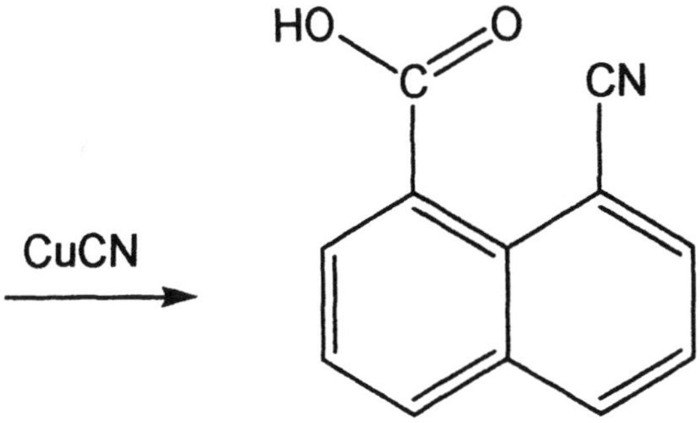

CuCN

Compound I:
naphthoic acid
MW = 197, NE = 197

Hydrolysis of 8-cyano-1-naphthoic acid (Compound I) produced 1,8-naphthalenedicarboxylic acid (Compound V), which is heated to produce 1,8-naphthalenedicarboxylic anhydride (Compound VI).

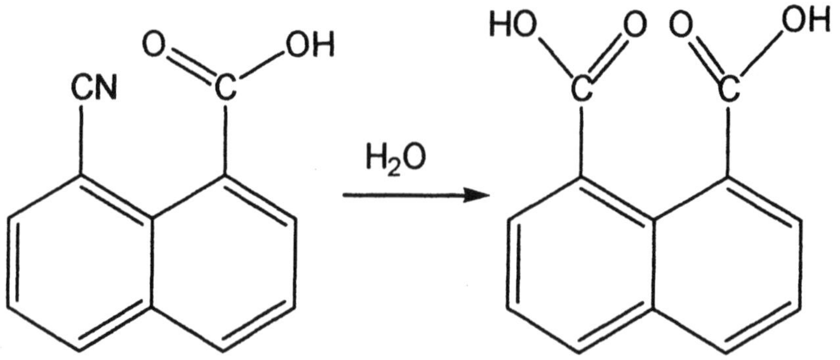

Compound I:
8-cyano-1-naphthoic acid

Compound V:
1,8-naphthalenedicarboxylic acid
MW = 216, NE = 108

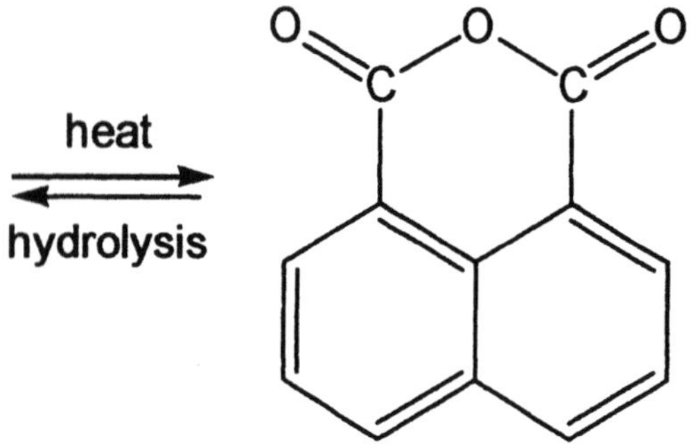

Compound VI:
1,8-napthalic
anhydride

Oxidation of 8-amino-1-naphthoic acid (Compound IV) produced another acid, 1,2,3-benzenetricarboxylic acid (Compound VII).

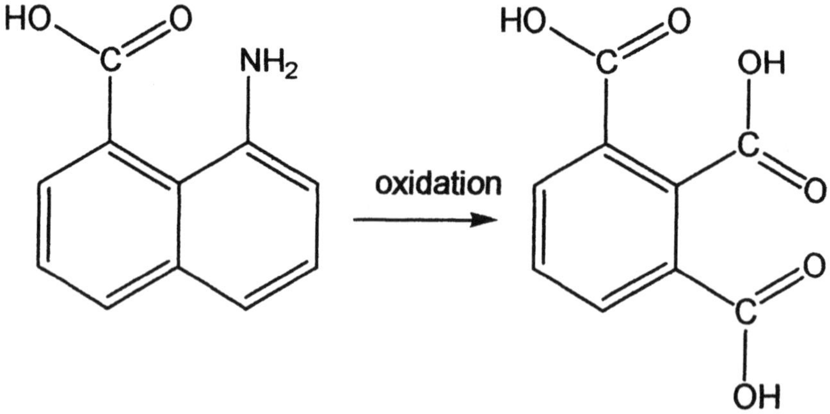

Compound IV:
8-amino-1-naphthoic acid
MW = 187, NE = 187

Compound VII:
1,2,3-benzenetricarboxylic acid
MW = 210, NE = 70

The ¹H NMR spectrum matches the structure for Compound I, with the aromatic peaks at δ 7.6, δ 7.9, and δ 8.28 and the –OH of the acid at δ 11.8.

2. The solubility class for this compound is A₁, A₂, or B, consisting of carboxylic acids, phenols, enols, oximes, imides, sulfonamides, thiophenols, nitro compounds, β-diketones, sulfonamides, amines, or anilines. Acetic anhydride reacts with active hydrogens such as those on amines and alcohols. Since nitrogen was evolved when the compound was treated with nitrous acid, a primary amine must be present. From these tests, the original compound (Compound I) must contain a –COOH and a –NH₂ group, which would also explain the inability to obtain a satisfactory neutralization equivalent.

    The ¹H NMR spectrum of the oxidation product indicates the presence of a *p*-disubstituted benzene ring (4H) with doublets at δ 7.60 and δ 7.95, and a broad singlet indicating a –OH peak for the carboxylic acid group at δ 12.18. The only bromine-containing compound that matches the ¹H NMR spectrum and the neutralization equivalent of 201 is 4-bromobenzoic acid.

    Compounds I, II, and III must also be disubstituted with carbon-containing groups in the position *para* to the bromine. By using the neutralization equivalent of 270 ± 2 for compound II, the structure of compound I can be determined.

$$270 \pm 2 \quad - \quad (76 \quad + \quad 80 \quad + \quad 15 \quad + \quad 45 \quad + \quad 43)$$

| | *p*-disubstituted ring | Br | –NH | –COOH | CH₃CO– |

$$= 11 \pm 2, \text{ which is CH}$$

Thus Compound I must be amino-(4-bromophenyl)acetic acid, which reacts with acetic anhydride to produce acetylamino-(4-bromophenyl)acetic acid (Compound II).

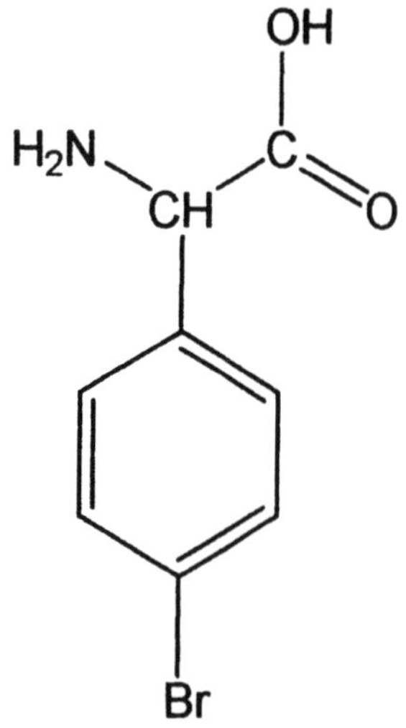

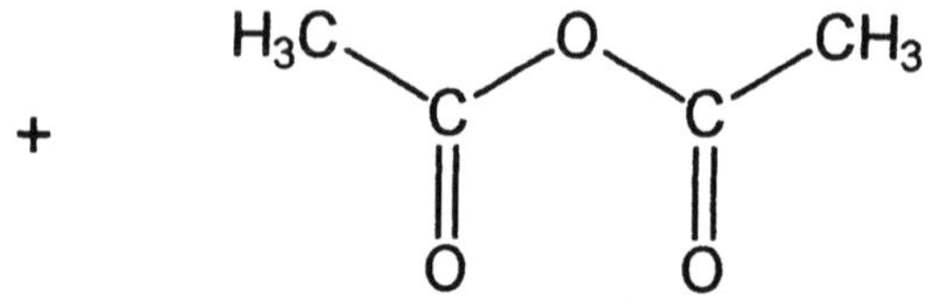

Compound I:
amino-(4-bromophenyl)acetic acid

acetic anhydride

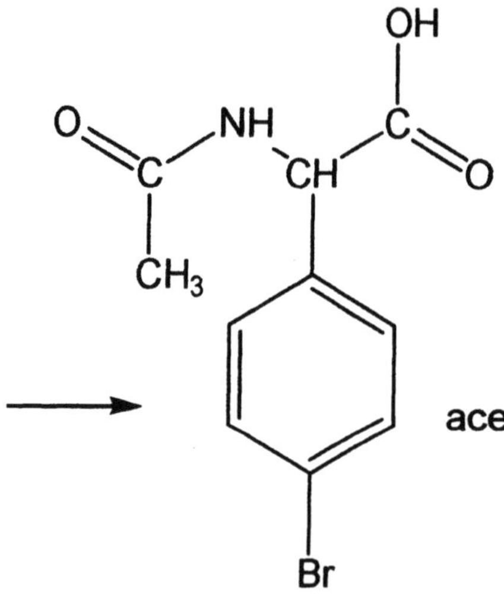

Compound II:
acetylamino-(4-bromophenyl)-
acetic acid
MW = 272, NE = 272

2-Amino-2-(4-bromophenyl)ethanoic acid (Compound I) is treated with nitrous acid to produce 2-(4-bromophenyl)-2-hydroxy-ethanoic acid (Compound III).

Compound I:
amino-(4-bromophenyl)acetic acid

Compound III:
(4-bromophenyl)hydroxyacetic acid

Vigorous oxidation of amino-(4-bromophenyl)acetic acid (Compound I), acetylamino-(4-bromophenyl)acetic acid (Compound II), or (4-bromophenyl)hydroxyethanoic acid (Compound III) produces 4-bromobenzoic acid.

I, II, or III

vigorous
oxidation

4-bromobenzoic acid
MW = 201, NE = 201

3. Esters hydrolyze to acids and alcohols. Compound II contains an active hydrogen (positive acetyl chloride test) and is a phenol (positive ferric chloride test). The ^{1}H NMR spectrum for Compound II indicates an isolated $-CH_3$ at δ 2.25, an $-OH$ at δ 5.67, and a disubstituted aromatic ring (4H) at δ6.5–δ7.3. Therefore, Compound II is a methylphenol. The ^{1}H NMR spectrum for Compound III shows a ratio of 2:1 between the aromatic signals

at δ 7.4–δ 7.9 and the –COOH groups at δ 12.08. Therefore, Compound III is 1,2-benzenedicarboxylic acid (phthalic acid), since it undergoes a reversible dehydration.

The original ester (Compound I) is di-3-methylphenyl 1,2-benzenedicarboxylate, which is hydrolyzed to 1,2-benzenedicarboxylic acid (phthalic acid) (Compound III) and 3-methylphenol (Compound II).

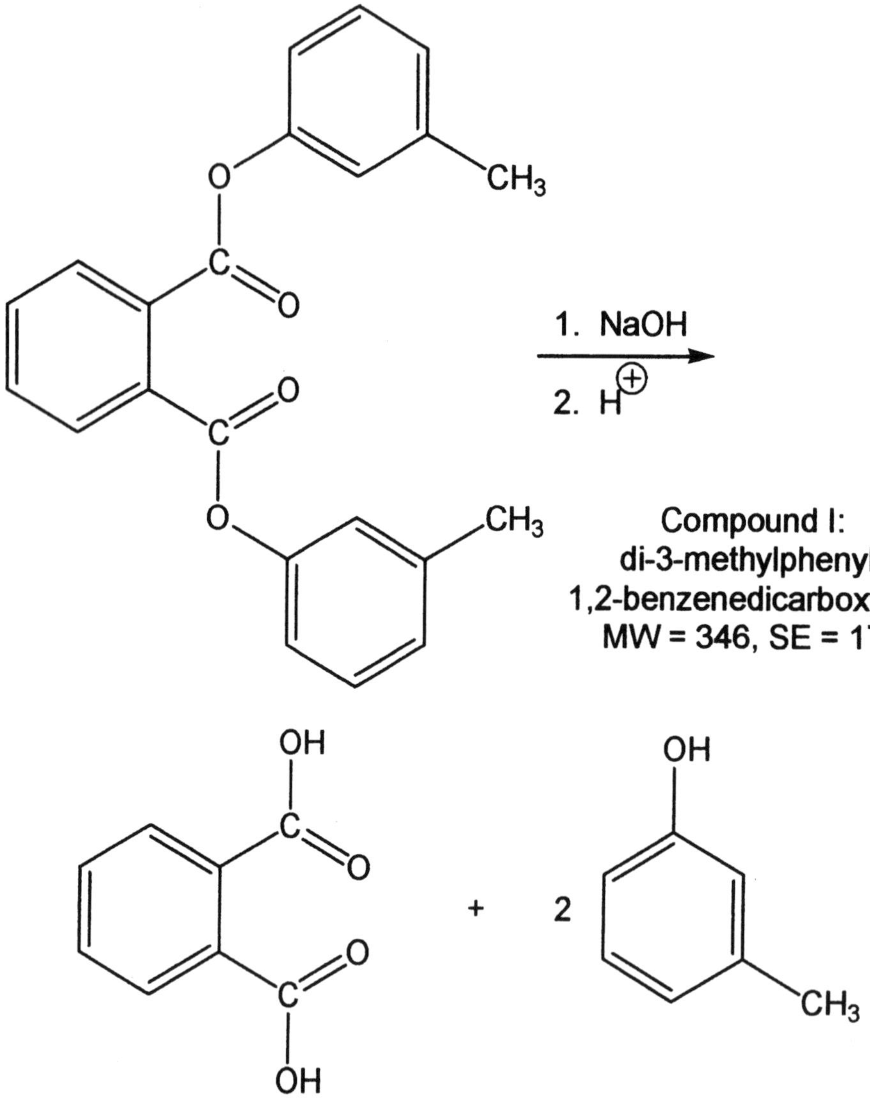

Compound I:
di-3-methylphenyl-
1,2-benzenedicarboxylate
MW = 346, SE = 173

Compound III: phthalic acid
MW = 166, NE = 83

Compound II:
3-methylphenol

When heated, phthalic acid (Compound III) loses water and becomes phthalic anhydride (Compound IV).

Compound III:	Compound IV:
phthalic acid	phthalic anhydride

4. Solid I is in the solubility class of S_2, consisting of salts of an organic acid, amine hydrochlorides, amino acids, or polyfunctional compounds. Treatment of an amine salt of a carboxylic acid with base liberates the salt of a carboxylic acid and an amine (Compound II). The reaction of the amine (Compound II) with benzenesulfonyl chloride produces a benzenesulfonamide. The amine must be a secondary amine since the benzenesulfonamide is insoluble in base.

   The salt of the carboxylic acid from the cold alkali reaction is acidified to liberate the carboxylic acid (Compound III). The acid must also contain a nitro group, since it reacts with zinc dust and ammonium chloride to form a product (hydrazine, hydroxylamine, or aminophenol) that reduces Tollens reagent.

   The ^{1}H NMR spectrum of the amine (Compound II) corresponds to dibutylamine. The amino is at δ 0.53 (1H, s); the two methyls are at δ 0.90 (6H, t); the four methylenes on carbons two and three are at δ 1.2–δ 1.7 (8H, m); and the two methylenes on carbon one are at δ 2.52 (4H, t). The ^{1}H NMR spectrum of the carboxylic acid (III) is a nitrobenzoic acid; it is *meta* because of the splitting pattern in the aromatic region (δ 7.77 - δ 8.96, 4H). The peak at δ 12.0 is the –OH of the acid. The original compound (Compound I) is dibutylammonium 3-nitrobenzoate, which is treated with base to form sodium 3-nitrobenzoate and dibutylamine (Compound II).

Compound I: dibutylammonium 3-nitrobenzoate

sodium 3-nitrobenzoate   Compound II: dibutylamine

The dibutylammonium 3-nitrobenzoate (Compound I) is acidified to give 3-nitrobenzoic acid (Compound III), which is reduced to 3-hydroxyaminobenzoic acid (Compound IV).

Compound I: dibutylammonium 3-nitrobenzoate

Compound III:
3-nitrobenzoic acid
MW = 167, NE = 167

Compound IV:
3-hydroxyaminobenzoic acid

5. From the formula of $C_5H_{10}O$, it is obvious that the compound contains a double bond. Additional proof is seen in the decolorization of bromine and potassium permanganate solutions. The compound is not an aldehyde or a ketone (negative phenylhydrazine test). A negative Lucas test indicates that the compound is not a tertiary or secondary alcohol. The compound is not a methyl ketone nor does it have a methyl next to a carbon with a hydroxy group (negative iodoform test). The compound contains an active hydrogen (positive acetyl chloride test).

The original compound is oxidized with potassium permanganate to a carboxylic acid, with a neutralization equivalent of 59 ± 1. For this acid to be able to lose carbon dioxide upon heating and increase in neutralization equivalent from 59 ± 1 to 73 ± 1, it must be a *gem* dicarboxylic acid. The second acid has a neutralization equivalent of 73 ± 1. The original compound must be 2-methyl-3-buten-1-ol, which is oxidized to 2-methylpropanedioic acid. When heated, the acid is converted to propanoic acid and carbon dioxide.

2-methyl-3-buten-1-ol

2-methylpropanedioic acid
MW = 118, NE = 59

propanoic acid
MW = 74, NE = 74

6. Compound I is in the solubility class of A_2 or B, which include phenols, enols, oximes, imides, sulfonamides, thiophenols, β-diketones, nitro compounds, amines, anilines, or ethers. Compound I was treated with excess acetic anhydride to produce compound II, which is in solubility class of MN, N, or I. Compound I is a phenol since it decolorized bromine water. Compound I reacts with nitrous acid to form an *N*-nitroso (Compound III) without the evolution of nitrogen, indicating the presence of a secondary amine.

From the ^{1}H NMR spectrum of compound I, a $-CH_3$ group is present at $\delta 3.65$, a *p*-disubstituted benzene is indicated from the two doublets in the aromatic region (4H) at $\delta 7.12$ and $\delta 7.49$, and the broad singlets corresponding to the amino and hydroxy groups are at $\delta 3.31$ and $\delta 5.51$.

From the spectrum and the classification tests, compound I is *N*-methyl-4-aminophenol. 4-Acetoxy-*N*-methylacetanilide (Compound II) is the diacetyl derivative.

Compound I:
N-methyl-4-aminophenol

Compound II:
4-acetoxy-*N*-methylacetanilide

Treatment of *N*-methyl-4-aminophenol (Compound I) with nitrous acid yielded 4-hydroxy-*N*-methyl-*N*-nitrosoaniline (Compound III).

Compound I:
N-methyl-4-aminophenol

Compound III:
4-hydroxy-*N*-methyl-*N*-nitrosoaniline

Problem Set 12

1. From the formula the compound must be aromatic with an aliphatic ring, in order to obtain the correct number of carbons and hydrogens. As Compound I ($C_9H_6O_2$) is oxidized, the aliphatic ring opens to form the potassium salt of a carboxylic acid (Compound II) ($C_9H_8O_3$). Compound I must contain a double bond, since it accepts bromine to form a dibromo structure (Compound III) ($C_9H_8O_2Br_2$). When Compound I was heated with sodium hydroxide, followed by acidification, Compound IV ($C_9H_6O_3$) was formed, which involved the loss of both bromine atoms and the formation of a carboxylic acid. The best place to start is to look in any handbook that lists formulas and structures and draw out a few possibilities for compounds I and IV. After scrutinizing the various choices, the only one choice that meets the criteria is coumarin for Compound I.

   Coumarin (Compound I) ($C_9H_6O_2$) is oxidized to the potassium salt of 3-(2-hydroxyphenyl)propenoic acid (Compound II) ($C_9H_8O_3$).

| Compound I: coumarin | potassium salt of Compound II: potassium 3-(2-hydroxyphenyl)-2-propenoate |

   Treatment of coumarin (Compound I) with bromine yields the 3,4-dbromocoumarin (Compound III) ($C_9H_8O_2Br_2$). This product undergoes reaction with base, followed by acid to yield coumarilic acid (Compound IV) ($C_9H_6O_3$).

Compound II:
coumarin

Compound III:
3,4-dibromocoumarin

Compound IV:
benzofuran-2-carboxylic acid

The ^1H NMR spectrum of 3-(2-hydroxyphenyl)-2-propenoic acid (Compound II) gives an integration ratio of 3:1 between the aromatic and alkene protons at $\delta 6.4–\delta 8.2$ and the –OH protons at $\delta 10.3$.

2. From the formula, the structure must be aromatic. For the nitrogen to be lost upon hydrolysis with no loss of carbon(s), a primary amide must be present. For the acid to be optically active, the benzene ring must be monosubstituted. 2-Bromo-2-phenylethanamide (C_8H_8ONBr) reacts with potassium hydroxide to produce the potassium salt of 2-hydroxy-2-phenylethanoic acid ($C_8H_8O_3$).

2-bromo-2-phenylethanamide

potassium salt of 2-hydroxy-
2-phenylethanoic acid

The ^{1}H NMR spectrum of the 2-hydroxy-2-phenylethanoic acid indicates the singlet –CH group at δ 5.22 and the aromatic ring and the –OH groups (7H) at δ 7.2–δ 7.7.

3.   This compound must be aromatic as deduced from the formula. This is a very difficult problem to solve. When Compound I ($C_{11}H_{12}O_4$) is treated with hot sodium hydroxide solution, the ketone group is reduced and the aldehyde group is oxidized to form Compound II ($C_{11}H_{13}O_5Na$). Compound II is protonated to form Compound III ($C_9H_8O_4$), along with the loss of two carbons. In the last step, compound III dimerizes to form compound IV ($C_{18}H_{12}O_6$). The *o, m,* or *p* structures are all possible answers.

hot

NaOH

Compound I:
dimethoxymethylphenyl-
2-oxoethanol

H_2SO_4

-2 CH_3OH

Compound II:
sodium dimethoxymethylphenyl-
2-oxoethanoic acid

Compound III:
carboxyphenyl-2-hydroxy-
ethanoic acid

Compound IV:
3,6-di(carboxyphenyl)-
1,4-dioxane-2,5-dione

4. From the formula of $C_5H_8O_2$, Compound I must contain two double bonds, two rings, or one double bond and a ring. Hydrogen chloride is added across a double bond or a ring producing Compound II ($C_5H_9O_2Cl$). With potassium hydroxide, the chlorine in compound II is replaced by –OH and the salt of the carboxylic acid is formed to yield compound III ($C_9H_5O_3K$). In the presence of potassium permanganate, compound III is oxidized from a secondary alcohol to a ketone and the anion of the acid is protonated to produce compound IV ($C_5H_8O_3$). Hypochlorite ion oxidizes the methyl ketone in compound IV to give Compound V. With a formula of $C_4H_6O_4$ and a neutralization equivalent of 58 ± 1, Compound V must be a dicarboxylic acid. The dicarboxylic acid (Compound V) cyclizes with loss of water to yield Compound VI ($C_4H_4O_3$).

With all of these reactions in mind, the original Compound (I)
must be either 4-pentenoic acid or *γ*-methyl-*γ*-butyrolactone.

Compound I:
4-pentenoic acid

Compound I:
γ–methyl-γ-butyrolactone
(lactone of 4-hydroxybutanoic acid)

Compound II: 4-chloropentanoic acid

Compound III:
potassium 4-hydroxypentanoate

Compound IV:
4-oxopentanoic acid

NaOCl →

Compound V:
butanedioic acid
MW = 118, NE = 59

heat →

Compound VI:
butanedioic anhydride

The ^1H NMR spectrum of butanedioic acid (Compound V) agrees with the structure of butanedioic acid. The integration ratio of 2:1 for the signal at $\delta 2.43$ compared to the $\delta 11.8$ correlates to the ratio of $-CH_2$ to $-OH$ peaks.

5. The singlet at $\delta 3.89$ indicates an isolated methyl group. The pair of doublets, with four protons, at $\delta 6.91$ and $\delta 8.12$ indicates a *p*-disubstituted aromatic ring. The structure must be 4-nitroanisole $(C_7H_7NO_3)$.

CH_3O—⟨benzene⟩—NO_2 4-nitroanisole

6. The compound must be aromatic as indicated from the formula of $C_{16}H_{13}N$. The ^1H NMR spectrum shows a broad singlet at $\delta 5.61$ which indicates a $-NH$. The remainder of the peaks are aromatic (12H) ($\delta 6.60–\delta 7.55$, $\delta 7.80$). Two structures are possible, 1-phenylaminonaphthalene or 2-phenylaminonaphthalene.

1-phenylaminonaphthalene   2-phenylaminonaphthalene

7.   Potassium permanganate oxidizes alkenes to 1,2-diols, which can be further oxidized to ketones. Potassium permanganate also oxidizes alkynes to carboxylic acids. Sodium tests for the presence of an active hydrogen attached to oxygen, nitrogen, sulfur, or a carbon-carbon triple bond. From the formula, the compound must be aromatic. From drawing various structures with a formula of $C_{14}H_{10}O$, two benzene rings must be present in the structure and these rings are not fused together. Possible structures are 2-phenoxyphenyl-ethyne, 3-phenoxyphenylethyne, or 4-phenoxyphenylethyne.

2-phenoxyphenylethyne, 3-phenoxy-
phenylethyne, or 4-phenoxyphenylethyne

Oxidation of 2-phenoxyphenylethyne, 3-phenoxyphenylethyne, or 4-phenoxyphenylethyne ($C_{14}H_{10}O$) produces 2-phenoxybenzoic acid, 3-phenoxybenzoic acid, or 4-phenoxybenzoic acid ($C_{13}H_{10}O$).

2-phenoxyphenylethyne, 3-phenoxy-
phenylethyne, or 4-phenoxyphenylethyne

2-phenoxybenzoic acid, 3-phenoxy-
benzoic acid, or 4-phenoxybenzoic acid

Treatment of 2-phenoxyphenylethyne, 3-phenoxyphenylethyne, or 4-phenoxyphenylethyne ($C_{14}H_{10}O$) with sodium produces the sodium salt of 2-phenoxyphenylethyne, 3-phenoxyphenylethyne, or 4-phenoxyphenylethyne ($C_{14}H_9ONa$).

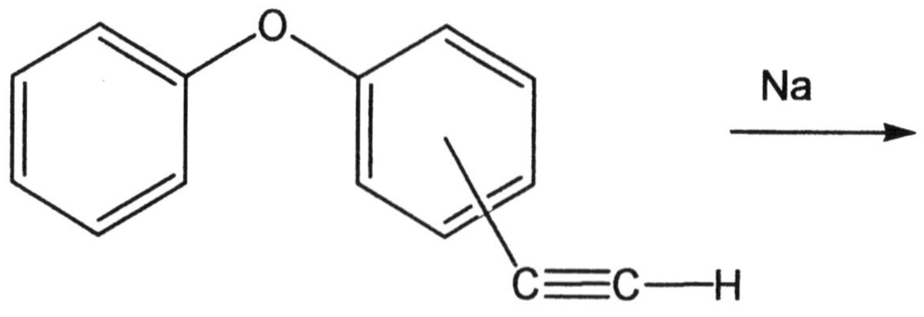

2-phenoxyphenylethyne, 3-phenoxy-
phenylethyne, or 4-phenoxyphenylethyne

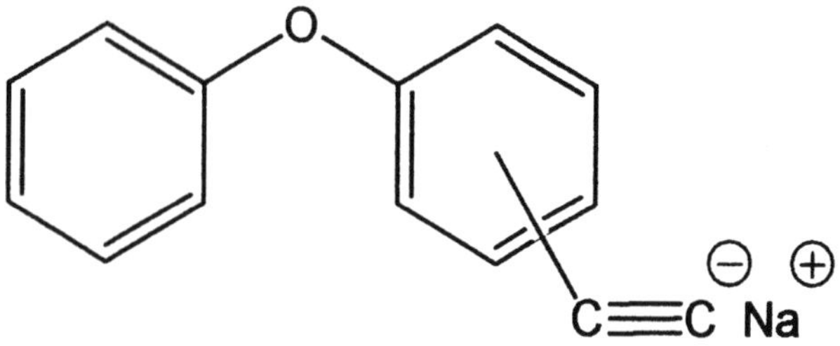

sodium salt of 2-phenoxyphenylethyne, 3-phenoxy-
phenylethyne, or 4-phenoxyphenylethyne

8. This structure cannot have double bonds that can accept bromine. A possible structure would be 1,2,3,4,5,6-benzenehexacarboxylic acid.

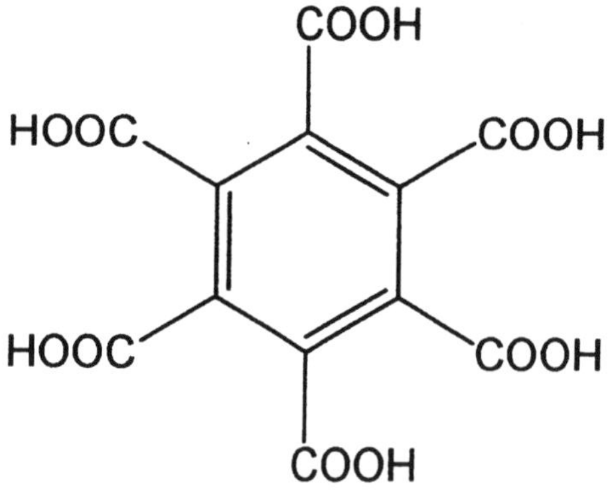

1,2,3,4,5,6-benzene-
hexacarboxylic acid
MW = 342, NE = 57

Problem Set 13

1.   Compound I ($C_5H_4O_2$) must have a double bond (positive potassium permanganate test) and is an aldehyde (positive Tollens test). Compound I is dimerized with heat and alkali cyanide to produce Compound II. Compound I reacts with excess ethanol to form an acetal (Compound III) ($C_9H_{14}O_3$). Because of the formula, Compound I must be cyclic. The original compound (Compound I) must be furfural. Furfural (Compound I) ($C_5H_4O_2$) is dimerized with heat and alkali cyanide to form furoin (Compound II).

Compound I: furfural   Compound II: furoin

   Furfural (Compound I) ($C_5H_4O_2$) reacts with excess ethanol to form the acetal (Compound III) ($C_9H_{14}O_3$) of furfural.

Compound I: Compound III:
furfural   acetal of furfural

   The ^1H NMR spectrum of furfural (Compound I) ($C_5H_4O_2$) indicates a conjugated system. The three protons in the range of δ 6.63–δ 7.72 are the hydrogens attached to the carbons in the ring. The singlet at δ 9.67 is indicative of an isolated aldehyde. Thus the protons are identified as follows: δ 6.63, 1H, d of d (**a**); δ 7.28, 1H, d (**b**); δ 7.72, 1H, m (**c**); and δ 9.67, 1H, s (**d**).

furfural

2.   Compound I is in solubility class A$_1$ or A$_2$, consisting of carboxylic acids, phenols, β-diketones, enols, oximes, imides, sulfonamides, thiophenols, or nitro compounds. From the formula of C$_4$H$_4$O$_4$, the compound must contain three double bonds, three rings, two double bonds and one ring, two rings and a double bond, a triple bond and one ring, or a triple bond and a double bond. Treatment of Compound I with bromine yields Compound II (C$_4$H$_4$O$_4$Br$_2$), thus indicating the presence of one carbon-carbon double bond. Compound I is regenerated by the reaction of compound II with zinc dust, confirming the presence of a carbon-carbon double bond in I.

   Compound I is hydrogenated to produce Compound III. Compound III has a neutralization equivalent of 59 and a molecular weight of 118 (probably C$_4$H$_6$O$_4$), thus indicating the presence of two –COOH groups. Compound III lost a molecule of water when heated, forming a cyclic anhydride (Compound IV). The anhydride (Compound IV) reacted with aluminum chloride and benzene in a Friedel-Craft acylation to form a ketone (Compound V) (C$_{10}$H$_{10}$O$_3$). The presence of a ketone is confirmed by a positive test with phenylhydrazine and vigorous oxidation to produce benzoic acid.

   The original compound (Compound I) must be *cis* or *trans* butenedioic acid (C$_4$H$_4$O$_4$). Butenedioic acid (Compound I) (C$_4$H$_4$O$_4$) is brominated to yield 2,3-dibromobutanedioic acid (Compound II) (C$_4$H$_4$Br$_2$O$_4$), which is debrominated with zinc.

Compound I:
butenedioic acid

Compound II:
2,3-dibromobutanedioic acid

Butenedioic acid (Compound I) ($C_4H_4O_4$) is hydrogenated to yield butanedioic acid (Compound III), which loses water to produce butanedioic anhydride (Compound IV). Butanedioic anhydride (Compound IV) reacts with aluminum chloride and benzene to yield 4-oxo-4-phenylbutanedioic acid (Compound V) ($C_{10}H_{10}O_3$), which is oxidized to benzoic acid.

Compound I: butenedioic acid

Compound III: butanedioic acid
MW = 118, NE = 59

Compound IV: butanedioic anhydride

Compound V:
4-oxo-4-phenylbutanoic acid

benzoic acid

The ^1H NMR spectrum of 4-oxo-4-phenylbutanoic acid (Compound V) ($C_{10}H_{10}O_3$), shows two triplets of two hydrogens each at δ 2.8 and δ 3.3 indicating the two $-CH_2-$ groups adjacent to each other. Additionally, a monosubstituted benzene (5H) is indicated at δ7.2–δ8.0, and the broad singlet at δ11.7 is the –OH.

3.   From the formula of $C_{11}H_{10}N_2$, the original compound must be aromatic. Possible answers would be a dipyridylmethane oxidizing to a dipyridyl ketone. The methylene group may be attached to the pyridine ring in the 2, 3, or 4 positions of either ring.

dipyridylmethane

dipyridyl ketone

4.   The formula, $C_5H_{10}O$, indicates the presence of a ring, a carbon-carbon double bond, a carbon-oxygen double bond, or an aliphatic ring. A compound that decolorizes a potassium permanganate solution, but is not affected by bromine in carbon tetrachloride contains a carbonyl group.
      The only aldehyde that has a boiling point of 75°C and has a formula of $C_5H_{10}O$ is 2,2-dimethylpropanal.

2,2-dimethylpropanal
(pivaldehyde)

      The following fragments can be assigned in the mass spectrum.

m/z	fragment	
86	 $H_3C - \overset{\overset{\displaystyle CH_3}{\vert}}{\underset{\underset{\displaystyle CH_3}{\vert}}{C}} - \overset{\displaystyle H}{\underset{\underset{\displaystyle O}{\parallel}}{C}} \overset{+}{\cdot}$	
57	$H_3C - \overset{\overset{\displaystyle CH_3}{\vert}}{\underset{\underset{\displaystyle +}{}}{C}} - CH_3$	
55	$H_2C - \overset{\overset{\displaystyle CH_2}{\parallel}}{\underset{\underset{\displaystyle +}{}}{C}} - CH_3$	(due to rearrangement)
41	$H_2C = \underset{\underset{\displaystyle +}{}}{C} - CH_3$	(due to rearrangement)
29	$CH_3CH_2{}^+$	(due to rearrangement)
27	$H_2C = CH^+$	(due to rearrangement)

5. From the formula of $C_{14}H_{12}O$, the structure must be aromatic. To produce a carboxylic acid from chromic acid oxidation, the original compound must be a primary alcohol, an aldehyde, or an alkyl group attached to an aromatic ring. By calculating the difference in the molecular weight of the original compound ($C_{14}H_{12}O$) and the neutralization equivalent of the product ($226 - 196 = 30$), a value of 30 is obtained, indicating that the original compound must be gaining two oxygens and losing two hydrogens upon oxidation.

In the ^{1}H NMR spectrum of the oxidation product, the aromatic peaks are seen as nine hydrogens in the range of δ 7.45–δ 8.19. The –OH of the carboxylic acid appears as a singlet at δ 10.8. This compound must consist of two nonfused benzene rings. Possibilities for this structure include 2, 3, or 4-benzoylbenzoic acid. The original compound must be 2, 3, or 4-methylbenzophenone.

2, 3, or 4-methylbenzophenone

2, 3, or 4-benzoylbenzoic acid
MW = 228, NE = 226

6.   Compound I must be highly conjugated because of the formula $C_{10}H_6O_4$ and the ^{1}H NMR spectrum. Based upon this information, compound I would be 1,2-difuryl-1,2-ethanedione. This compound, when treated with sodium hydroxide, can rearrange in a manner similar to the benzoin rearrangement to benzilic acid to yield 2,2-difuryl-2-hydroxyethanoic acid (Compound II).

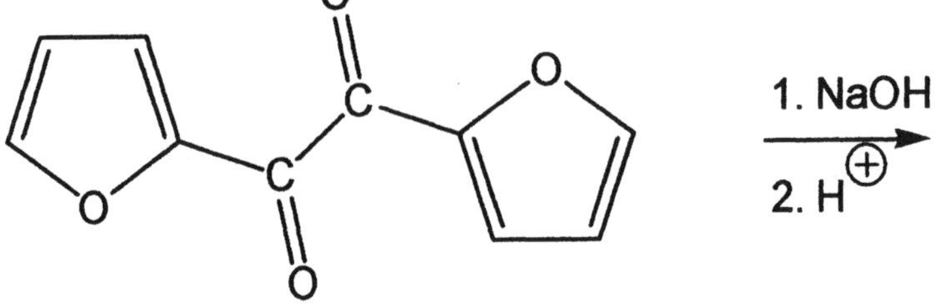

Compound I: 1,2-difuryl-1,2-ethanedione

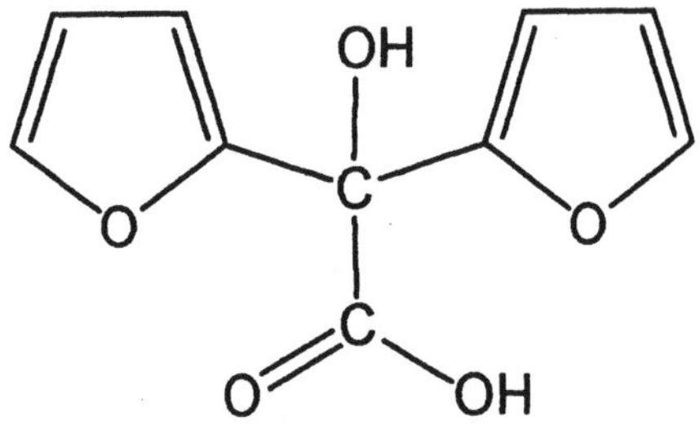

Compound II: 2,2-difuryl-2-hydroxyethanoic acid
MW = 208, NE = 208

Treatment of 1,2-difuryl-1,2-ethanedione (Compound I) ($C_{10}H_6O_4$) with hydrogen peroxide can cause cleavage between the two carbonyl groups to produce two molecules of furoic acid (Compound III).

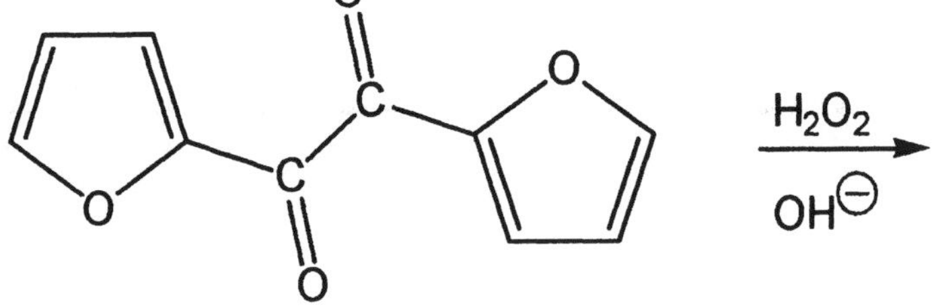

Compound I: 1,2-difuryl-1,2-ethanedione

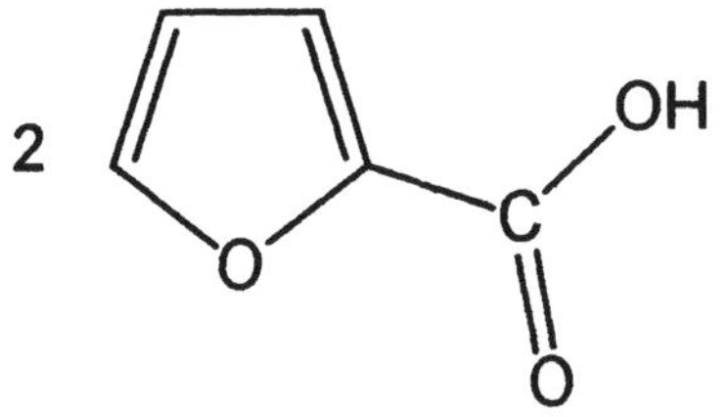

Compound III:
furoic acid
MW = 112, NE = 112

The ^1H NMR spectrum of 1,2-difuryl-1,2-ethanedione (Compound I) ($C_{10}H_6O_4$) indicates a conjugated system. The six hydrogens in the range of $\delta 6.64$–$\delta 7.78$ are the hydrogens attached to the carbons in the ring. The protons are identified as follows: δ 6.64, 2H, d of d (*a*); $\delta 7.63$, 2H, d (*b*); and $\delta 7.78$, 2H, d (*c*).

7.   The reaction of compound I with iron and hydrochloric acid reduced the nitro group in structure I to an amino group in structure II. In both compounds, the rest of the formula, $C_{10}H_7$, must be a naphthalene ring. Thus compound I is 1-nitro-naphthalene and compound II is 1-aminonaphthalene.

Compound I:
1-nitronaphthalene

Compound II:
1-aminonaphthalene

Both 1-nitronaphthalene (Compound I) ($C_{10}H_7NO_2$) and 2-aminonaphthalene (Compound II) ($C_{10}H_9N$) are oxidized, with loss of two carbons each to form 1,2-benzenedicarboxylic acids ($C_8H_5O_6N$ and $C_8H_6O_4$). In 1-nitronaphthalene (Compound I) ($C_{10}H_7NO_2$), the nitro group makes that ring less reactive than the other ring, so carbons five and eight become the –COOH groups. In 2-aminonaphthalene (Compound II) ($C_{10}H_9N$), the amino group makes that ring more reactive than the other ring, so carbons one and four become the –COOH groups.

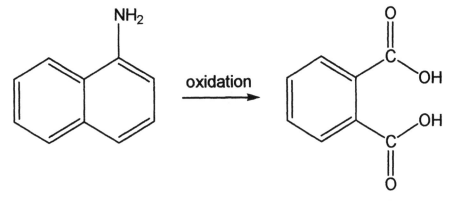

Compound I:
1-nitronaphthalene

3-nitro-1,2-benzene-
dicarboxylic acid

Compound II:
1-aminonaphthalene

1,2-benzenedicarboxylic acid
(phthalic acid)

heat

−H₂O

1,2-benzenedicarboxylic acid
anhydride
(phthalic anhydride)

The ^{1}H NMR spectrum of any of these compounds will only show protons in the aromatic region.

Problem Set 14

1. Compound I ($C_{10}H_{10}O_2$) does not have an active hydrogen (negative acetyl chloride test) and is not an aldehyde or a ketone (negative phenylhydrazine test). Compound II also does not have an active hydrogen (negative acetyl chloride test) and is not an aldehyde or a ketone (negative phenylhydrazine test). Both compounds contain a multiple bond because they decolorized solutions of bromine and of potassium permanganate. Compound III is an aldehyde or a ketone (positive phenylhydrazine test). The formula ($C_8H_6O_4$) for compound IV indicates that it is a substituted benzoic acid.

    The naturally occurring compound that fits this data is safrole (Compound I) which isomerizes to isosafrole (Compound II). Isosafrole (Compound II) is then converted to piperonal (Compound III). Isosafrole (Compound II) and piperonal (Compound III) are both oxidized with alkaline potassium permanganate to piperonylic acid (Compound IV).

Compound I:
safrole,
1-allyl-3,4-methylene-
dioxybenzene

Compound II:
isosafrole,
1-propenyl-3,4-methylene-
dioxybenzene

Compound III:
piperonal,
3,4-methylene-
dioxybenzaldehyde

II or III $\xrightarrow{\text{KMnO}_4}$

Compound IV:
piperonylic acid,
3,4-methylene-
dioxybenzoic acid

The structure for safrole (Compound I) $(C_{10}H_{10}O_2)$ is supported by the following ^{1}H NMR spectrum: the methylene outside of the ring is at δ 3.30, 2H, d of m (**a**); the alkene protons are at δ 4.90, 1H, m (**b**); δ 5.15, 1H, m (**c**); and δ 5.6–δ 6.2, 1H, m (**e**); the methylene that is between two oxygens is at δ 5.88, 2H, s (**d**); and the aromatic protons are located at δ 6.67, 3H, s (**f**).

Compound I:
safrole,
1-allyl-3,4-methylene-
dioxybenzene

2. From the formula $(C_8H_5ClO_2)$, compound I has to be aromatic. Treatment of compound I with ethanol resulted in the addition of three molecules of ethanol to form compound II. Thus, an aldehyde, ketone, or acid halide must be present.

Compound II was oxidized with potassium permanganate to yield a dicarboxylic acid ($C_8H_6O_4$), which was then treated with excess thionyl chloride to produce a diacid chloride (Compound III) ($C_8H_4Cl_2O_2$).

The reaction of compound I with aniline yielded a compound ($C_{20}H_{16}N_2O$) that had close to two aniline molecules added; aniline adds to aldehydes, ketones, and acid halides.

Since all of the ^1H NMR signals for compound III are in the aromatic range of δ 7.72–δ 8.82, the functional groups must be attached to a benzene ring. The benzene ring is disubstituted since there are only 4H. Since there are only three sets of equivalent hydrogens, the groups must be *meta* to each other. If the groups were *para*, then all the aromatic protons would be equivalent and only one aromatic signal would be present. If the groups were *ortho*, there would be two sets of aromatic protons, each with two protons.

Therefore, Compound III must be 1,3-benzenedicarboxylic dichloride. Working backwards, Compound I must be 2-formyl-benzoyl chloride. 2-Formylbenzoyl chloride (Compound I) ($C_8H_5ClO_2$) is treated with excess ethanol to form the acetal (Compound II) ($C_{14}H_{20}O_4$). The acetal (Compound II) ($C_{14}H_{20}O_4$) is oxidized to 1,3-benzenedicarboxylic acid (isophthalic acid) ($C_8H_6O_4$), which is converted with thionyl chloride to 1,3-benzenedicarboxylic chloride (Compound III) ($C_8H_4Cl_2O_2$).

excess
CH₃CH₂OH

Compound I:
3-formylbenzoyl chloride

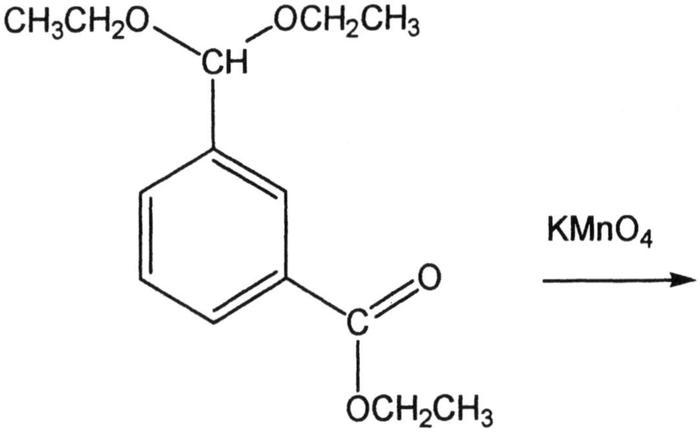

Compound II:
ethyl 3-diethoxymethylbenzoate

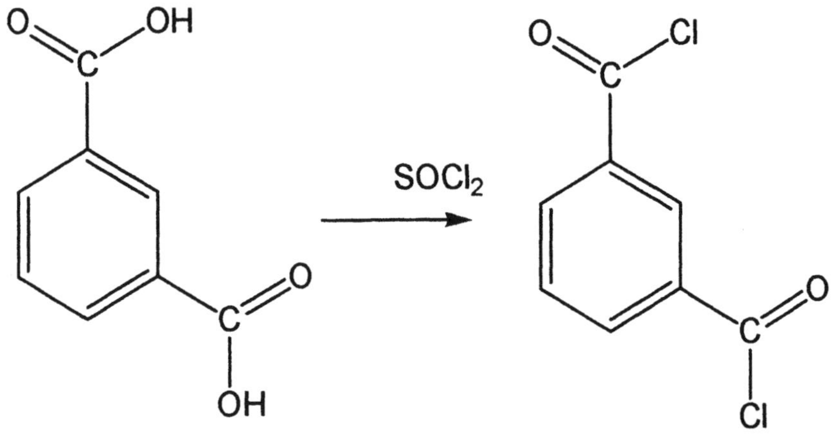

KMnO₄ →

1,3-benzenedicarboxylic acid,
isophthalic acid

SOCl₂ →

Compound II,
1,3-benzenedicarboxylic dichloride
isophthaloyl dichloride

2-Formylbenzoyl chloride (Compound I) ($C_8H_5ClO_2$) is reacted with excess aniline to produce *N*-phenyl-3-phenyliminomethyl-benzamide ($C_{20}H_{16}N_2O$).

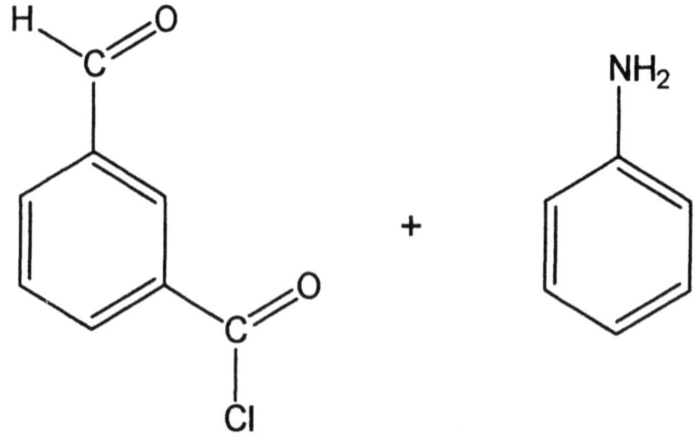

Compound I:
2-formylbenzoyl chloride

excess aniine

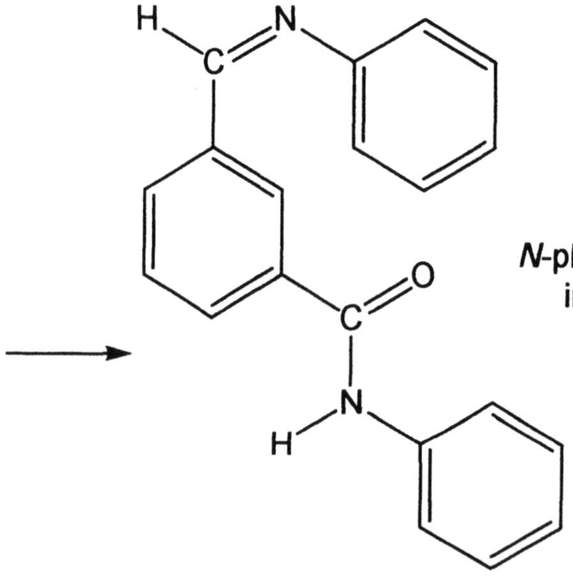

+

N-phenyl-3-phenyl-
iminomethyl-
benzamide

3. From the formula of $C_8H_{14}O_3$, the compound contains two rings, two double bonds, one triple bond, or a double bond and a ring. The compound is in the solubility class of A_1 or A_2, indicating a carboxylic acid, phenol, β-diketone, enol, oxime, imide, sulfonamide, thiophenol, or nitro compound.

　　Compound III gave a positive test with semicarbazide indicating an aldehyde or a ketone, but a negative test with iodoform indicating that a methyl ketone or methyl secondary alcohol is not present. From the ^1H NMR spectrum of the semicarbazone of carbonyl compound III, evidence is seen that compound III is 3-pentanone.

Working backwards, compound I ($C_8H_{14}O_3$) is ethyl 2-methyl-3-oxopentanoate. When ethyl 2-methyl-3-oxopentanoate (Compound I) is treated with phenylhydrazine, the initial phenylhydrazone is formed. The reaction continues as the ethoxy group is lost and the chain cyclizes to form ethyl 2-methyl-3-(phenylhydrazono)pentanoate (Compound II) ($C_{12}H_{14}N_2O$).

Compound I:
ethyl 2-methyl-3-oxopentanoate

phenylhydrazine

ethyl 2-methyl-3-(phenylhydrazono)pentanoate

Compound II:
5-ethyl-4-methyl-2-phenyl-3-pyrazolone

β-Keto acids undergo decarboxylation to yield ketones. Since ethyl 2-methyl-3-oxopentanoate (Compound I) $(C_8H_{14}O_3)$ is an β-keto ester, it hydrolyzes to produce a β-keto acid. The acid then decarboxylates to give 3-pentanone (Compound III) $(C_5H_{10}O)$.

Compound I: ethyl 2-methyl-3-oxopentanoate

2-methyl-3-oxopentanoic acid

Compound III: 3-pentanone

3-Pentanone (Compound III) ($C_5H_{10}O$) reacts with semicarbazide to form the semicarbazone.

Compound III: 3-pentanone

semicarbazide

semicarbazone
of 3-pentanone

    The ^1H NMR spectrum of the semicarbazone shows as a triplet with six hydrogens at δ 1.09, indicating a CH_3 adjacent to a CH_2; a quartet with four hydrogens at δ 2.26, indicating a CH_2 adjacent to a CH_3; a broad singlet with two hydrogens at δ 5.89, indicating a NH_2; and a broad singlet with one hydrogen at δ 8.59, referring to the NH.

4.   From the formula of C_9H_7N, a fused ring system is indicated, with nitrogen being a member of one of the rings. Two possibilities of a structure with this formula are quinoline or isoquinoline. However, later oxidation steps eliminate quinoline as a possibility.

Isoquinoline (Compound I) (C$_9$H$_7$N), is catalytically reduced to 1,2,3,4-tetrahydroisoquinoline (Compound II) (C$_9$H$_{11}$N). 1,2,3,4-Tetrahydroisoquinoline (Compound II) (C$_9$H$_{11}$N) is treated with excess methyl iodide to form a quaternary ammonium salt, *N,N*-dimethyl-1,2,3,4-tetrahydroisoquinolinium iodide. With silver oxide, this salt is changed to the hydroxide salt, *N,N*-dimethyl-1,2,3,4-tetrahydroiso-quinolinium hydroxide. This hydroxide salt, when heated, decomposes to water and 2-(*N,N*-dimethylaminomethyl)styrene (Compound III) (C$_{11}$H$_{15}$N). Vigorous oxidation of 2-(*N,N*-dimethylaminomethyl)-styrene (Compound III) (C$_{11}$H$_{15}$N) produces 1,2-benzenedicarboxylic acid (phthalic acid) (Compound IV) (C$_8$H$_6$O$_4$).

Compound I:
isoquinoline

catalytic
reduction

Compound II:
1,2,3,4-tetrahydroisoquinoline

CH$_3$I

Ag$_2$O

N,N-dimethyl-1,2,3,4-tetrahydro-
isoquinoline iodide

heat
−H$_2$O

N,N-dimethyl-1,2,3,4-tetrahydro-
isoquinoline hydroxide

Compound III:
2-(*N,N*-dimethylamino-
methyl)styrene

Compound IV:
1,2-benzenedicarboxylic acid
(phthalic acid)

Ozonolysis of (*N,N*-dimethylaminomethyl)styrene (Compound III) ($C_{11}H_{15}N$), followed by hydrolysis oxidized the alkene group to form 2-(*N,N*-dimethylaminomethyl)benzaldehyde (Compound V) ($C_{10}H_{13}NO$). 2-(*N,N*-Dimethylaminomethyl)benzaldehyde (Compound V) ($C_{10}H_{13}NO$), when treated with a solution of potassium cyanide, dimerized to form 2,2'-(*N,N*-dimethylaminomethyl)benzoin (Compound VI) ($C_{20}H_{26}N_2O_2$), which is then oxidized back to 1,2-benzene-dicarboxylic acid (Compound IV) ($C_8H_6O_4$).

1. O_3
2. H_2O

Compound III:
2-(*N,N*-dimethylamino-
methyl)styrene

Compound V:
2-(*N,N*-dimethylamino-
methyl)benzaldehyde

KCN

Compound VI:
2,2'-(*N,N*-dimethylaminomethyl)benzoin

vigorous
⟶
oxidation

Compound IV:
1,2-benzenedicarboxylic acid
(phthalic acid)

The ^{1}H NMR spectrum agrees with the structure of isoquinoline (Compound I) (C_9H_7N), with seven protons in the aromatic region (δ 7.25–δ 9.26).

5. The very last product with the formula of $C_8H_6O_4$ has a ^{1}H NMR spectrum that shows only two singlets at δ 8.08 and δ 11.00 with an integration ratio of 2:1. The structure of this compound is 1,4-benzenedicarboxylic acid (terephthalic acid) and the ratio agrees with the comparison of aromatic protons to –OH protons. This is the *para* product, since *ortho* and *meta* would produce more splitting in the aromatic range. Compound I must have an active hydrogen since it reacted with sodium.

Sulfuric acid and mercuric sulfate react with terminal alkynes to form methyl ketones. Methyl ketones are oxidized to carboxylic acids with sodium hypochlorite. Since treatment of compound III with hydrogen bromide resulted in the loss of a carbon, the CH_3OCH_2 group must be attached to the benzene ring. Therefore, Compound I must be 4-methoxymethylphenyl-ethyne ($C_{10}H_{10}O$).

Treatment of 4-methoxymethylphenylethyne (Compound I) ($C_{10}H_{10}O$) with sodium produced the sodium salt of 4-methoxymethylphenylethyne ($C_{10}H_9ONa$), which gave 4-methoxymethylphenylethyne (Compound I) when treated with water.

Compound I:
4-methoxymethylphenylethyne

sodium salt of 4-methoxy-
methylphenylethyne

Sulfuric acid and mercuric sulfate reacted with 4-methoxymethylphenylethyne (Compound I) ($C_{10}H_{10}O$), to yield 4-methoxymethylacetophenone (Compound II) ($C_{10}H_{12}O_2$). Sodium hypochlorite oxidized 4-methoxymethylacetophenone (Compound II) ($C_{10}H_{12}O_2$) to 4-methoxymethylbenzoic acid (Compound III) ($C_9H_{10}O_3$). Treatment of 4-methoxymethylbenzoic acid (Compound III) ($C_9H_{10}O_3$) with hydrogen bromide produced 4-bromomethylbenzoic acid (Compound IV) ($C_8H_7BrO_2$), which was oxidized to 1,4-benzenedicarboxylic acid (Compound V) ($C_8H_6O_4$).

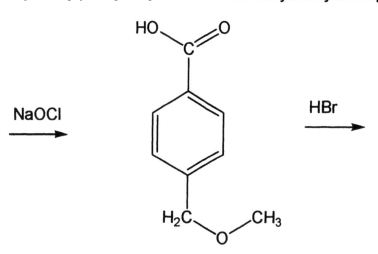

Compound I:
4-methoxymethylphenylethyne

Compound II:
4-methoxymethylacetophenone

Compound III: 4-methoxymethylbenzoic acid

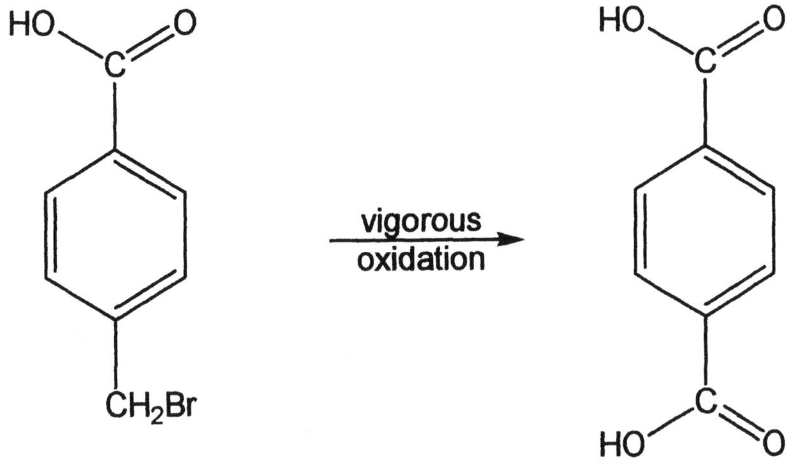

Compound IV:
4-bromomethylbenzoic acid

vigorous
oxidation →

Compound V:
1,4-benzenedicarboxylic acid,
terephthalic acid

6.   From the formula of $C_{14}H_{12}$, the compound obviously contains two
benzene rings that are not fused together. Inferred from the oxidation
reaction is that both benzene rings are attached to the same carbon.
The only possible structure for $C_{14}H_{12}$ is 1,1-diphenylethene. This
compound is then oxidized to benzo-phenone ($C_{13}H_{10}O$).
Benzophenone cannot be oxidized any further.

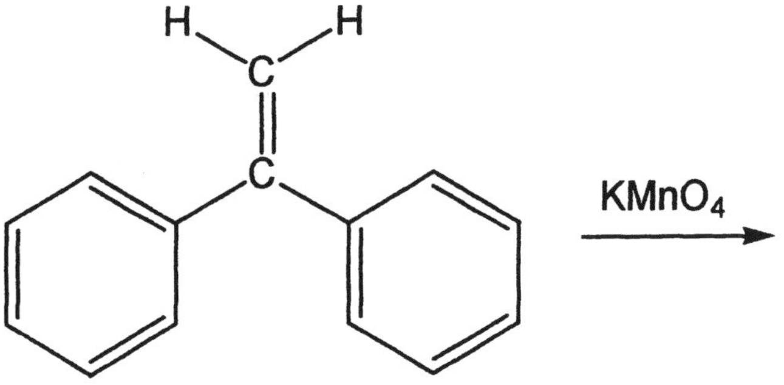

1,1-diphenylmethane

$KMnO_4$ →

benzophenone

The ^{1}H NMR spectrum of 1,1-diphenylmethane ($C_{14}H_{12}$) shows the $-CH_2$ peak at δ 5.35 and ten aryl protons at δ 7.21.

7.   In the ^{1}H NMR spectrum of the isomer, the singlet at δ 0.95 indicates a $-NH_2$ group, the singlet at δ 2.30 indicates an isolated $-CH_3$, the multiplet at δ 2.5–δ 3.0 signifies two $-CH_2$ groups, and the four protons at δ 7.05 indicates a p-disubstituted benzene. The structure that matches this ^{1}H NMR spectrum is α-methyl-4-methylbenzylamine. The only optically active p-disubstituted compound that agrees with the formula ($C_9H_{13}N$) is 4-(aminoethyl)toluene. Vigorous oxidation of either amine produces 1,4-benzenedicarboxylic acid.

α-methyl-4-methyl-
benzylamine
(original compound)

1,4-benzenedicarboxylic
acid (phthalic acid)
MW = 166, NE = 83

4-(aminoethyl)-
toluene
(isomer)

8. The compound is not a phenol (negative bromine water test), but has two active hydrogens (positive sodium test to give a disodium compound). The first carboxylic acid contains only one –COOH group, since the neutralization equivalent of 152 is equal to the molecular weight (probably 152 from $C_8H_8O_3$). The second carboxylic acid must be a dicarboxylic acid, since the neutralization equivalent of 82 ± 2 is half of the molecular weight (166, calculated from $C_8H_6O_4$). Since there is only one aromatic peak on the ^{1}H NMR spectrum, all protons on the ring must be equivalent. Thus, the second acid must be 1,4-benzenedicarboxylic acid (terephthalic acid). Working backwards, the original compound must be 4-hydroxymethyl-benzoic acid.

Thus, 4-hydroxymethylbenzoic acid reacts with sodium to give the disodium salt of 4-hydroxymethylbenzoic acid ($C_8H_6O_3Na_2$). 4-Hydroxymethylbenzoic acid is vigorously oxidized to 1,4-benzene-dicarboxylic acid (terephthalic acid) ($C_8H_6O_4$).

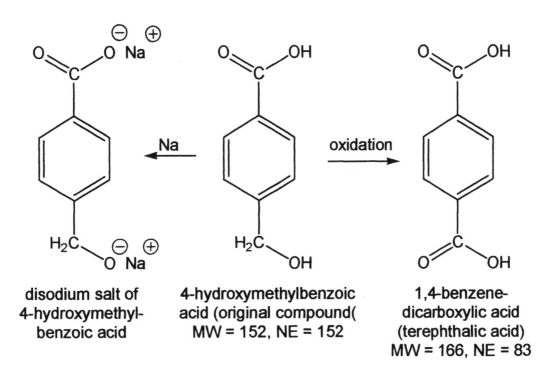

| disodium salt of 4-hydroxymethyl-benzoic acid | 4-hydroxymethylbenzoic acid (original compound(MW = 152, NE = 152 | 1,4-benzene-dicarboxylic acid (terephthalic acid) MW = 166, NE = 83 |

Problem Set 15

1. Solid I is in solubility class of S_2 and thus is a salt of an organic acid, an amine hydrochloride, an amino acid, or a polyfunctional compound. Evolution of a gas, probably nitrogen, from a reaction with nitrous acid (sodium nitrite and hydrochloric acid) indicates a primary amine; however, another part of this gas also dissolved in potassium hydroxide solution which shows that an acid is present. The second gas is carbon dioxide. The presence of sodium nitrate in the residue shows that the nitrate ion must have been in the original compound.

   When the original compound is treated with dilute sodium hydroxide solution, ammonia is evolved. Upon acidification of this solution, another gas (probably carbon dioxide) is evolved. The only possibility for solid I is urea nitrate.

   Treatment of urea nitrate (Compound I) with sodium nitrite and hydrochloric acid produces carbon dioxide, nitrogen, and inorganic salts. The carbon dioxide is soluble in potassium hydroxide.

Compound I:
urea nitrate
MW = 123, NE = 123

$$CO_2 \xrightarrow{\text{KOH}} CO_3^{-2}$$

carbon dioxide   carbonate ion

   Treatment of urea nitrate (Compound I) with sodium hydroxide produced carbonate ion, ammonia, and inorganic salts. Carbonate ion reacts with hydrochloric acid to produce carbon dioxide.

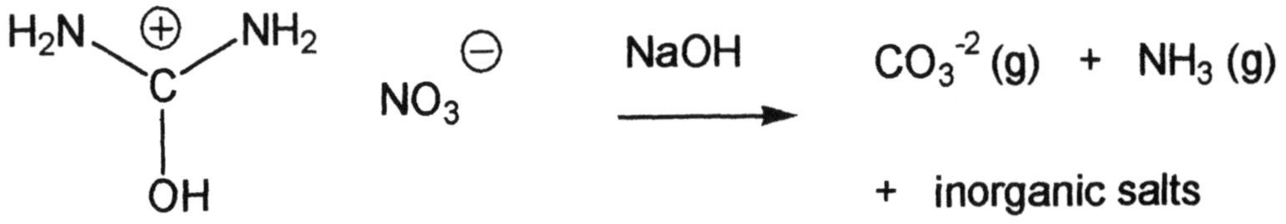

Compound I:
urea nitrate
MW = 123, NE = 123

$$CO_3^{-2} \xrightarrow{HCl} CO_2\,(g)$$

carbonate ion — carbon dioxide

Urea (mp 135°C) is the neutralized compound.

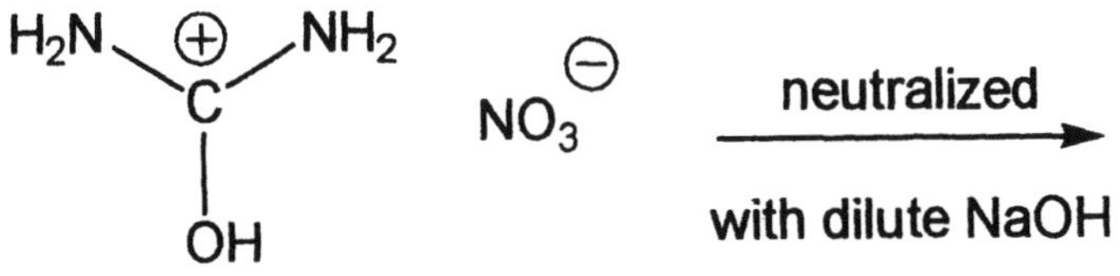

Compound I:
urea nitrate
MW = 123, NE = 123

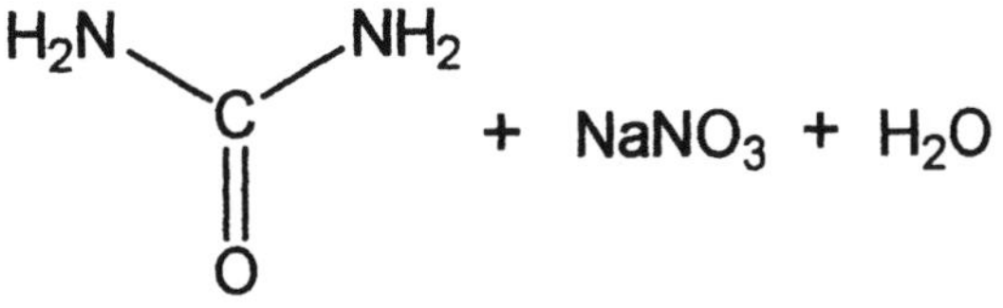

urea

The mass spectrum can be interpreted as follows:

m/e	fragment
60	
44	

In the IR spectrum, the doublet at 3350 and 3450 cm^{-1} indicates the presence of N–H stretching for a primary amine. The signals at 1640 and 1690 cm^{-1} indicate the C=O stretching of the carbonyl group.

2. From the solubilities given, the compound is in the solubility class of N or I. This means that the compound could be an alcohol, aldehyde, ketone, ester, ether, epoxide, alkene, alkyne, aromatic, saturated hydrocarbon, haloalkane, or aryl halide. Saturated hydrocarbon, haloalkane, and aryl halide are eliminated as possibilities since the compound contains carbon, hydrogen, oxygen, and nitrogen. The compound does not contain an active hydrogen (negative acetyl chloride test) and it is not an aldehyde or ketone (negative phenylhydrazine test).

    From a positive iodoform test, the alkali distillate compound is either a methyl ketone or has a methyl adjacent to a hydroxyl group. A negative reaction with zinc chloride and hydrochloric acid indicates that this compound is not a secondary or tertiary alcohol. The alkali distillate could be ethanol, since it gives a negative Lucas test. The ^1H NMR spectrum showed a peak that was concentration dependent, which indicates a –OH.

    The alkali residue, after acidification and distillation, yielded a volatile carboxylic acid. This volatile acid reduced potassium permanganate which indicates the presence of a multiple bond, a primary alcohol, a secondary alcohol, or an aldehyde. Analysis of the ^1H NMR spectrum for the volatile acid shows a multiplet in the range of δ 5.8–δ 6.75 with only three hydrogens and a singlet with one proton at δ 12.4. The multiplet must be an H₂C=CH– group and the singlet would be from a –COOH group, which would mean that the volatile acid is propenoic acid.

The neutralized residue from the second distillation is a non-volatile carboxylic acid. The non-volatile acid contains a *para* disubstituted aromatic ring, as indicated from the pair of doublets (4H) at δ 6.75 and δ 7.83. The singlet (3H) at δ 6.55 is the –COOH and the –NH₂ protons. Thus, the nonvolatile acid is 4-aminobenzoic acid.

The original compound must have been an ester, since esters hydrolyze slowly with base. Based upon the above information, the original compound must be ethyl 4-[*N*-(1-oxo-2-propenyl)]amino-benzoate.

Ethyl 4-[*N*-(1-oxo-2-propenyl)]aminobenzoate reacts with hot aqueous alkali, followed by acidification, to produce 4-[*N*-(1-oxo-2-propenyl)]aminobenzoic acid and ethanol.

ethyl 4[*N*-(1-oxo-2-propenyl)]amino benzoate

4-[*N*-1-oxo-2-propenyl)]-aminobenzoic acid (residue)

ethanol (distillate)

The original alkaline solution containing the 4-[*N*-(1-oxo-2-propenyl)]aminobenzoic acid was acidified and distilled to yield 4-aminobenzoic acid and propenoic acid.

4-[*N*-1-oxo-2-propenyl)]-   4-aminobenzoic acid
amino benzoic acid   (nonvolatile acid)
                                MW = 137, NE =137

propenoic acid
(volatile acid)

3.   Compound I ($C_6H_8O_4$) must be a diacid, since it lost 2 –OH groups and gained two –Cl groups from the treatment of I with phosphorus pentachloride to produce the diacid chloride (Compound II) ($C_6H_6O_2Cl$). The diacid chloride (Compound II) yielded a diketone (Compound III) ($C_{18}H_{16}O_2$) from the Friedel-Craft acylation reaction. The structure is confirmed as a diketone (negative potassium permanganate test, positive bromine test).   The formation of a dioxime is in agreement that III is a diketone. For the last reaction to occur, compound IV ($C_{18}H_{18}O_2N_2$) must be able to be easily hydrolyzed to a diacid. Thus, compound IV ($C_{18}H_{18}O_2N_2$) is a diamide.
        From the formula of $C_6H_8O_4$ for compound I, a hexenedioic acid or a cyclobutanedicarboxylic acid can be drawn. However, a double bond is absent as indicated from the negative reaction of potassium permanganate solution with III. Thus, compound I must be a cyclobutanedicarboxylic acid.

In the ^{1}H NMR spectrum of compound I, a multiplet of four protons exists in the range of δ 2.2–δ 2.6 indicating that the carboxylic acid groups are *trans* and are on the ring on adjacent carbons; these peaks are the –CH$_2$ protons. The triplet at δ 3.90 indicates the protons on the same carbons as the –COOH. The broad singlet at δ 11.9 indicates the –OH of the acid. Therefore, compound I (C$_6$H$_8$O$_4$) is *trans*-1,2-cyclobutanedicarboxylic acid.

Trans-1,2-cyclobutanedicarboxylic acid (Compound I) (C$_6$H$_8$O$_4$) is converted with phosphorus pentachloride to the *trans*-1,2-cyclobutanedicarboxylic dichloride (Compound II) (C$_6$H$_6$O$_2$Cl$_2$), which then undergoes a Friedel-Crafts acylation with benzene and aluminum chloride to produce *trans*-1,2-dibenzoylcyclobutane (Compound III) (C$_{18}$H$_{16}$O$_2$). *Trans*-1,2-dibenzoylcyclobutane (Compound III) (C$_{18}$H$_{16}$O$_2$) reacts with hydroxylamine to form *trans*-1,2-cyclobutanedicarboxylic acid dioxime, which rearranges in the presence of phosphorus pentachloride to yield *trans*-N,N-diphenyl-1,2-cyclobutanedicarboxylic acid diamide (Compound IV) (C$_{18}$H$_{18}$O$_2$N$_2$). *Trans*-N,N-diphenyl-1,2-cyclobutanedicarboxylic acid diamide (Compound IV) (C$_{18}$H$_{18}$O$_2$N$_2$) hydrolyzes to *trans*-1,2-cyclobutanedicarboxylic acid (Compound I) (C$_6$H$_8$O$_4$).

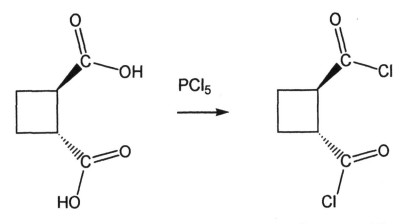

Compound I:
trans-cyclobutane-
1,2-dicarboxylic acid

Compound II:
trans-cyclobutane-
1,2-dicarboxylic dichloride

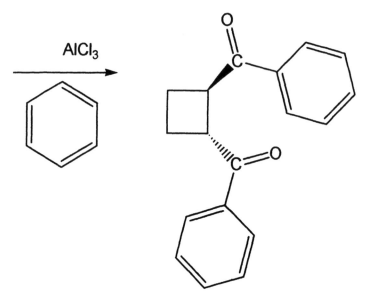

Compound III: *trans*-1,2-dibenzoylcyclobutane

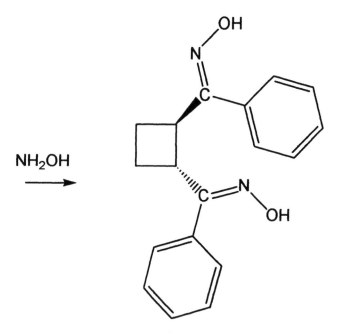

trans-1,2-cyclobutanedicarboxylic acid dioxime

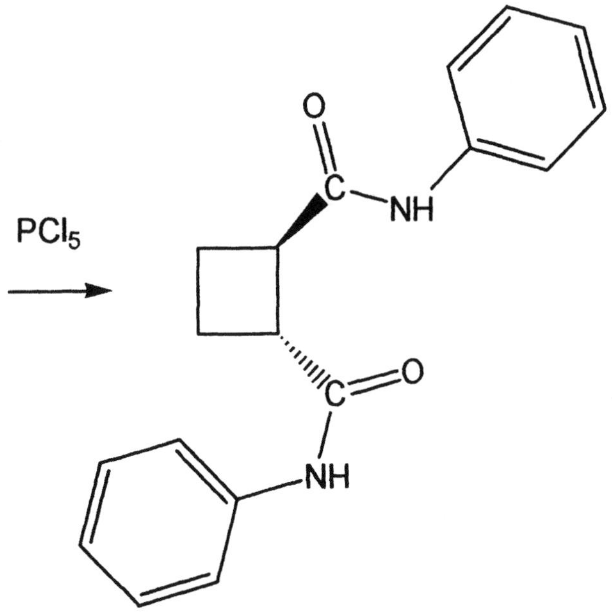

PCl$_5$

Compound IV:
trans-N,N-diphenyl-1,2-cyclobutanedicarboxylic acid diamide

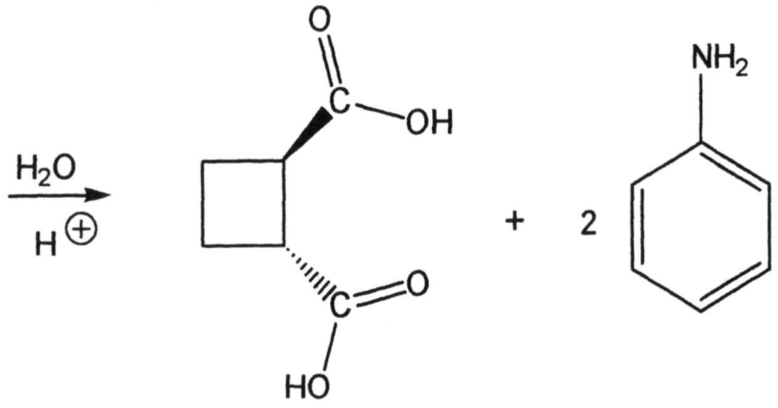

$\xrightarrow[\text{H} \oplus]{\text{H}_2\text{O}}$

+ 2

Compound I:
trans-1,2-cyclobutanedicarboxylic acid

aniline

4.   Compound I is in solubility class S$_2$, indicating a salt of an organic acid, an amine hydrochloride, an amino acid, or a polyfunctional compound. A reaction with silver nitrate shows that the chlorine is labile. When compound I was exactly neutralized to form Compound

II, the chlorine was released which means that chloride ion was initially present.

The reaction of Compound II with acetic anhydride signifies the presence of a primary or secondary amine. The reaction of Compound II with benzenesulfonyl chloride and with nitrous acid without the evolution of gas confirms the identity of this compound as a secondary amine.

Compound III is a carboxylic acid since it gave a neutralization equivalent. Since Compounds II and III were soluble in base, but not soluble in acid, the presence of a carboxylic acid group is confirmed.

The oxidation of Compounds I, II, or III to produce a nitrogen-free acid, which is insoluble in water, shows that the nitrogen is not directly attached to the aromatic ring.

The ^1H NMR spectrum of the nitrogen-free acid shows three sets of aromatic signals (4H) at $\delta 7.62$, $\delta 8.25$, and $\delta 8.70$ indicating *meta* disubstitution. The two protons in a singlet at δ 12.28 correspond to the –OH of the two –COOH. Therefore, the final product is 1,3-benzenedicarboxylic acid (isophthalic acid). Thus, working backwards, Compound I must be 3-carboxy-*N*-methylbenzyl-ammonium chloride.

3-Carboxy-*N*-methylbenzylammonium chloride (Compound I) is reacted with sodium hydroxide to produce an amine, 3-carboxy-*N*-methylbenzylamine (Compound II). Treatment of 3-carboxy-*N*-methylbenzylamine (Compound II) with acetic anhydride yields an amide, 3-carboxy-*N*-acetyl-*N*-methylbenzylamine (Compound III).

Compound I:
3-carboxy-*N*-methyl-
benzyl ammonium chloride

Compound II:
3-carboxy-*N*-methyl-
benzylamine

Compound III:
3-carboxy-*N*-acetyl-*N*-methylbenzylamine
MW = 207, NE = 207

Oxidation of 3-carboxy-*N*-methylbenzylammonium chloride (Compound I), 3-carboxy-*N*-methylbenzylamine (Compound II), or 3-carboxy-*N*-acetyl-*N*-methylbenzylamine (Compound III) produces a dicarboxylic acid, 1,3-benzenedicarboxylic acid (isophthalic acid).

Compounds I, II, or III   <u>oxidation</u>

1,3-benzenedicarboxylic acid,
isophthalic acid
MW = 166, NE = 166

5.   Compound I ($C_{10}H_6O_3$) contains a multiple bond (positive potassium permanganate test) and is an aldehyde or a ketone (positive hydroxylamine test). Compound II ($C_9H_6O_2$) contains an active hydrogen (positive sodium test). Compound III ($C_8H_6O_4$) is a

carboxylic acid which decarboxylates with soda lime to form Compound IV ($C_7H_6O_2$). Compound IV ($C_7H_6O_2$) is treated with hydrochloric acid under pressure to yield a weakly acidic compound ($C_6H_6O_2$).

The ^1H NMR spectrum of Compound IV indicates an isolated singlet indicating a $-CH_2-$ at $\delta 5.90$ and a disubstituted aromatic ring (4H) at $\delta 6.83$. For compound III, an isolated $-CH_2-$ is at $\delta 6.00$, a trisubstituted aromatic ring (3H) is in the range of $\delta 6.8-\delta 7.55$, and a $-COOH$ is present as a broad singlet at $\delta 7.6$.

The only structure that can be drawn for Compound IV that matches the formula of $C_7H_6O_2$ and the ^1H NMR spectrum is 1,2-methylenedioxybenzene. By working backwards, the other structures can be determined. Compound I is 3-(3,4-methylenedioxyphenyl)-propynal ($C_{10}H_6O_3$).

In the presence of heat, 3-(3,4-methylenedioxyphenyl)-propynal (Compound I) ($C_6H_{10}O_3$) is converted to 1-(3,4-methylene-dioxyphenyl)ethyne (Compound II) ($C_9H_6O_2$), which is then oxidized to 3,4-methylenedioxybenzoic acid (Compound III) ($C_8H_6O_4$). 3,4-Methylenedioxybenzoic acid (Compound III) ($C_8H_6O_4$) decarboxylates with heat and soda lime to produce 1,2-methylenedioxybenzene (Compound IV) ($C_7H_6O_2$), which is then decomposed by heating with hydrochloric acid and pressure to 1,2-dihydroxybenzene (weak acid).

Compound I:
3-(3,4-methylenedioxyphenyl)propynal

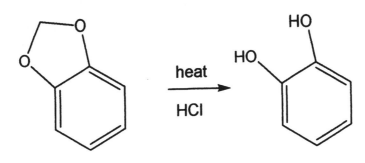

oxidation →

Compound II:
1-(3,4-methylenedioxyphenyl)ethyne

soda
lime →

Compound III:
3,4-methylenedioxybenzoic acid

heat
HCl →

Compound IV:
1,2-methylenedioxybenzene

1,2-dihydroxybenzene
(weak acid)

Problem Set 16

1. Compound A belongs to the solubility class of N which indicates that it is an alcohol, aldehyde, ketone, ester, ether, epoxide, alkene, alkyne, or aromatic. Compound A is not an aldehyde or a ketone (negative phenylhydrazine test) and does not contain an active hydrogen (negative acetyl chloride test). The volatile acid, from the base hydrolysis of compound A, must be acetic acid due to the low neutralization equivalent. Compound B contains an active hydrogen (positive acetyl chloride test), but is not an aldehyde or ketone (negative phenylhydrazine test). Compound C contains a labile halogen (positive silver nitrate test) and is a phenol (positive ferric chloride test and positive bromine water test). Compound D was esterified to produce compound E.

    The ^{1}H NMR spectrum of compound E can be interpreted to have a –CH$_3$ group adjacent to a –CH$_2$– at both δ 1.35 and δ 1.40. The signal at δ 3.8–δ 4.5 is two overlapping quartets and shows a –CH$_2$– adjacent to –CH$_3$ for each quartet. These two ethyl groups are in slightly different environments. The pair of doublets (4H) in the aromatic region at δ 6.79 and δ 7.89 indicate a *p*-disubstituted benzene ring.

    Compound E must be ethyl 4-ethoxybenzoate. This structure matches the ^{1}H NMR spectrum and is an ester. From this structure, the other structures can be determined.

    4-Ethoxybenzyl acetate (Compound A) is heated with alkali and neutralized to produce 4-ethoxybenzyl alcohol (Compound B) and acetic acid. 4-Ethoxybenzyl alcohol (Compound B) is treated with hydrogen bromide to produce 4-ethoxybenzyl bromide (Compound C).

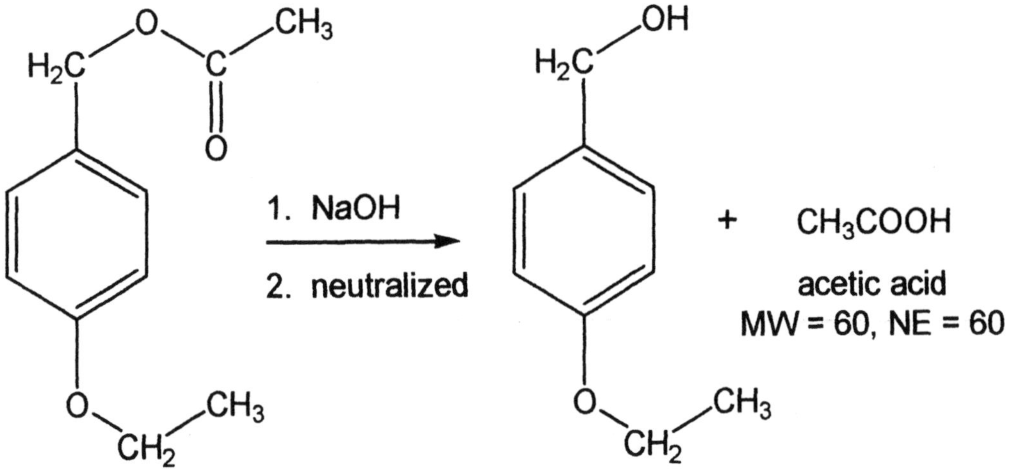

Compound A:
4-ethoxybenzyl acetate

1. NaOH
2. neutralized

Compound B:
4-ethoxybenzyl alcohol

+ CH₃COOH

acetic acid
MW = 60, NE = 60

HBr

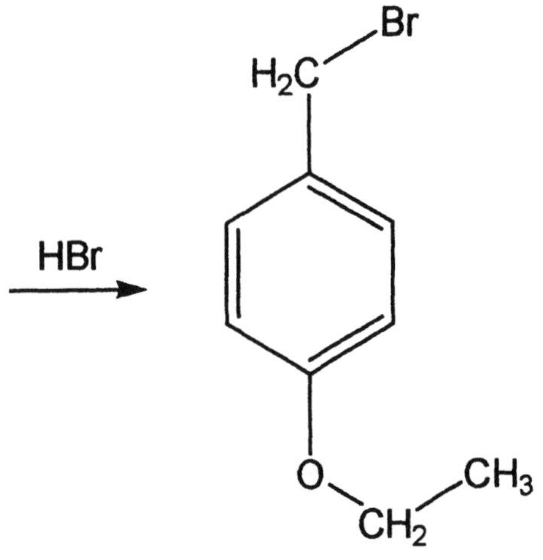

Compound C:
4-ethoxybenzyl bromide

Oxidation of 4-ethoxybenzyl alcohol (Compound B) yielded 4-ethoxybenzoic acid (Compound D), which is esterified to produce ethyl 4-ethoxybenzoate (Compound E).

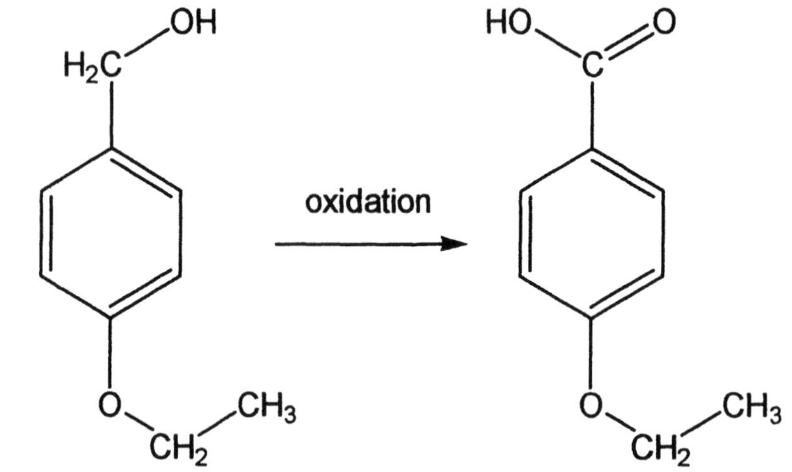

Compound B:
4-ethoxybenzyl alcohol

oxidation

Compound D:
4-ethoxybenzoic acid
ME = 166, NE = 166

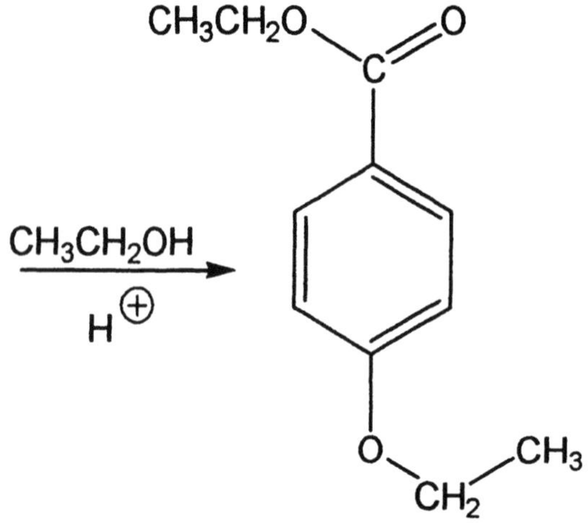

Compound E: 4-ethoxybenzoate

2. Compound I is in the solubility class of A_2, indicating that it is a phenol, enol, oxime, imide, sulfonamide, thiophenol, β-diketone, or nitro compound. Since compound I tests positive for nitrogen, phenol, enol, thiophenol, and β-diketone are eliminated as possibilities. Sulfonamide is also eliminated since sulfur is not present. Compound I contains an active hydrogen (positive acetyl chloride test) and the chlorine is not labile (negative silver nitrate

test). Since boiling alkali liberated ammonia from Compound I to produce carboxylic acid III, then compound I must be an imide.

The ^1H NMR spectrum of Compound IV indicates a benzenetricarboxylic acid. There are three protons present in the aromatic region at δ 7.57 and δ 8.18 and three carboxylic acid protons at δ 11.2. Since all of the protons are split, the answer must be 1,2,3-benzenetricarboxylic acid. 1,2,3-Benzenetricarboxylic acid (Compound IV) has a molecular weight of 210 and a neutralization equivalent of 70, matching the neutralization equivalent given. 1,2,3-Benzenetricarboxylic acid (Compound IV) was chlorinated to produce a compound with a neutralization equivalent of 81.5.

Working backwards, Compound I is 5-chloro-3-hydroxy-phthalimide. Chlorine can actually be on carbons 4, 5, or 6, based upon the information given. 5-Chloro-3-hydroxyphthalimide (Compound I) reacts with acetyl chloride to produce 5-chloro-3-hydroxymethylphthalimide acetate (Compound II).

Compound I:
5-chloro-3-hydroxy-
methylphthalimide
MW = 211.5, NE = 211.5

acetyl chloride

Compound II:
5-chloro-3-hydroxy-methylphthalimide acetate
MW = 233.5, NE = 233.5

5-Chloro-3-hydroxymethylphthalimide (Compound I) reacts with boiling alkali to liberate ammonia. Acidification of the residue produced a carboxylic acid, 5-chloro-3-hydroxymethyl-1,2-benzene-dicarboxylic acid (Compound III), which is oxidized to 5-chloro-1,2,3-benzenetricarboxylic acid.

Compound I:
5-chloro-3-hydroxymethylphthalimide

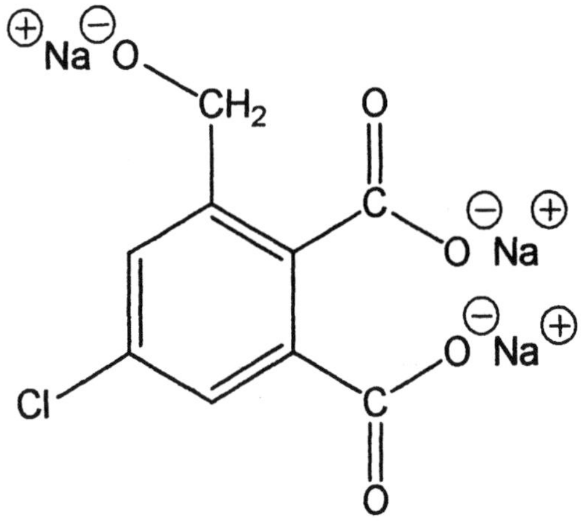

trisodium salt of 5-chloro-3-hydroxymethyl-
1,2-benzenedicarboxylic acid

+   NH$_3$

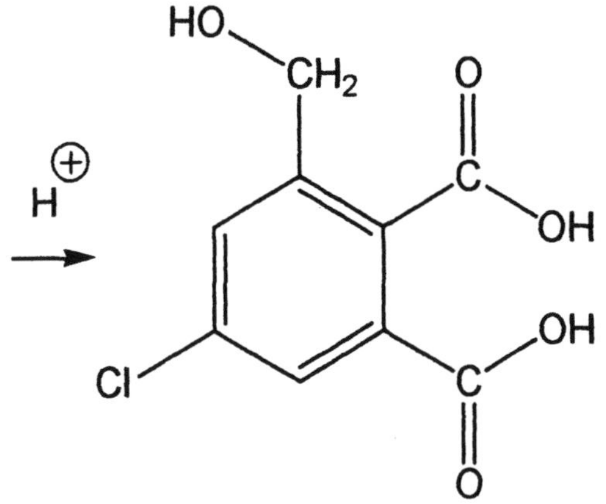

Compound III:
5-chloro-3-hydroxymethyl-
1,2-benzene-
dicarboxylic acid
MW = 231.5, NE = 115.5

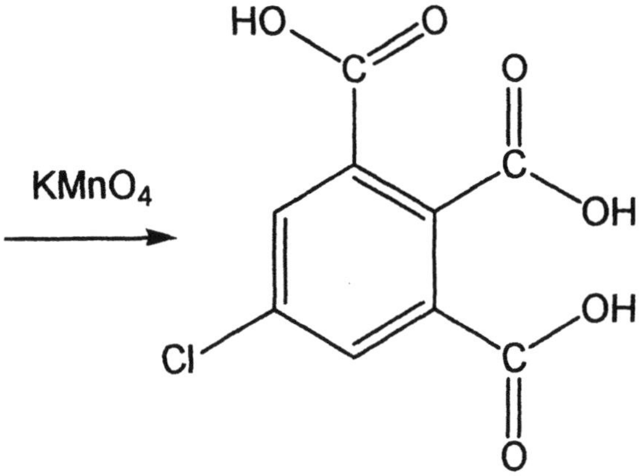

KMnO$_4$

5-chloro-1,2,3-
benzenetri-
carboxylic acid
MW = 244.5, NE = 81.5

1,2,3-Benzenetricarboxylic acid (Compound IV) is chlorinated to give 5-chloro-1,2,3-benzenetricarboxylic acid.

| Compound IV:
1,2,3-tricarboxylic acid
MW = 210, NE = 70 | 5-chloro-1,2,3-benzene-
tricarboxylic acid
MW = 244.5, NE = 81.5 |

The IR spectrum shows the C=O stretch of 5-chloro-3-hydroxyphthalimide at 1712 and 1719 cm^{-1}.

3.   Compound A is an aldehyde or a ketone (positive phenylhydrazine test) and contains an active hydrogen (positive acetyl chloride test). Compound B is an aldehyde or ketone (positive phenylhydrazine test), but does not have an active hydrogen (negative acetyl chloride test). Compound C is benzoic acid.
   The ^{1}H NMR spectrum of compound A shows ten hydrogens in the aromatic region of δ 7.2–δ 7.85, indicating two nonfused benzene rings. The singlet as δ 5.9 is a –CH and the broad singlet at δ 4.5 is an –OH. From this information, compound A is deduced to be benzoin.
   Benzoin (Compound A) is oxidized to benzil (Compound B). Vigorous oxidation of benzil (Compound B) produced benzoic acid (Compound C).

Compound A: benzoin

mild
───────▶
oxidation

Compound B: benzil

vigorous
───────▶
oxidation

Compound C: benzoic acid
MW = 122, NE = 122

4. Liquid I is in the solubility class N, which includes alcohols, aldehydes, ketones, esters, alkenes, alkynes, ethers, epoxides, or aromatic compounds. Liquid I does not contain a labile halogen (negative silver nitrate test), does not have an active hydrogen (negative acetyl chloride test), but could be an aldehyde or a ketone (positive phenylhydrazine test). A negative Tollens test indicates that liquid I is not an aldehyde. Liquid I has a saponification equivalent of 113 ± 1 and is probably an ester. Liquid I is heated with sodium hydroxide; the distillate from this reaction is a methyl ketone or has

a methyl adjacent to a secondary alcohol (positive iodoform test). The residue was acidified to produce a solid II with a neutralization equivalent of 156 ± 1 and an acidic filtrate with a neutralization equivalent of 61 ± 1, indicating possible carboxylic acids.

The ^{1}H NMR spectrum of Compound II indicates a possible *ortho* disubstituted benzene ring (4H) in the aromatic region of δ 7.1–δ 7.82 and a –COOH group at δ 9.0. 2-Chlorobenzoic acid matches the ^{1}H NMR spectrum and the neutralization equivalent for compound II. With a neutralization equivalent of 61 ± 1, the acidic filtrate is acetic acid. Working backwards, the initial compound is ethyl 3-(2-chlorophenyl)-3-oxopropanoate.

Ethyl 3-(2-chlorophenyl)-3-oxopropanoate (Compound I) is saponified to give ethanol (distillate) and a residue, which is acidified to yield 2-chlorobenzoic acid (Compound II) and acetic acid (filtrate).

Compound I:
ethyl 3-(2-chlorophenyl)-3-oxopropanoate
MW = 226.5, SE = 226.5

ethanol
(distillate)

2-chlorobenzoic acid
MW = 156.5, NE = 156.5

acetic acid (filtrate)
MW = 60, NE = 60

5. The unknown is an ester, since a saponification equivalent was determined. Saponification of an ester yields an alcohol and a carboxylic acid. The alcohol is a phenol (positive ferric chloride test). The acid is acetic acid, because of the neutralization equivalent of 60.

   The ^1H NMR spectrum indicated two isolated methyl groups in the same environment at $\delta 2.18$, the –OH group at $\delta 5.73$, and three aromatic ring protons at $\delta 6.33$ and $\delta 6.45$. The aromatic protons are all isolated, since they are singlets. The benzene ring must be trisubstituted, since only three aromatic protons are present. The original compound is 3,5-dimethylphenyl acetate. 3,5-Dimethylphenyl acetate is saponified to yield 3,5-dimethylphenol and acetic acid.

3,5-dimethylphenyl acetate
MW = 164, SE = 164

1. NaOH
2. HCl

3,5-dimethylphenol

acetic acid
MW = 60, NE = 60

Problem Set 17

1. Compound I is in the solubility class of S_2, indicating that this compound is an amino acid, a salt of an organic acid, an amine hydrochloride, or a polyfunctional compound. Treatment of Compound I with nitrous acid liberated a gas, indicating that this compound is a primary amine. With a neutralization equivalent of 37 ± 1, the structure is small. The ^1H NMR spectrum for Compound I shows isolated amino protons at δ 1.08, a methylene adjacent to a methylene at δ 2.75, and a methylene between two methylenes at δ 1.58. 1,3-Diaminopropane agrees with the neutralization equivalent and the ^1H NMR spectrum.

    Treatment of 1,3-diaminopropane (Compound I) with sodium nitrite and hydrochloride acid, followed by benzoic anhydride produced 1,3-propanediol dibenzoate (Compound III).

Compound I:
1,3-diaminopropane
MW = 74, NE = 37

1,3-propanediol

benzoic anhydride

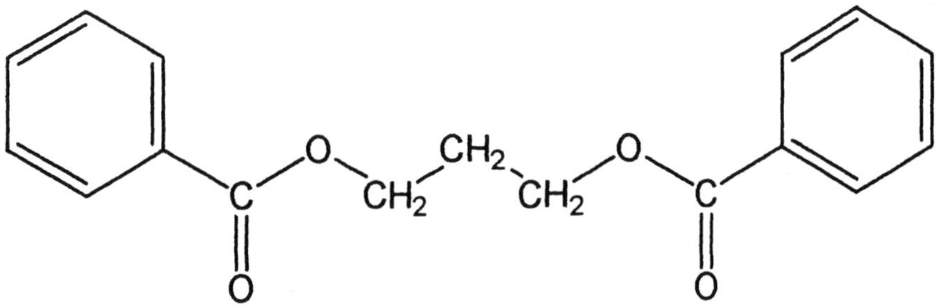

Compound III: 1,3-propanediol dibenzoate
MW = 284, SE = 142

Treatment of 1,3-diaminopropane (Compound I) with sodium hydroxide and benzoyl chloride yielded dibenzamide of 1,3-diaminopropane (Compound II).

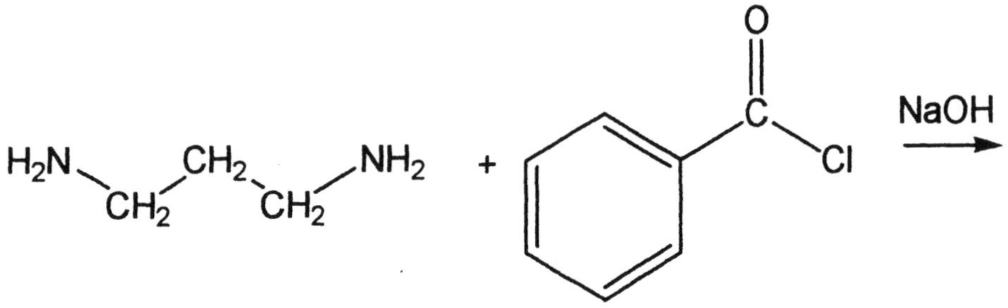

Compound I:
1,3-diaminopropane
MW = 74 , NE = 37

benzoyl chloride

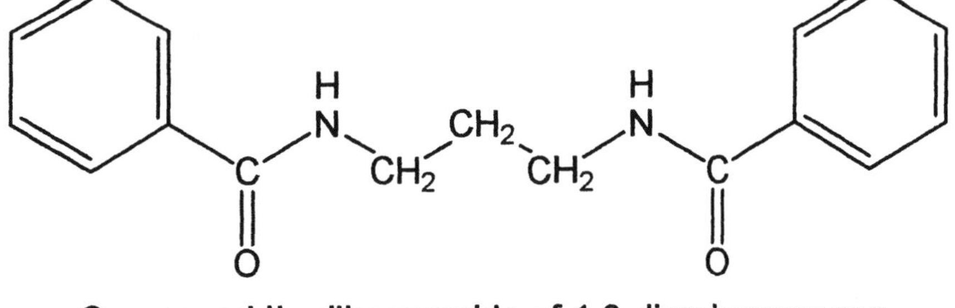

Compound II: dibenzamide of 1,3-diaminopropane
MW = 282

2.   Liquid I is in the solubility class of **N**, thus indicating an alcohol, an aldehyde, a ketone, an ester, an epoxide, an alkene, an alkyne, an aromatic compound, or an ether. This compound does not contain an active hydrogen (negative acetyl chloride test) and is not an aldehyde or a ketone (negative phenylhydrazine test). Compound I was treated with phosphoric acid to yield compound II as an oil and compound VI.

Compound II is an aldehyde or a ketone (positive phenylhydrazine and sodium bisulfite tests), but does not contain an active hydrogen (negative acetyl chloride test). Compound II was treated with strong alkali to yield compound III and the anion of compound IV.

Compound III contains an active hydrogen (positive acetyl chloride test), but is not an aldehyde or a ketone (negative phenylhydrazine test). Compounds IV and V are carboxylic acids.

Compound VI contains an active hydrogen (positive sodium and acetyl chloride tests), is a methyl ketone or has a methyl next to a –CHOH (positive iodoform test), and is not a tertiary or secondary alcohol (negative Lucas test). Compound VI is ethanol.

The ^{1}H NMR spectrum for Compound II shows an isolated methyl at δ 2.42, a *p*-disubstituted benzene ring (4H) with signals at δ 7.18 and δ 7.66, and an aldehyde signal at δ 9.81. From this spectrum, Compound II must be 4-methylbenzaldehyde. Since Compound VI is ethanol, Compound I is an acetal.

The acetal of 4-methylbenzaldehyde (Compound II) is hydrolyzed to 4-methylbenzaldehyde (Compound II) and ethanol (Compound VI). 4-Methylbenzaldehyde (Compound II) undergoes a Cannizzaro reaction to yield 4-methylbenzyl alcohol (Compound III) and sodium 4-methylbenzoate, which is acidified to yield 4-methylbenzoic acid (Compound IV). Oxidation of 4-methylbenzoic acid (Compound IV) produces 1,4-benzenedicarboxylic acid (terephthalic acid) (Compound V).

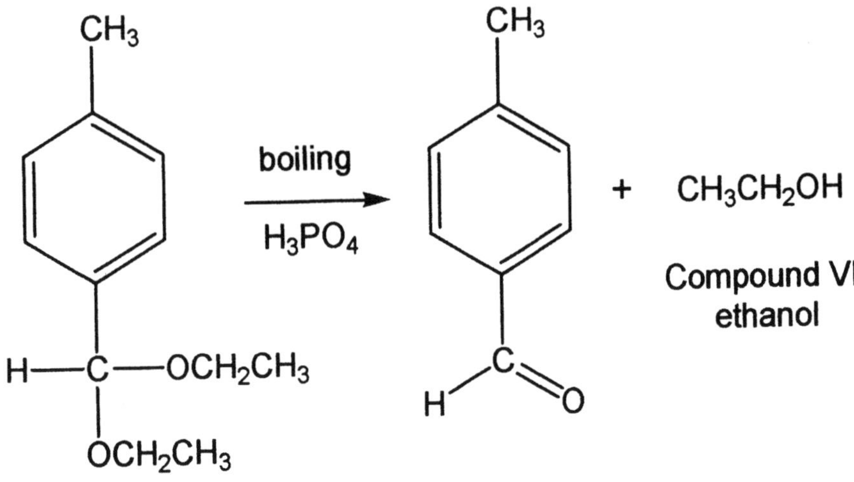

CH₃ group on benzene ring. Structure with H—C—OCH₂CH₃ and OCH₂CH₃

boiling
→
H₃PO₄

+ CH₃CH₂OH

Compound VI:
ethanol

Compound I:
acetal of
4-methylbenzaldehyde

Compound II:
4-methylbenzaldehyde
(oil)

NaOH
→

+

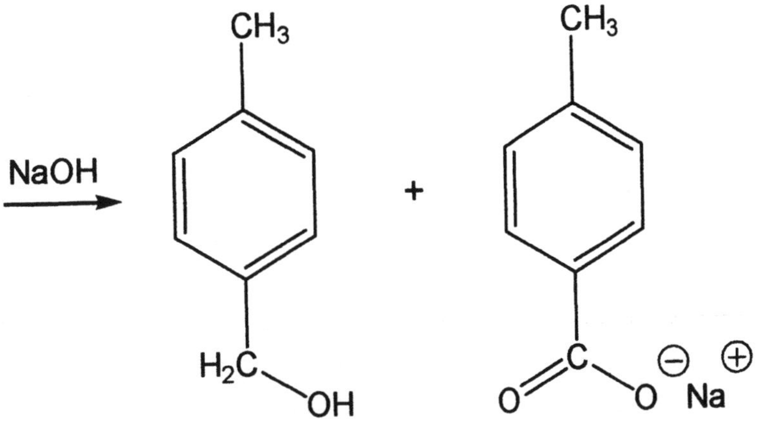

Compound III:
4-methylbenzyl
alcohol

sodium
4-methylbenzoate

Compound III:
4–methylbenzoic acid
MW = 136, NE = 136

Compound IV:
1,4-benzenedicarboxylic acid,
terephthalic acid
MW = 166, NE = 83

3. Liquid I is in either the S_A, S_B, or S_1 solubility class, indicating a monofunctional carboxylic acid with five carbons or fewer, an arylsulfonic acid; a monofunctional amine with six carbons or fewer; or an alcohol, an aldehyde, a ketone, an ester, a nitrile, or amide with five carbons or fewer. Arylsulfonic acids, amines, nitriles, and amides are eliminated as possibilities, since the compound does not contain sulfur or nitrogen. Liquid I does not contain an active hydrogen (negative sodium and acetyl chloride tests), is not an aldehyde or a ketone (negative phenylhydrazine test), and does not have a multiple bond (negative potassium permanganate and bromine tests).

Liquid I reacts with hydrobromic acid to yield oil II. Oil II is in the solubility class of I, which indicates a saturated hydrocarbon, an haloalkane, an aryl halide, a diaryl ether, or an aromatic compound. This compound contains a labile bromine (positive silver nitrate test).

Oil II undergoes a reaction with magnesium to produce a gas III, as well as a reaction with alcoholic potassium hydroxide to liberate a gas IV. Compounds III and IV both decolorize potassium permanganate and bromine solutions, thus indicating the presence of a multiple bond.

The ^{1}H NMR spectrum indicates only one signal at δ 3.69, showing that all of the protons are equivalent. One structure that fits the above criteria for compound I is 1,4-dioxane. The *m/z* value for the parent peak for 1,4-dioxane is 88.

1,4-Dioxane, I, is treated with excess hydrobromic acid to form 1,2-ethanediol (ethylene glycol), 2-bromoethanol (ethylene bromohydrin), and 1,2-dibromoethane (Compound II). 1,2-Ethanediol and 2-bromoethanol are infinitely soluble in the aqueous solution; 1,2-dibromoethane (Compound II) is only slightly soluble. Therefore, 1,2-dibromoethane (Compound II) will separate as an oil.

Compound I:
1,4-dioxane

excess HBr
heat

1,2-ethanediol
(ethylene glycol)
(water soluble)

+

2-bromoethanol
(ethylene bromohydrin)
(water soluble)

+

Compound II:
1,2-dibromoethane
(slightly water soluble)

Treatment of 1,2-dibromoethane (Compound II) with magnesium produces ethene (Compound III).

Mg

Compound II:
1,2-dibromoethane

Compound III:
ethene

The reaction of 1,2-dibromoethane (Compound II) with potassium hydroxide yields ethyne (acetylene) (Compound IV).

Br–CH₂–CH₂–Br $\xrightarrow{\text{KOH}}$ HC≡CH

Compound II: Compound IV:
1,2-dibromoethane   ethyne, acetylene

4. Compound I is in the solubility class of I, indicating a saturated hydrocarbon, a haloalkane, an aryl halide, a diaryl ether, or a deactivated aromatic compound. When Compound I was heated with hydrochloric acid, acid II was formed. Oxidation of Compound I with potassium dichromate in sulfuric acid yielded another acid III. The reaction of I with benzaldehyde produced a benzal derivative, thus indicating that I has an acidic proton. Since tin (II) chloride and hydrogen chloride reduces a nitro group to an amino group or a nitrile group to an aldehyde group, then compound I must have either a nitro group or a nitrile group and Compound IV has an amino group or an aldehyde group.

Using the ^{1}H NMR spectrum, Compound I is a disubstituted aromatic compound with four hydrogens at $\delta 7.5$–$\delta 8.25$. An isolated –CH$_2$– is also present as a singlet at δ 4.25. Compound II also contains the isolated –CH$_2$– group as a singlet at δ 4.02 and is a disubstituted aromatic compound with four hydrogens at $\delta 7.3$–$\delta 8.0$. A –COOH group is present as a broad singlet at $\delta 11.2$. Compound III contains a disubstituted aromatic ring since four hydrogens are present in the aromatic range of $\delta 7.5$–$\delta 8.0$. The broad singlet at δ 12.15 is due to the –COOH. Compound IV shows seven aromatic and alkene protons at $\delta 6.38$–$\delta 7.65$.

Based on the information above, compound I is 2-nitrobenzyl cyanide. 2-Nitrobenzyl cyanide (Compound I) is hydrolyzed to 2-nitrophenylacetic acid (Compound II).

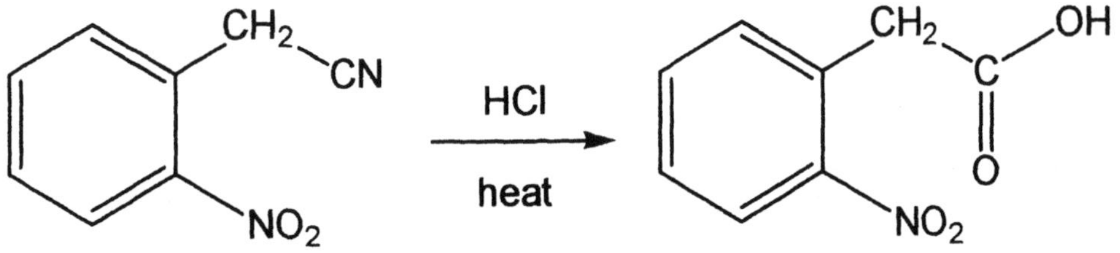

Compound I:
2-nitrobenzyl cyanide

HCl
heat

Compound II:
2-nitrophenylacetic acid
MW = 181, NE = 181
(solid)

2-Nitrobenzyl cyanide (Compound I) is oxidized to 2-nitrobenzoic acid (Compound III).

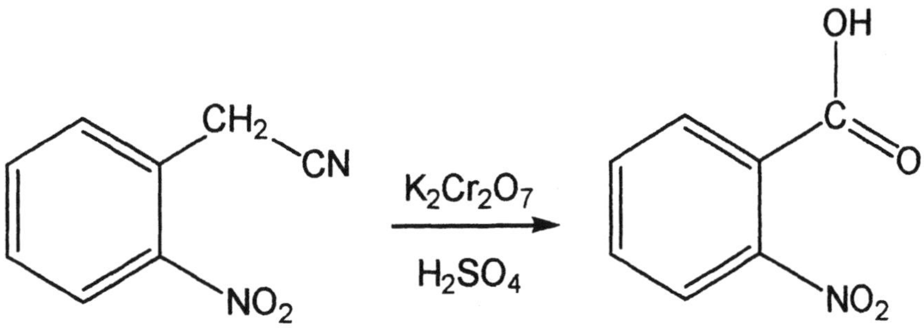

Compound I:
2-nitrobenzyl cyanide

$K_2Cr_2O_7$

H_2SO_4

Compound III: 2-nitrobenzoic acid
MW = 167, NE = 167 (solid)

2-Nitrobenzyl cyanide (Compound I) is treated with benzaldehyde to produce a benzal derivative, 2-(2-nitrophenyl)-3-phenylacrylonitrile

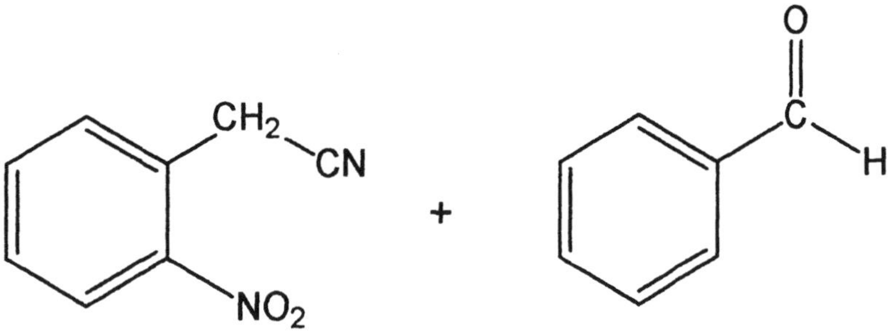

Compound I: 2-nitrobenzyl cyanide   benzaldehyde

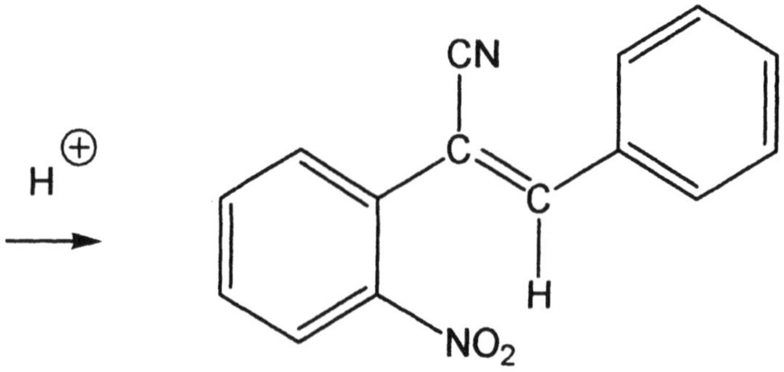

2-(2-nitrophenyl)-3-phenylacrylonitrile

2-Nitrobenzyl cyanide (Compound I) is treated with tin (Compound II) chloride and hydrochoric acid, which converts the nitrile group to an acid group and the nitro group to an amino group, to yield (2-aminophenyl)ethanal. Cyclization of this compound occurs to form indole (Compound IV) (C_8H_7N).

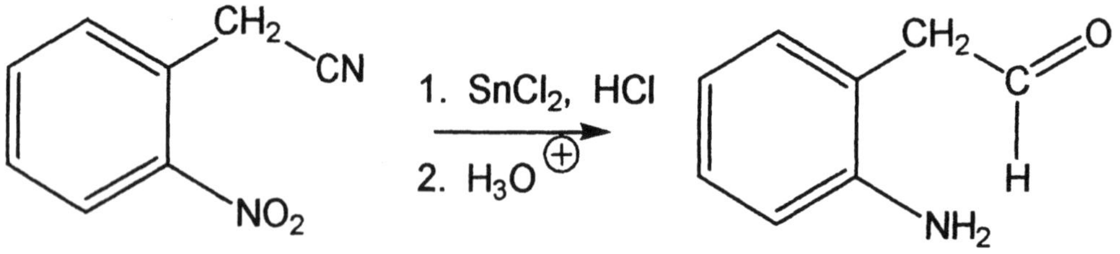

Compound I:
2-nitrobenzyl cyanide

(2-aminophenyl)ethanal

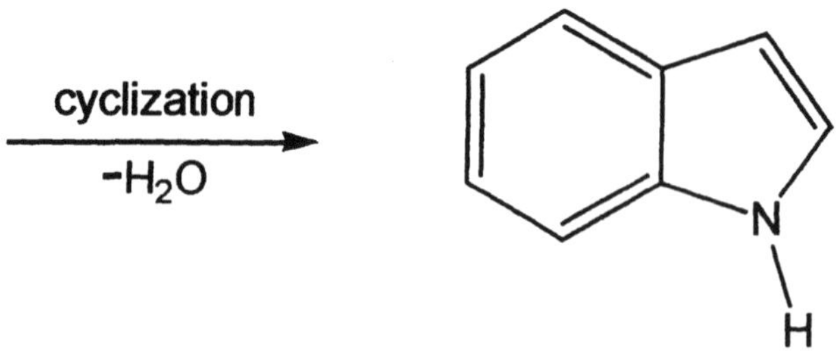

Compound IV: indole

Problem Set 18

1. Liquid I is an aldehyde or a ketone (positive phenylhydrazine test), does not contain an active hydrogen (negative acetyl chloride test), and has a multiple bond, is an alcohol, or is an aldehyde (positive potassium permanganate test). Compound I reacts with base in a Cannizzaro reaction to yield an alcohol II (positive acetyl chloride test) and an acid III. Upon heating, Compound III loses carbon dioxide to produce a compound with a formula of C_4H_4O. This compound contains a multiple bond, is an alcohol, or is an aldehyde (positive potassium permanganate test), does not have an active hydrogen (negative sodium test), and is not an aldehyde or a ketone (negative phenylhydrazine test). Compound I dimerized to produce Compound IV, which has an active hydrogen (positive acetyl chloride test) and is an aldehyde or a ketone (positive phenylhydrazine test).

   The only possible structure for the formula of C_4H_4O is furan. Thus Compound III must be 2-furancarboxylic acid, since carbon dioxide is lost from Compound III to form furan. Furfural (Compound I) reacts with base in a Cannizzaro reaction to yield 2-furylmethanol (Compound II) and 2-furancarboxylic acid (Compound III).

Compound I:
furfural

Compound II:
2-furylmethanol

Compound III:
2-furancarboxylic acid
furoic acid

2-Furancarboxylic acid (Compound III) decarboxylates to yield furan (C_4H_4O).

Compound III: furan
2-furancarboxylic acid

Furfural (Compound I) dimerizes to form 1,2-di(2-furyl)-1-oxo-2-ethanol (Compound IV) in a reaction similar to the benzoin condensation.

Compound I: Compound IV:
furfural   1,2-di(2-furyl)-1-oxo-2-ethanol

1,2-Di(2-furyl)-1-oxo-2-ethanol (Compound IV) reacts with periodic acid to form furfural (Compound I) and 2-furancarboxylic acid (Compound III).

Compound IV:
1,2-di(2-furyl)-1-oxo-2-ethanol

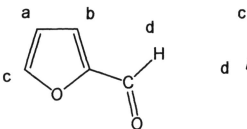

Compound I:
furfural

+

Compound III:
2-furancarboxylic acid

The ¹H NMR spectra of Compounds I, II, III and C_4H_4O do not contain any aromatic protons. The protons are identified in the structures below. Furfural (Compound I): δ 6.63, 1H, d of d (***a***); δ 7.28, 1H, d (***b***); δ 7.72, 1H, m (***c***); δ 9.67, 1H, s (***d***). 2-Furylmethanol (Compound II): δ 2.83, 1H, s (***a***); δ 4.57, 2H, s (***b***); δ 6.33, 2H, m (***c***); δ 7.44, 1H, m (***d***). 2-Furancarboxylic acid (Compound III): δ 6.56, 1H, m (***a***); δ 7.20, 1H, m (***b***); δ 7.65, 1H, m (***c***); δ 12.18, 1H, bs (***d***). Furan (C_4H_4O): δ 6.37, 2H, t (***a***); δ 7.42, 2H, t (***b***).

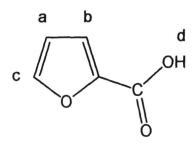

Compound I: furfural

Compound II: 2-furylmethanol

Compound III:
2-furancarboxylic acid (furan)

furan

2. Compound I contains an active hydrogen (positive acetyl chloride test) and is not an aldehyde or a ketone (negative phenylhydrazine test). Compound I is easily hydrolyzed to Compound II. Compound II does not have an active hydrogen (negative acetyl chloride test) and is a ketone (negative Tollens test, positive phenylhydrazine and sodium bisulfite tests). Compound III contains an active hydrogen, since it reacts with benzoyl chloride to produce an ester or an amide IV. The hydrolysis of Compound IV produces two compounds, a carboxylic acid V and Compound III. Compound VI is an amine salt, which is from a primary amine (liberates a gas with sodium nitrite).

Compound I is cyclopentanone oxime, which is hydrolyzed to cyclopentanone (Compound II).

Compound I:
cyclopentanone oxime

Compound II:
cyclopentanone

In the presence of phosphorus pentachloride, cyclopentanone oxime (Compound I) rearranges to pentanelactam (Compound III). Pentanelactam (Compound III) is treated with sodium hydroxide and benzoyl chloride to form (5-benzoylamino)pentanoic acid (Compound IV), which is hydrolyzed to benzoic acid (Compound V) and pentanelactam (Compound III). Pentanelactam (Compound III) is acidified to yield the hydrochloride salt of 5-aminopentanoic acid (Compound VI), which gives off a gas when treated with nitrous acid.

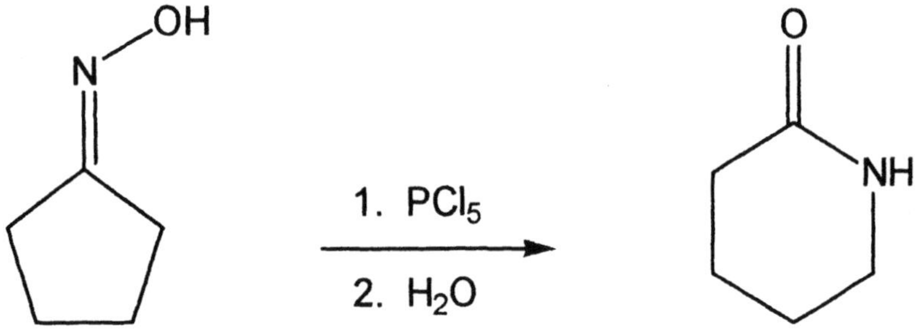

Compound I:
cyclopentanone oxime

1. PCl$_5$

2. H$_2$O

Compound III:
pentanelactam

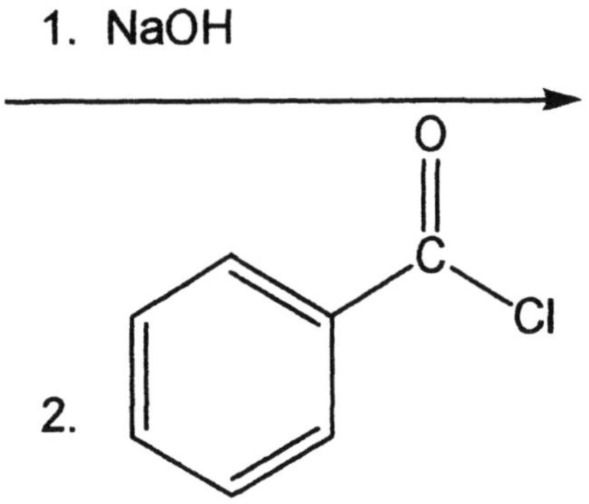

1. NaOH

2. benzoyl chloride

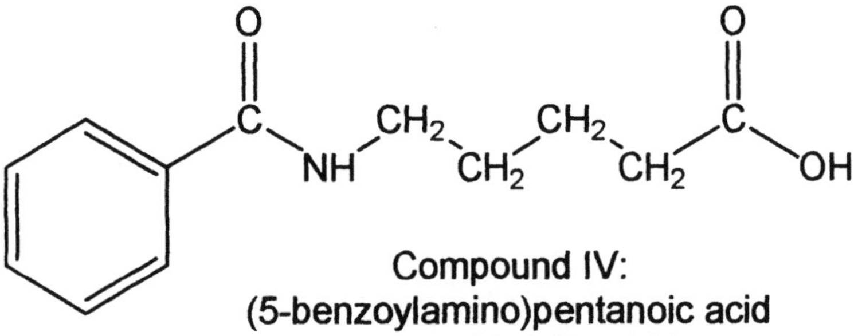

Compound IV:
(5-benzoylamino)pentanoic acid
MW = 221, NE = 221

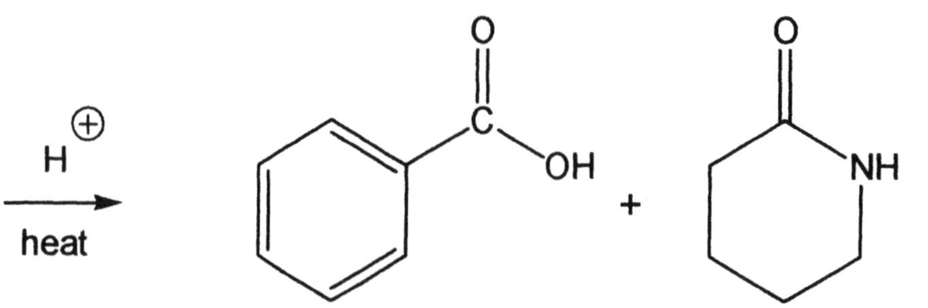

Compound V:
MW = 122, NE = 122

+

Compound III:
pentanelactam

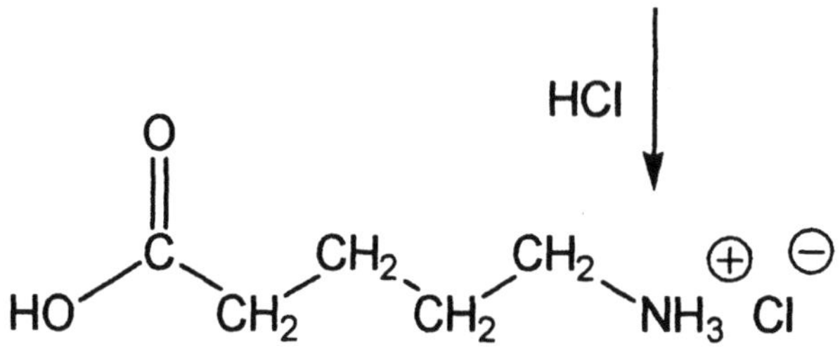

Compound VI:
hydrochloride salt of 5-aminopentanoic acid

The ^{1}H NMR spectrum for cyclopentanone oxime (Compound I) shows an –OH at δ 9.12. For pentanelactam (Compound III), the –NH is seen at δ 7.60. The protons are identified in the structures below. Cyclopentanone oxime (Compound I): δ 1.77, 4H, m (**a**); δ 2.40, 4H, m (**b**); δ 9.12, 1H, bs (**c**). Pentanelactam (Compound III): δ 1.80, 4H, m (**a**); δ 2.38, 2H, m (**b**); δ 3.32, 2H, m (**c**); δ 7.60, 1H, bs (**d**).

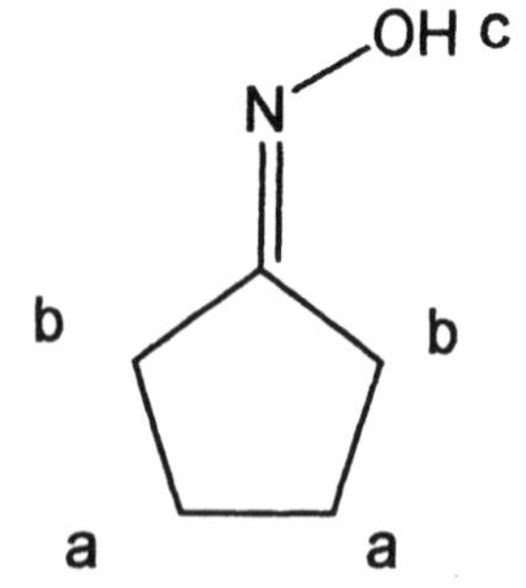

Compound I:
cyclopentanone oxime

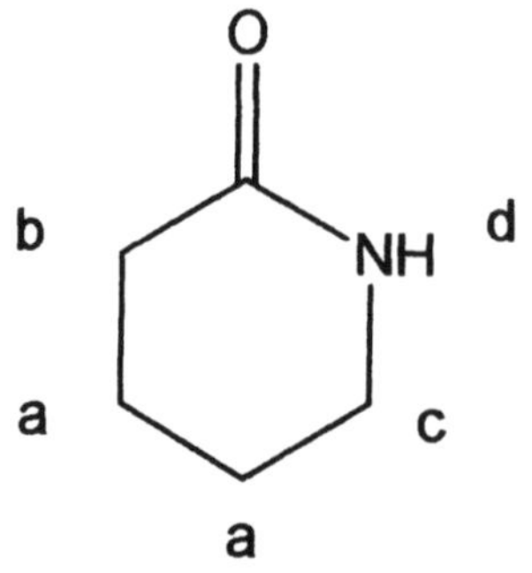

Compound III:
pentanelactam

3. Compound I is in the solubility class of S_2, which includes salts of organic acids, amine hydrochlorides, amino acids, or polyfunctional structures. Compound II is in the solubility class of I, which includes saturated hydrocarbons, haloalkanes, aryl halides, diaryl ethers, or deactivated aromatic compounds. Both Compounds I and II contain a labile halogen (positive silver nitrate test). Since Compound III did not give a benzenesulfonamide derivative, but did give a nitroso derivative, it must be a tertiary aniline.

   The ^{1}H NMR spectrum of I shows a ratio of 9:5 between methyl groups at δ 3.70 (s) and the aromatic range at δ 7.4 (m), indicating three methyls and a monosubstituted aromatic ring. This information corresponds to an N,N,N-trimethylanilinium halide. The ^{1}H NMR spectrum for II shows only a methyl at δ 2.20. For compound III, two methyls are seen at δ 2.85 (s) and the monosubstituted ring protons at δ 6.4–δ 7.10 (m). This NMR spectrum corresponds to N,N-dimethylaniline.

Compound I: trimethylanilinium halide

heat

CH_3X +

Compound II: methyl halide

Compound III: N,N-dimethylaniline

4. Compound I contains chlorine, bromine, and iodine. Reaction of compound I with silver nitrate to produce a white solid indicates the presence of a labile chlorine. Compound I contains an aldehyde or a ketone (positive phenylhydrazine test), but does not contain an active hydrogen (negative acetyl chloride test). Compound II contains bromine and iodine and is a carboxylic acid. Since Compound II is produced from the hydrolysis of compound I with the loss of chlorine,

then I must have been an acid chloride. Compound I is subjected to hot alkali, followed by acidification to yield Compound III as a carboxylic acid. Compound III contains only iodine, indicating the loss of chlorine and bromine. Bromine could be lost from the hydrolysis of an alkyl bromide. Compound III is an aldehyde or a ketone (positive phenylhydrazine test) and has an active hydrogen (positive acetyl chloride test).

Based upon the information given, Compound I is a trisubstituted benzene ring, but positions of the various groups on the ring cannot be determined. The example below is given with the groups at the 1, 3, and 5 positions.

3-Bromoacetyl-5-iodobenzoyl chloride (Compound I) is treated with cold alkali, followed by acidification, to yield 3-bromoacetyl-5-iodobenzoic acid (Compound II). The acid chloride is hydrolyzed to a carboxylic acid.

Compound I:
3-bromoacetyl-5-
iodobenzoyl chloride

Compound II:
3-bromoacetyl-5-
iodobenzoic acid
MW = 369, NE = 369

3-Bromoacetyl-5-iodobenzoyl chloride (Compound I) is treated with hot alkali, followed by acidification, to yield 3-hydroxyacetyl-5-iodobenzoic acid (Compound III). In this reaction, both the chloride and bromide ions are cleaved.

Compound I:
3-bromoacetyl-5-
iodobenzoyl chloride

Compound III:
3-hydroxyacetyl-5-iodobenzoic acid
MW = 306, NE = 306

With sodium hypochlorite, 3-bromoacetyl-5-iodobenzoyl chloride (Compound I) or 3-bromoacetyl-5-iodobenzoic acid (Compound II) is oxidized to 5-iodo-1,3-benzenedicarboxylic acid.

Compounds I or II

1. NaOCl
2. H$^{\oplus}$

5-iodo-1,3-benzenedicarboxylic acid
MW = 292, NE = 146

5. Compound A does not contain a labile bromine (negative silver nitrate test), an active hydrogen (negative acetyl chloride test), or a multiple bond (negative bromine test). Compound A is not an aldehyde or a ketone (negative phenylhydrazine test). Compound A was treated with alkaline solution, then acidified to produce Compounds B, C, and

the precursor to D. Compounds B and C are carboxylic acids. The precursor to Compound D was treated with alkaline, then treated with benzoyl chloride to yield Compound D.

From the ^{1}H NMR spectrum, Compound B contains a *p*-disubstituted benzene ring with 4 hydrogens as doublets at δ 7.71 and δ 7.90. The hydrogen from the –COOH group appears as a broad singlet at δ 7.3. Since Compound B also contains bromine, Compound B must be 4-bromobenzoic acid.

The ^{1}H NMR spectrum of Compound C indicates an aliphatic structure. A CH$_3$– adjacent to a –CH$_2$– is indicated by a triplet with three hydrogens at δ 0.93. Two –CH$_2$– are seen at δ 1.2–δ 1.8, and a –CH$_2$– adjacent to a –CH$_2$– is seen at δ 2.31. At δ 11.7, the hydrogen from the –COOH is indicated. Compound C has a neutralization equivalent of 102 ± 1. From this information, Compound C must be pentanoic acid.

The precursor to Compound D must be a diol, since two carboxylic acids have already been identified. This diol undergoes reaction with benzoyl chloride to yield Compound D, a diester with a saponification equivalent of 135 ± 1. Compound D must be the dibenzoyl ester of 1,2-ethanediol.

Compound A must be 4-bromo-2-pentanoyloxyethyl benzoate, which is hydrolyzed to 4-bromobenzoic acid (Compound B), pentanoic acid (Compound C), and 1,2-ethanediol.

Compound A:
4-bromo-2-pentanoyloxyethyl benzoate

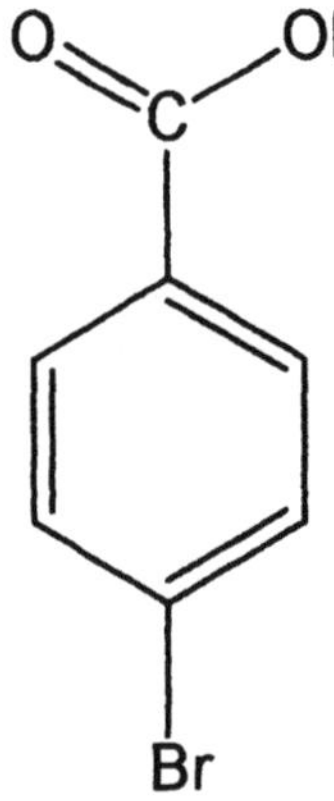

Compound B:
4-bromobenzoic acid
MW = 201, NE = 201

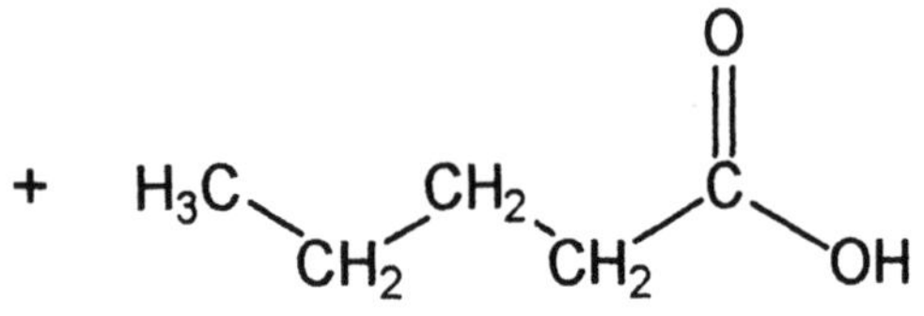

Compound C:
pentanoic acid
MW = 102, NE = 102

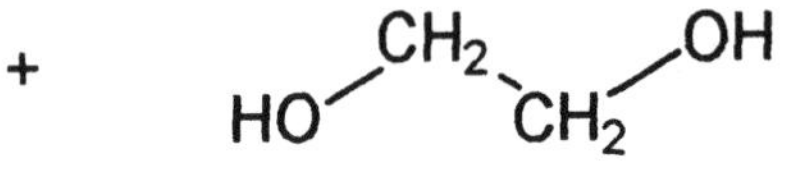

1,2-ethanediol
(precursor to
Compound D)

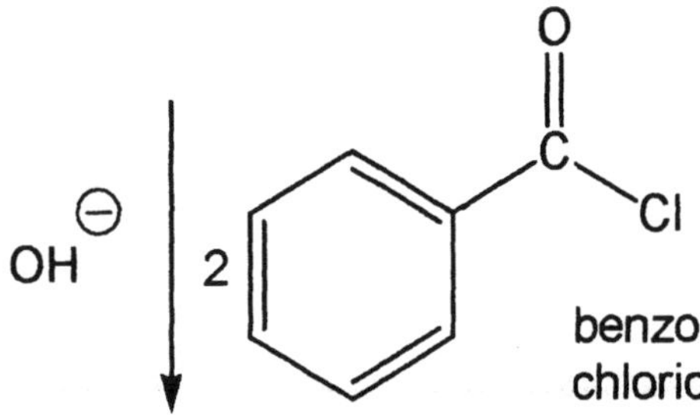

OH⊖ 2 benzoyl
chloride

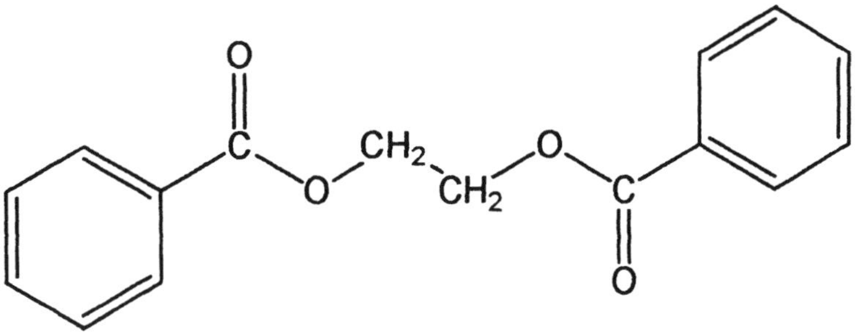

Compound D:
1,2-dibenzoyl 1,2-ethanediol
MW = 270, SE = 135

Problem Set 19

1. Compound I was hydrolyzed into two acids, Compounds II and III. Since Compound I also has a neutralization equivalent, then Compound I contains both an ester group and a carboxylic acid group. Compound II is a phenol, because of the positive test with both bromine water and ferric chloride.

    Since all four of the protons are split in the range of δ 6.7 to δ 7.75 in the ^{1}H NMR spectrum of Compound I, it is an *o*-disubstituted benzene ring. Two hydroxy protons are present at δ 11.55. With a neutralization equivalent of 138 ± 1, Compound II is salicylic acid (2-hydroxybenzoic acid). With a neutralization equivalent of 60 ± 1, Compound III is acetic acid. Therefore, Compound I is acetyl salicylic acid (2-acetyl benzoic acid).

    Acetyl salicylic acid (2-acetyl benzoic acid) (Compound I) is hydrolyzed to salicylic acid (2-hydroxybenzoic acid) (Compound II) and acetic acid (Compound III).

Compound I:
acetylsalicylic acid, 2-acetylbenzoic acid
MW = 180, NE = 180

Compound II:
salicylic acid,
2-hydroxybenzoic acid
MW = 138, NE = 138

Compound III:
acetic acid
MW = 60, NE = 60

2. Compound I is in the B solubility class, which contains amines, anilines, and ethers. A primary amine is indicated since the product from treatment with benzenesulfonyl chloride is soluble in acid. Compound I is not an aldehyde or ketone (negative phenylhydrazine test).

Compound II contains an active hydrogen (positive acetyl chloride and sodium test) and is a primary amine, since the product from the reaction with benzenesulfonyl chloride is soluble in acid. Compound IV is the hydrochloride salt of Compound II.

Compound V is an aniline derivative, since it reacts with nitrous acid without the evolution of nitrogen and this solution reacts with 2-sodium naphthoxide to form a red solution. The ^{1}H NMR spectrum of Compound V indicates a *p*-disubstituted benzene ring from the doublets with four protons at δ 6.76 and δ 7.83 and three hydroxy and/or amino protons at δ 6.55. Compound V is 4-aminobenzoic acid.

Methylation of Compound V followed by mild reduction produces 4-(*N,N*-dimethylamino)benzyl alcohol, Compound II. Compound III is the sodium salt of Compound V. Compound I is 4-(*N,N*-dimethyl)aminobenzyl 4-aminobenzoate.

4-(*N,N*-Dimethyl)aminobenzyl 4-aminobenzoate (Compound I) is decomposed with hot sodium hydroxide to yield 4-(*N,N*-dimethylamino)benzyl alcohol (Compound II) and sodium benzoate (Compound III).

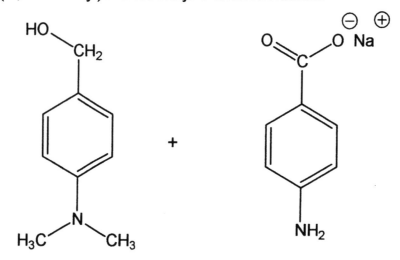

Compound I:
4-(*N,N*-dimethyl)aminobenzyl 4-aminobenzoate

+

Compound II:
4-(*N,N*-dimethylamino)benzyl alcohol

Compound III:
sodium benzoate

4-(*N,N*-Dimethyl)aminobenzyl alcohol (Compound II) is treated with hydrochloric acid to yield *N,N,N*-trimethyl-4-hydroxymethyl-anilinium chloride (Compound IV).

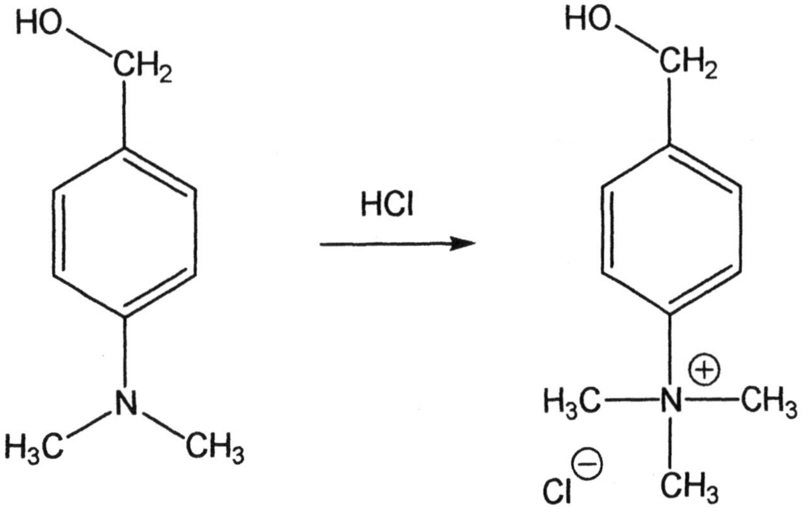

Compound II:
4-(*N*,*N*-dimethylamino)-
benzyl alcohol

Compound IV:
N,*N*,*N*-trimethyl-4-
hydroxymethyl anilinium chloride
MW = 201.5, NE = 201.5

Sodium 4-aminobenzoate (Compound III) is acidified to yield 4-aminobenzoic acid (Compound V), which is methylated twice to yield 4-(*N*,*N*-dimethylamino)benzyl alcohol (Compound II).

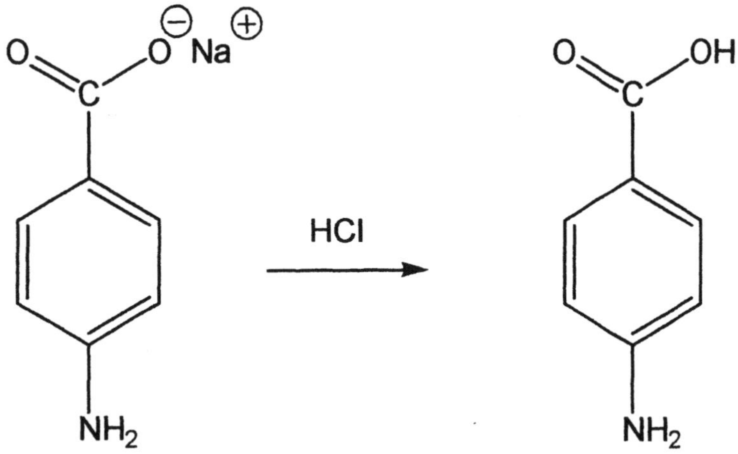

Compound III:
sodium 4-aminobenzoate

Compound V:
4-aminobenzoic acid
MW = 137, NE = 137

HO
CH₂

1. CH₃Cl
2. mild reduction

Compound II:
4-(*N,N*-dimethyl-
amino)benzyl alcohol

N
H₃C   CH₃

3.   Compound I is in the solubility class of S_2, which includes salts of organic acids, amine hydrochlorides, amino acids, and polyfunctional compounds.

Compound II is a primary amine since it reacts with benzenesulfonyl chloride and base to yield a soluble Compound III, which precipitates with acidification. Compound II does not contain a labile halogen (negative silver nitrate test), contains an activating group on the aromatic ring (positive bromine water test), and contains a multiple bond or is a phenol or an aniline (positive potassium permanganate test).

Compound I reacts with nitrous acid without the evolution of gas, followed by treatment with copper cyanide to produce Compound IV. This type of reaction is indicative of the diazotization of an aromatic amine with a subsequent Sandmeyer reaction resulting in a nitrile group attached to the benzene ring. Hydrolysis of the nitrile group to a carboxylic acid group produces Compound V. Compound V does not have a labile halogen (negative silver nitrate test) or non-aromatic multiple bond (negative potassium permanganate test).

The ^{1}H NMR spectrum of Compound II indicates a *p*-disubstituted benzene ring with doublets at δ6.57 and δ7.21. The amino protons appear as a broad singlet at δ 3.53. Since this structure also contains bromine, the structure must be 4-bromo-aniline. Compound III is the benzenesulfonamide derivative of Compound II.

Compound V must be 4-bromobenzoic acid, since it is an acid with a neutralization equivalent of 200 ± 2. Working backwards, Compound IV is 4-bromobenzonitrile. Using the neutralization equivalent and eliminating other possibilities, Compound I is 4-bromoanilinium sulfate.

4-Bromoanilinium sulfate (Compound I) was treated with alkali to yield 4-bromoaniline (Compound II), which is then reacted with benzenesulfonyl chloride and alkali to produce *N*-bromophenyl-benzenesulfonamide (Compound III).

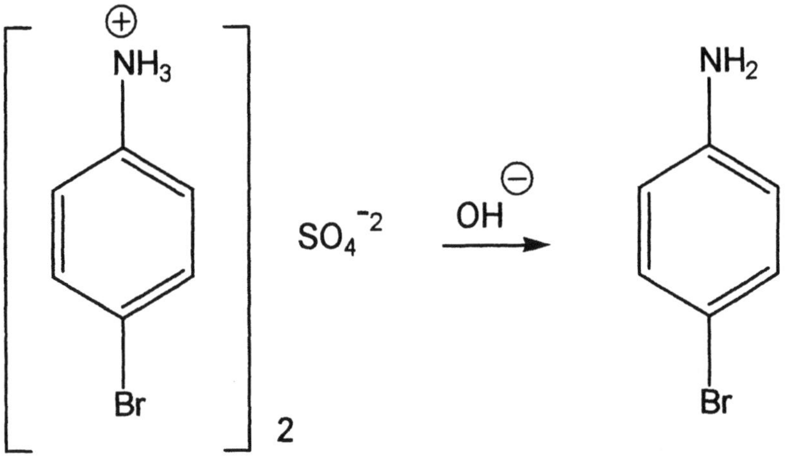

Compound I:
4-bromoanilinium sulfate
MW = 442, NE = 221

Compound II:
4-bromoaniline

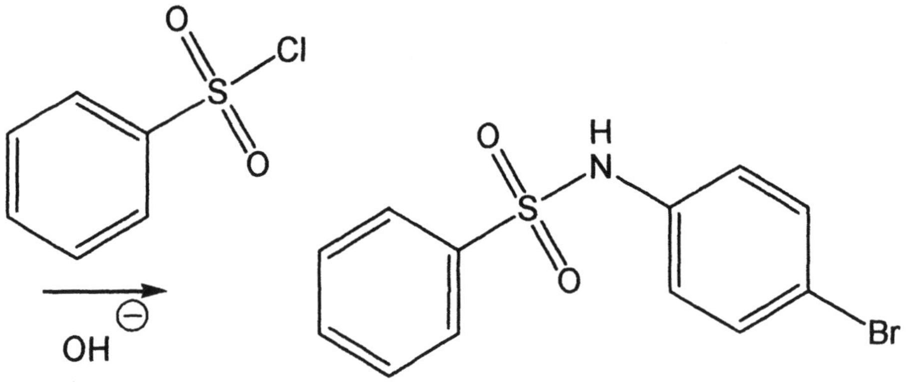

benzenesulfonyl
chloride

Compound III:
N-bromophenylbenzenesulfonamide

Treatment of 4-bromoanilinium sulfate (Compound I) with nitrous acid produced the diazonium salt, which reacted with copper cyanide to produce 4-bromobenzonitrile (Compound IV). Acid hydrolysis of 4-bromobenzonitrile (Compound IV) yielded 4-bromobenzoic acid (Compound V).

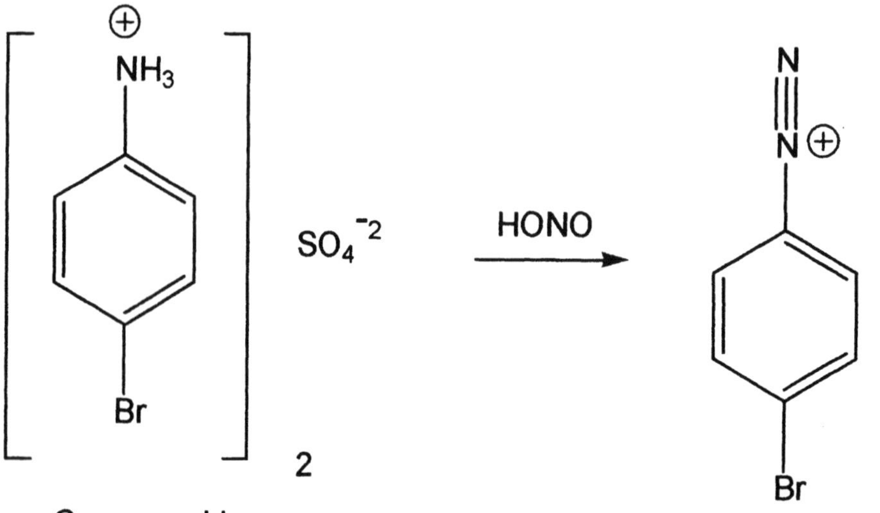

Compound I:
4-bromoanilinium sulfate
MW = 442, NE = 221

diazonium salt of
4-bromoaniline

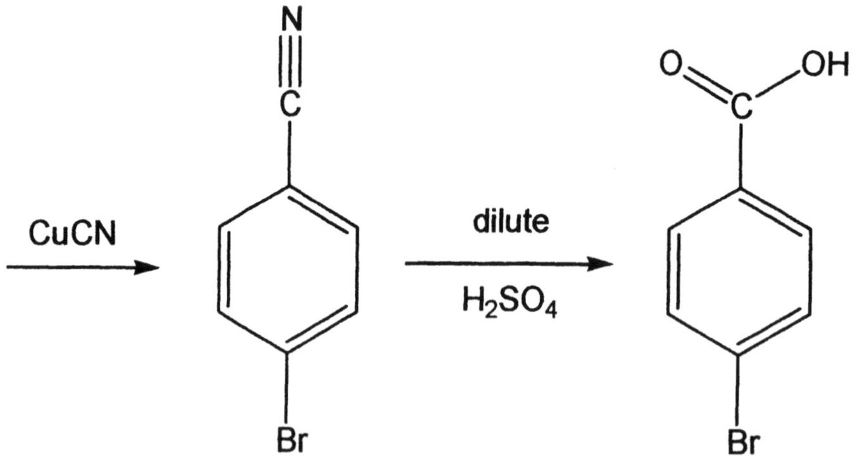

HONO

CuCN

dilute
H_2SO_4

Compound IV:
4-bromobenzonitrile

Compound V:
4-bromobenzoic acid
MW = 201, NE = 201

4. Compound I is in the solubility class of I, indicating that it is a saturated hydrocarbon, a haloalkane, an aryl halide, a diaryl ether, or a deactivated aromatic compound. Compound I is not an aldehyde or a ketone (negative phenylhydrazine test), does not have an active hydrogen (negative acetyl chloride test), and does not have a multiple bond (negative potassium permanganate test). A positive silver nitrate test was given after heating, thus showing that the halogen is not very labile and probably attached to an aromatic ring. Compound I reacts with zinc and ammonium chloride and the filtrate from this reaction reduces Tollens reagent, thus indicating the presence of a nitro group.

Compound I was vigorously oxidized to produce Compound II. Compound II still contains a bromine group and a nitro group. Compound II is also aromatic and a carboxylic acid. To come up with the neutralization equivalent of 145 ± 1, Compound II must be a bromonitro-1,2-benzenedicarboxylic acid. Compound I is a bromo-nitronaphthalene, with both the bromo and nitro group on the same ring.

Nitro groups are reduced to amino groups with tin and hydrochloric acid, thus Compound III contains an amino group. Compound III is an aniline or phenol (positive bromine water test). Additionally, Compound III contains an active hydrogen (positive acetyl chloride test) and the amine is primary (reaction with benzenesulfonyl chloride, and soluble in base).

Oxidation of Compound III produced Compound IV, which does not contain bromine or nitrogen and has a neutralization equivalent of 82 ± 1. A compound that fulfills these criteria is a benzene-dicarboxylic acid. Since Compound IV loses water to produce an anhydride, the only possible structure for this compound is 1,2-benzenedicarboxylic acid (phthalic acid). Compound III is an aminobromonaphthalene, with both the amino and the bromo group on the same ring.

The ^1H NMR spectrum of Compound III concurs with the structure of 1-amino-4-bromonaphthalene, with a broad singlet with two hydrogens for the amino group at $\delta 3.96$ and six hydrogens in the aromatic region in the range of δ 6.45–δ 8.2. Thus, the other structures can be determined.

1-Bromo-4-nitronaphthalene (Compound I) is vigorously oxidized to 3-bromo-6-nitro-1,2-benzenedicarboxylic acid (Compound II).

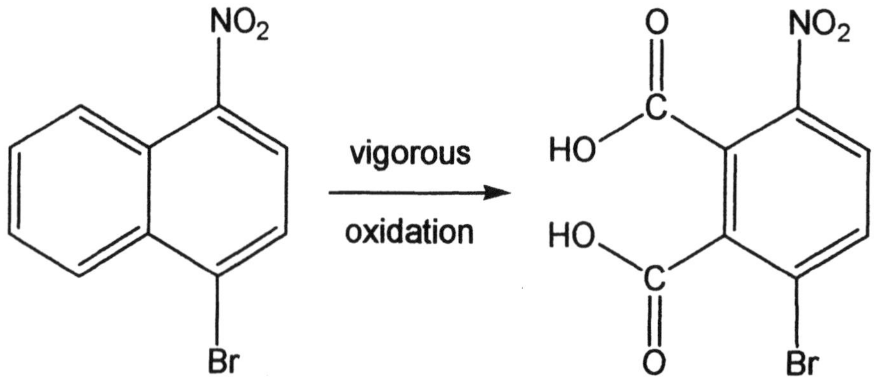

Compound I:
1-bromo-4-nitronaphthalene

Compound II:
3-bromo-6-nitro-1,2-
benzenedicarboxylic acid
MW = 290, NE = 145

Treatment of 1-bromo-4-nitronaphthalene (Compound I) with tin and hydrochloric acid reduces the nitro group to an amino group to produce 1-amino-4-bromonaphthalene (Compound III), which is vigorously oxidized to 1,2-benzenedicarboxylic acid (phthalic acid) (Compound IV).

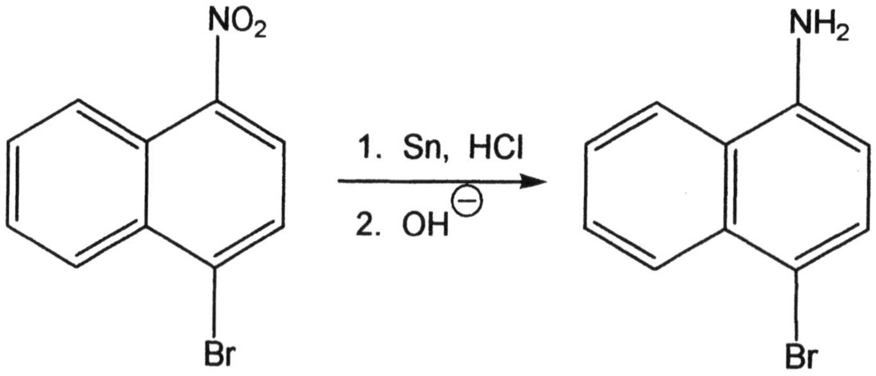

Compound I:
1-bromo-4-nitronaphthalene

Compound III:
1-amino-4-bromonapthalene

vigorous
⟶
oxidation

O
‖
C
‑OH

‑OH
C
‖
O

Compound IV:
1,2-benzenedicarboxylic acid.,
phthalic acid
MW = 166, NE = 83

5. Solid (Ccompound I) does not contain a labile chlorine (negative silver nitrate test), an active hydrogen (negative acetyl chloride test), or a multiple bond (negative bromine test). However, it is an aldehyde or a ketone (positive phenylhydrazine test).

Compound II contains chlorine and is a carboxylic acid.

Compound III is not a phenol (negative bromine water test), does not have a multiple bond (negative bromine in carbon tetrachloride test), and is not an aldehyde or a ketone (negative phenylhydrazine test).

Compound IV, which is produced from the oxidation of Compound I, is a carboxylic acid and contains chlorine. The appearance of two doublets with a total of four protons in the aromatic region of δ 7.37–δ 7.81 indicates a *p*-disubstituted benzene ring. The signal at δ 7.45 is due to the –COOH. Compound IV contains chlorine and has a neutralization equivalent of 156 ± 1. The only possible structure for Compound IV is 4-chlorobenzoic acid.

The neutralization equivalents for Compounds II and III are considerably higher than the neutralization equivalent for Compound IV, thus indicating that these two compounds may be twice the size of Compound IV. Since Compound II reacts with acetic anhydride to yield Compound III, then Compound II must be an alcohol. Compound II contains chlorine and is a carboxylic acid. From deductive reasoning, Compound II must be 4,4'-dichlorobenzilic acid and Compound I is 4,4'-dichlorobenzil.

In the presence of base, 4,4'-dichlorobenzil (Compound I) undergoes a benzilic acid rearrangement to form 4,4'-dichlorobenzilic acid (Compound II), which reacts with acetic anhydride to yield *α*-acetyl-4,4'-dichlorobenzilic acid (Compound III).

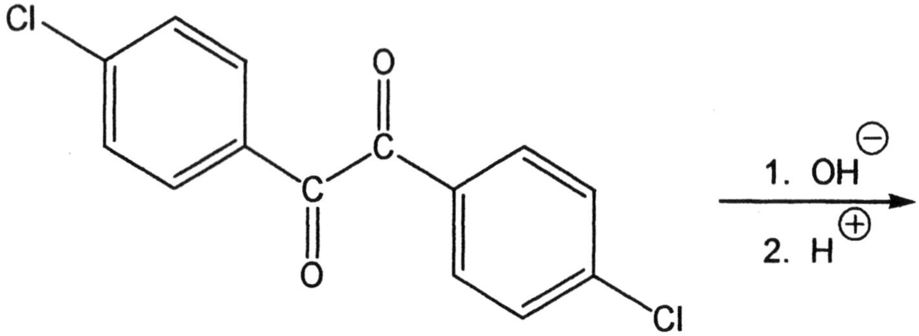

Compound I: 4,4'-dichlorobenzil

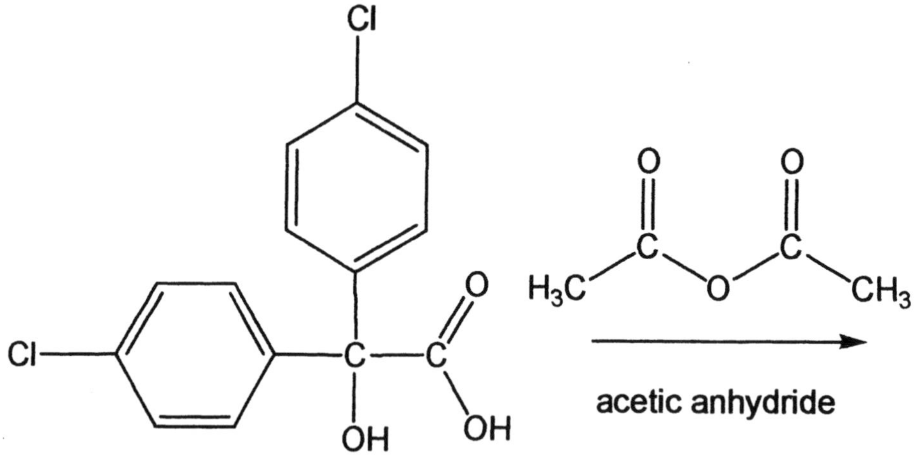

Compound II:
4,4'-dichlorobenzilic acid
MW = 297, NE = 297

acetic anhydride

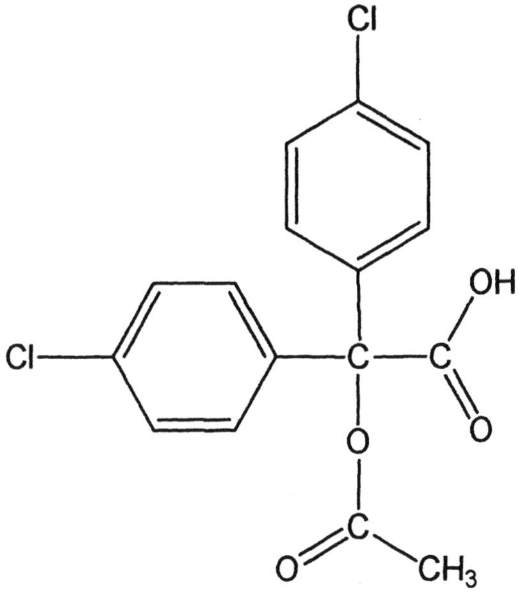

Compound III:
α-acetyl-4,4'-dichloro-
benzilic acid
MW = 339, NE = 339

4,4'-Dichlorobenzil (Compound I) is oxidized to 4-chloro-
benzoic acid (Compound IV).

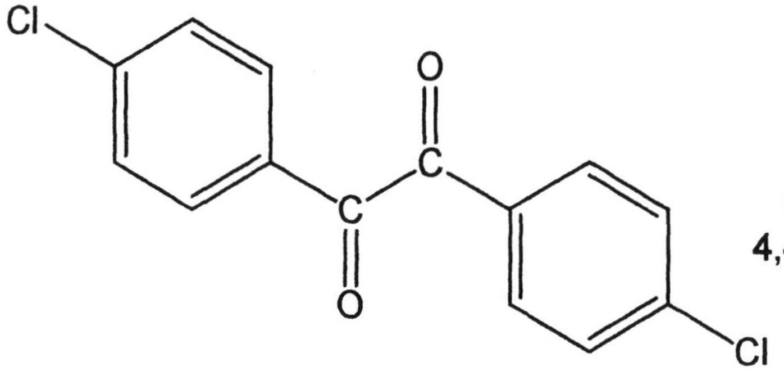

Compound I:
4,4'-dichlorobenzil

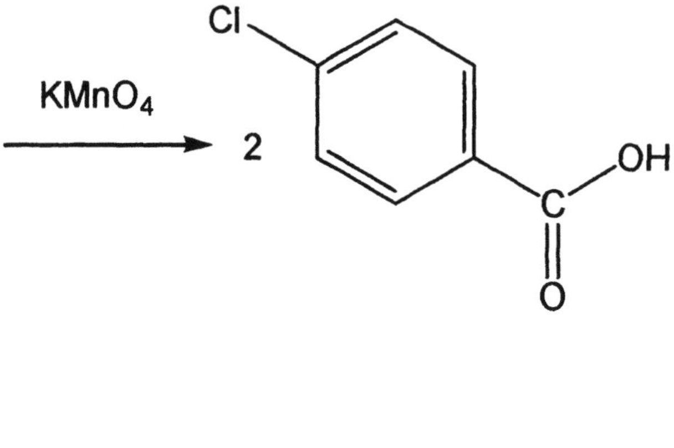

Compound IV:
4-chlorobenzoic acid
MW = 156.5, NE = 156.5

Problem Set 20

1. Compound I undergoes reaction with acetyl chloride to produce Compound II, indicating that Compound I has an active hydrogen. Compound I is oxidized to Compound III, which contains an aldehyde or ketone group (positive phenylhydrazine test). Vigorous oxidation of I, II, or III yields an acid with a neutralization equivalent of 122 ± 1, which corresponds to benzoic acid. The difference in the neutralization equivalents of Compounds I and III is 28 (150–122), which is a carbon and an oxygen. Thus Compound I is 2-phenyl-2-hydroxyethanoic acid (mandelic acid) and Compound III is 2-phenyl-2-oxoethanoic acid. Compound II is the acetyl derivative of Compound I.

    2-Phenyl-2-hydroxyethanoic acid (mandelic acid) (Compound I) is reacted with acetyl chloride to yield 2-acetyl-2-phenyl-2-hydroxyethanoic acid (Compound II).

Compound I:
2-phenyl-2-hydroxyethanoic acid,
mandelic acid
MW = 152, NE = 152

Compound II:
2-acetyl-2-phenyl-2-hydroxyethanoic acid
MW = 194, NE = 194

    2-Phenyl-2-hydroxyethanoic acid (mandelic acid) (Compound I) is oxidized to 2-phenyl-2-oxoethanoic acid (Compound III).

KMnO$_4$

Compound I:
2-phenyl-hydroxyethanoic acid,
mandelic acid
MW = 152, NE = 152

Compound III:
2-phenyl-2-oxoethanoic acid
MW = 150, NE = 150

Vigorous oxidation of 2-phenyl-2-hydroxyethanoic acid (mandelic acid) (Compound I), 2-acetyl-2-phenyl-2-hydroxyethanoic acid (Compound II), or 2-phenyl-2-oxoethanoic acid (Compound III) produces benzoic acid (Compound IV).

Compounds I, II, or III $\xrightarrow[\text{oxidation}]{\text{vigorous}}$

Compound IV:
benzoic acid
MW = 122, NE = 122

The ^{1}H NMR spectrum of 2-phenyl-2-hydroxyethanoic acid (mandelic acid) (Compound I) shows the isolated methine proton at δ 5.22 and the aromatic and hydroxy protons at δ 6.93–δ 7.88, with an integration of seven protons. The IR spectrum has a signal for the carbonyl stretching of the ketone at 1706 cm^{-1} and the –OH stretching of the carboxylic acid at 2400–3333 cm^{-1}.

2.   Since Compounds I, II, and III have neutralization equivalents, they are probably carboxylic acids. Zinc and hydrochloric acid reduces carbonyl groups in aldehydes and ketones to methylene groups and reduces nitro groups to amino groups.

      The ^{1}H NMR spectrum of compound II indicates a disubstituted benzene ring in the aromatic region of $\delta 7.72$–$\delta 8.71$ and the hydroxy peak of the carboxylic acid at δ 12.98. The groups are probably *meta*, due to the splitting pattern of the aromatic protons. Compound II is 3-nitrobenzoic acid, and Compound I is then 4-(3-nitrophenyl)-2-oxobutanoic acid.

      4-(3-Nitrophenyl)-2-oxobutanoic   acid   (Compound   I)   is vigorously oxidized to 3-nitrobenzoic acid (Compound II).

Compound I:
4-(3-nitrophenyl)-2-oxobutanoic acid
MW = 223, NE = 223

Compound II:
3-nitrobenzoic acid
MW = 167, NE = 167

With zinc and hydrochloric acid, the nitro and the carbonyl groups are reduced in 4-(3-nitrophenyl)-2-oxobutanoic acid (Compound I) to yield 4-(3-aminophenyl)butanoic acid (Compound III).

Compound I: 4-(3-nitrophenyl)-2-oxobutanoic acid
MW = 223, NE = 223

Compound III: 4-(3-aminophenyl)butanoic acid
MW = 179, NE = 179

3. The original sulfur-containing compound is in the solubility class of A_2, which includes phenols, enols, oximes, imides, sulfonamides, thiophenols, β-diketones, and nitro compounds. Oxidation of the original compound produces a sulfur-containing acid with a neutralization equivalent of 102 ± 1. The sulfur-containing acid was treated with superheated steam to yield a sulfur-free acid.

The ^{1}H NMR spectrum of the sulfur-free acid shows five aromatic protons at δ 7.00–δ 7.75, indicating a monosubstituted benzene. A singlet at δ 8.00 corresponds to the –OH of the carboxylic acid. Therefore, this spectrum corresponds to benzoic acid. The ^{1}H NMR spectrum of the original compound indicates an isolated methyl at δ 2.29, an isolated proton at δ 3.37, and a disubstituted benzene ring at δ 7.02. Since all of the aromatic protons are split, the compound is either *ortho* or *meta* disubstituted. The methyl group is probably attached to the benzene ring, and oxidation would convert the methyl to a carboxylic acid group to produce the sulfur-containing acid. By adding up the weights of C_6H_4 (a disubstituted benzene ring, –COOH, and –SO$_3$H), a value of 202 is obtained. Since it contains two acid groups, the neutralization equivalent is equal to 101.

In the second compound, the neutralization equivalent is much lower, indicating the presence of two acid groups. If the sulfur is in a sulfonic acid group, then the neutralization equivalent works out. Thus, the sulfur-containing acid is a carboxybenzenesulfonic acid. The original compound is a methylthiophenol and the final compound is benzoic acid. The answers are drawn for the *meta* disubstituted compounds.

Vigorous oxidation of 3-methylphenol (original compound) produced 3-carboxybenzenesulfonic acid (sulfur-containing acid), which is treated with steam to yield benzoic acid (sulfur-free acid).

3-methylthiophenol
(original acid)

3-carboxybenzenesulfonic acid
(sulfur-containing acid)
MW = 202, NE = 101

$$\xrightarrow{\text{steam}}$$

benzoic acid
(sulfur-free acid)

4. Compound I contains a labile halogen (positive silver nitrate test), is a methyl ketone or has a methyl group next to a hydroxy on a CH (positive iodoform test), and has an active hydrogen (positive acetyl chloride test). The hydrolysis of compound I yielded a chlorine-free compound (Compound II), which must be symmetrical since it was oxidized by periodic acid to a single compound (Compound III). For oxidation by periodic acid to occur, the compound must have a hydroxy group next to a hydroxy group, a hydroxy group next to a carbonyl group, or a carbonyl next to a carbonyl. Since Compound I was hydrolyzed to Compound II and it is symmetrical, then the hydroxy groups must be on adjacent carbons. Compound III contains a methyl ketone or a methyl group next to a hydroxy on a CH (positive iodoform test). Several structures are possible.

The chlorine group is replaced by –OH in 4-chloro-5-hydroxy-2,7-octadione (Compound I) to yield 4,5-dihydroxy-2,7-octadione (Compound II). Treatment of 4,5-dihydroxy-2,7-octadione (compound II) with periodic acid produces 3-oxobutanal (Compound III). Another set of answers could be 3-chloro-2-butanol (compound I), 2,3-butanediol (Compound II), and ethanal (compound III).

Compound I:
4-chloro-5-hydroxy-2,7-octadiene

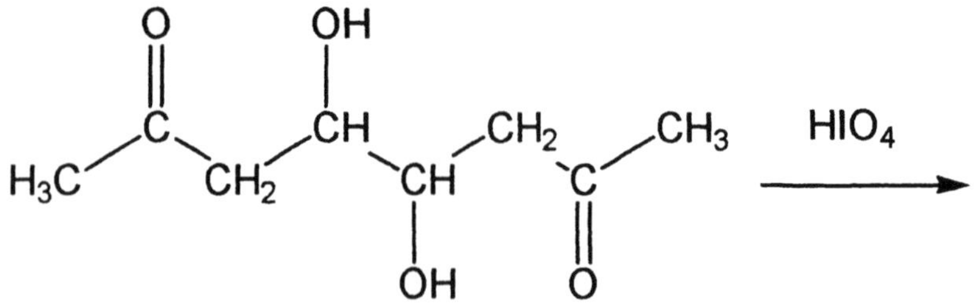

Compound II: 4,5-dihydroxy-2,7-octadiene

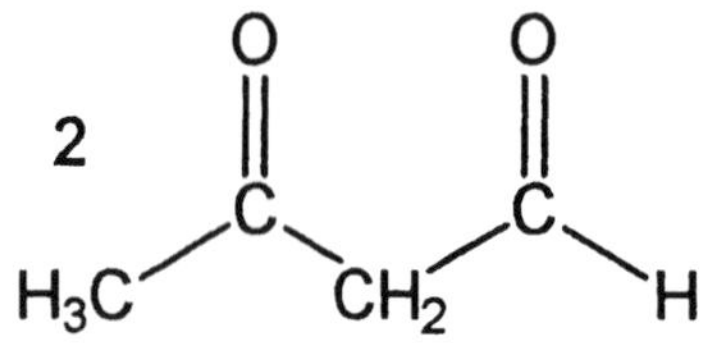

Compound III:
3-oxobutanal

OR

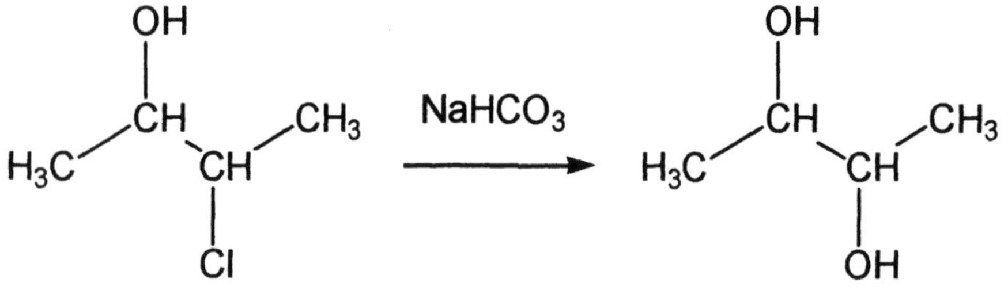

Compound I:
3-chloro-2-butanol

Compound II:
2,3-butanediol

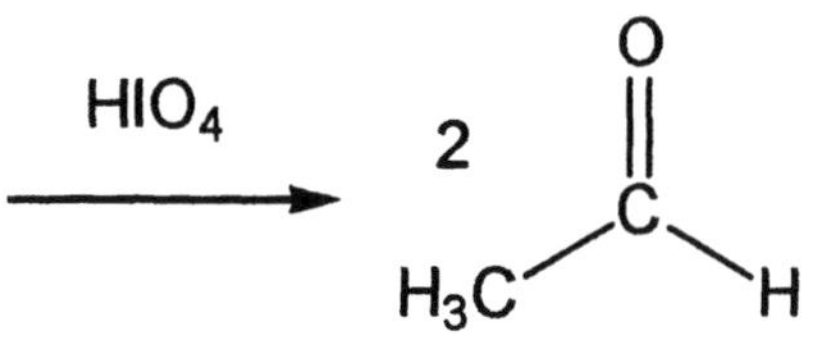

Compound III:
ethanal

5.   Compound A contains an active hydrogen (positive acetyl chloride test), but is not an aldehyde or a ketone (negative phenylhydrazine test). Compound B is an aldehyde (positive Tollens reagent), but is not an aldose, an α-hydroxyaldehyde, an α-hydroxyketone, or an α-ketoaldehyde (negative Fehling's test). Compound C contains an active hydrogen (positive acetyl chloride test) and is an aldehyde or a ketone (positive phenylhydrazine test).

The ^{1}H NMR spectrum of compound B indicates an isolated methyl group at δ 2.42, a *p*-disubstituted benzene ring (4H) with doublets at δ 7.18 and δ 7.56, and an isolated aldehyde group at δ 9.81. From this data, compound B is 4-methylbenzaldehyde. Compound A must be 1,2-di(4-methylphenyl)-1,2-ethanediol.

Treatment of 1,2-di(4-methylphenyl)-1,2-ethanediol (Compound A) with periodic acid produced 4-methylbenzaldehyde (Compound B). The reaction of 4-methylbenzaldehyde (Compound B) with potassium cyanide results in a benzoin condensation to form 4,4'-dimethylbenzoin (Compound C), which is oxidized to 4,4'-dimethylbenzil (Compound D). Oxidation of 4,4'-dimethylbenzil (Compound D) produced 4-methylbenzoic acid (Compound E).

Compound A:
1,2-di(4-methylphenyl)-1,2-ethanediol

Compound B:
4-methylbenzaldehyde

Compound C: 4,4'-dimethylbenzoin

Compound D:
4,4'-dimethylbenzil

Compound E:
4-methylbenzoic acid

Treatment of 4,4'-dimethylbenzil (Compound D) with *o*-phenylenediamine produced 2,3-di(4-tolyl)quinoxaline.

Compound D: 4,4'-dimethylbenzil

o-phenylenediamine

2,3-di(4-tolyl)quinoxaline

Catalytic hydrogenation of 4,4'-dimethylbenzoin (Compound C) or 4,4'-dimethylbenzil (Compound D) yielded 1,2-di(4-methylphenyl)-1,2-ethanediol (Compound A).

Compounds C or D

$\xrightarrow[\text{hydrogenation}]{\text{catalytic}}$

Compound A:
1,2-di(4-methylphenyl)-1,2-ethanediol

6. This problem is a very involved problem. The best way to figure out the structures is to list the characteristics of each compound and to list the types of reactions that are occurring.

Compound A does not have an active hydrogen (negative acetyl chloride test) or a multiple bond (negative bromine test). It is not a methyl ketone, nor does it have a methyl adjacent to an –CHOH (negative iodoform test). Compound A is not an aldehyde (negative Tollens test) and is probably a diaryl ketone, since it reacts slowly with 2,4-dinitrophenylhydrazine. A nitro group is present (positive zinc and ammonium chloride test, with product giving a positive Tollens test).

Compound A undergoes reaction with hydroxylamine hydrochloride in pyridine, thus converting the carbonyl to an oxime in Compound B. Phosphorus pentachloride rearranges the oxime group in Compound B to an amide in compound C. Boiling with acid hydrolyzed the amide in Compound C to an acid, Compound D, and an amine, compound E.

Acid D undergoes reaction with thionyl chloride and ammonium hydroxide to form an amide. The amide was subjected to the conditions of a Hofmann rearrangement, bromine and base, to produce an amine, Compound F. Compound F is an aniline derivative, since diazotization of this compound followed by addition of 2-naphthol produces a red color. Treatment of Compound F with tin and hydrochloric acid reduces the nitro group to an amine to form Compound G.

Compound E is an aniline derivative, since diazotization of this compound followed by addition of 2-naphthol produces a red color. When the diazotized solution is treated with base, a phenol derivative, Compound H, is formed. Oxidation of either Compounds E or H produced an acid with a neutralization equivalent of 83, which is easily converted to the anhydride. An acid that fulfills this criteria is 1,2-benzenedicarboxylic acid. Compound H does not contain an amino group, since it does not undergo coupling with diazonium solutions.

From the ^{1}H NMR spectrum, Compound F is a *p*-disubstituted benzene ring, with doublets (4H) at δ 6.58 and δ 7.88. From above, compound F contains both an amino group and a nitro group. The amino group appears at δ 5.68. The structure of Compound F must be 4-nitroaniline.

In the ^{1}H NMR spectrum of Compound H, six aromatic protons are present in the range of δ 7.1–δ 8.2. An isolated methyl is present at δ 2.32 and an –OH appears at δ 5.12. Therefore, Compound H is 1-hydroxy-2-methylnaphthalene and Compound A is 2-methyl-1-(4-nitrobenzoylphenyl)naphthalene.

2-Methyl-1-(4-nitrobenzoylphenyl)naphthalene (Compound A) is treated with hydroxylamine hydrochloride to form 2-methyl-1-(4-nitrobenzoylphenyl)naphthalene oxime (Compound B). 2-Methyl-1-(4-nitrobenzoylphenyl)naphthalene oxime (Compound B) undergoes a rearrangement in the presence of phosphorus pentachloride to yield 2-methyl-1-(4-nitrobenzoyl)aminonaphthalene (Compound C), which is hydrolyzed to 4-nitrobenzoic acid (Compound D) and the hydro-chloride salt of 1-amino-2-methylnaphthalene (Compound E).

Compound A:
2-methyl-1-(4-nitrobenzoylphenyl)naphthalene

Compound B:
2-methyl-1-(4-nitrobenzoylphenyl)naphthalene oxime

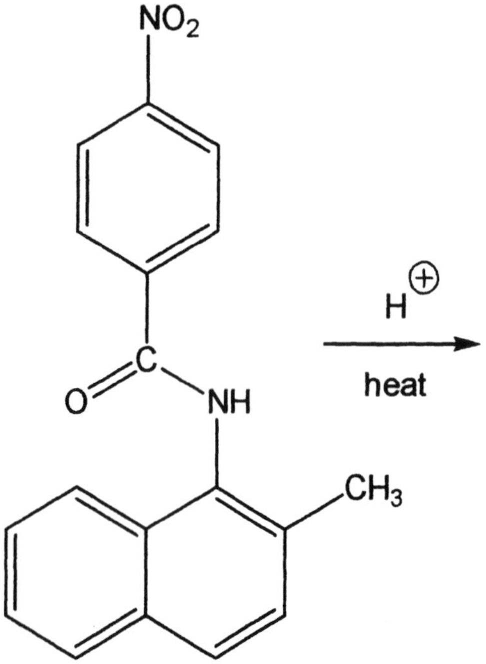

Compound C:
2-methyl-1-(4-nitrobenzoyl)aminonaphthalene

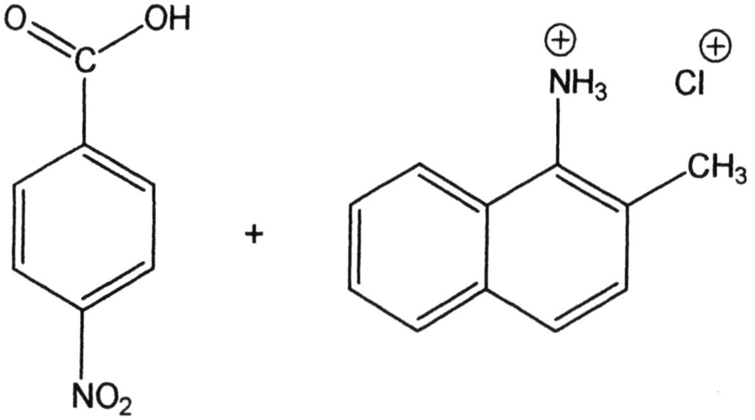

+

Compound D:
4-nitrobenzoic acid
MW = 167, NE = 167

Compound E:
hydrochloride salt of
1-amino-2-methylnaphthalene
MW = 193.5, NE = 193.5

4-Nitrobenzoic acid (Compound D) was treated with thionyl chloride to yield 4-nitrobenzoyl chloride, which was reacted with ammonia to form 4-nitrobenzamide. By reacting 4-nitrobenzamide with bromine and sodium hydroxide, a Hofmann rearrangement occurred, thus producing 4-nitroaniline (Compound F). The nitro group in 4-nitroaniline (Compound F) was reduced to an amino group to form 1,4-diaminobenzene (Compound G).

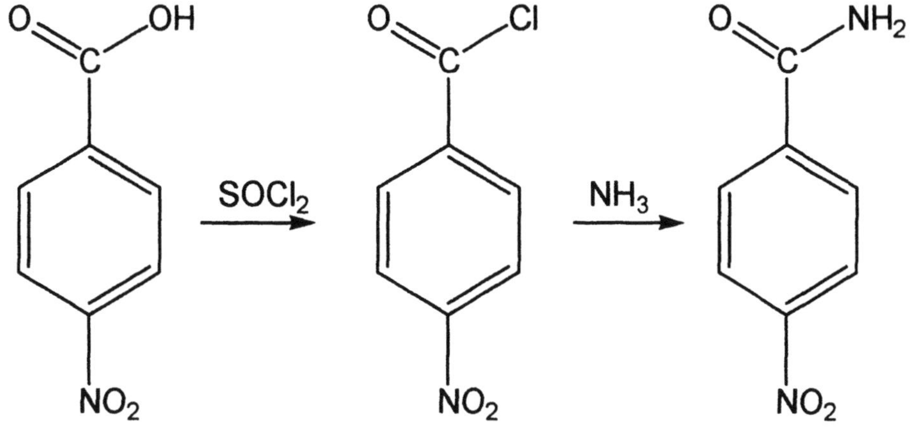

Compound D:
4-nitrobenzoic acid
MW = 167, NE = 167

4-nitrobenzoyl
chloride

4-nitrobenzamide

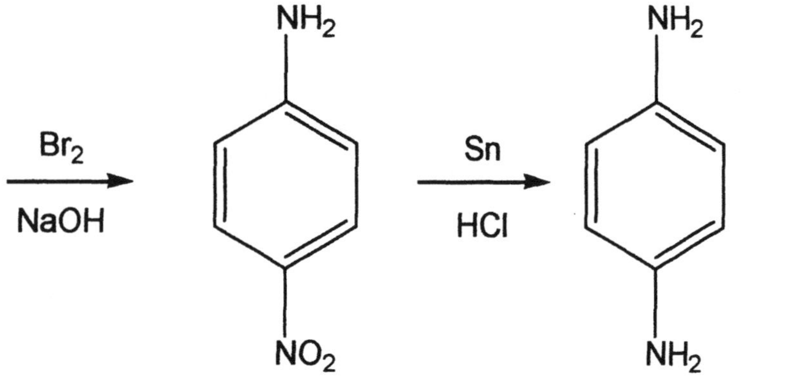

Compound F:
4-nitroaniline

Compound G:
1,4-diaminobenzene

The hydrochloride salt of 1-amino-2-methylnaphthalene (Compound E) is treated with nitrous acid to yield 1-hydroxy-2-methylnaphthalene (Compound H).

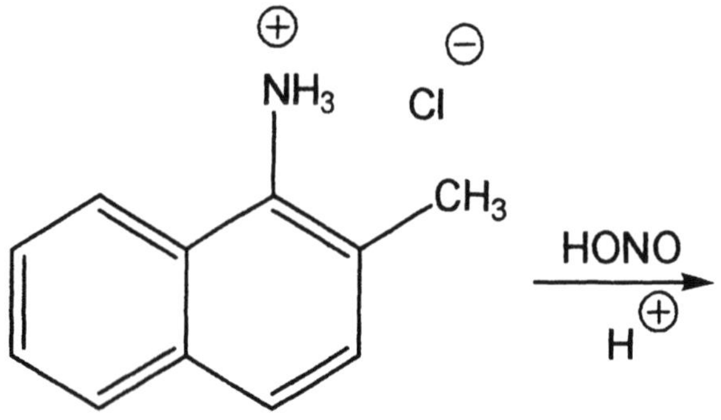

Compound E:
hydrochloride salt of
1-amino-2-methylnaphthalene
MW = 193.5, NE = 193.5

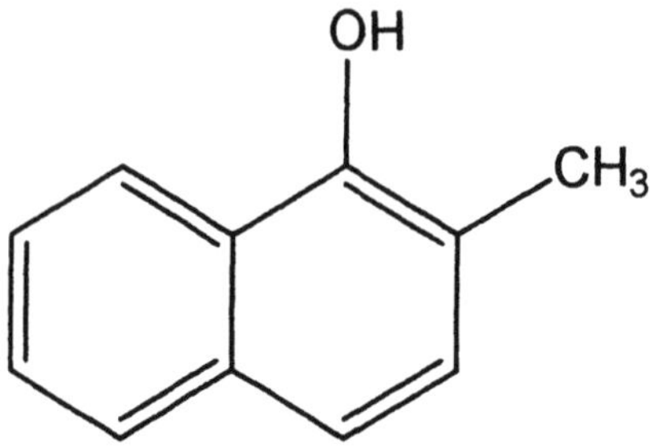

Compound H:
1-hydroxy-2-methylnaphthalene

The hydrochloride salt of 1-amino-2-methylnaphthalene (Compound E) and 1-hydroxy-2-methylnaphthalene (Compound H) are oxidized to 1,2-benzenedicarboxylic acid (phthalic acid), a compound that is readily converted to 1,2-benzenedicarboxylic anhydride (phthalic anhydride).

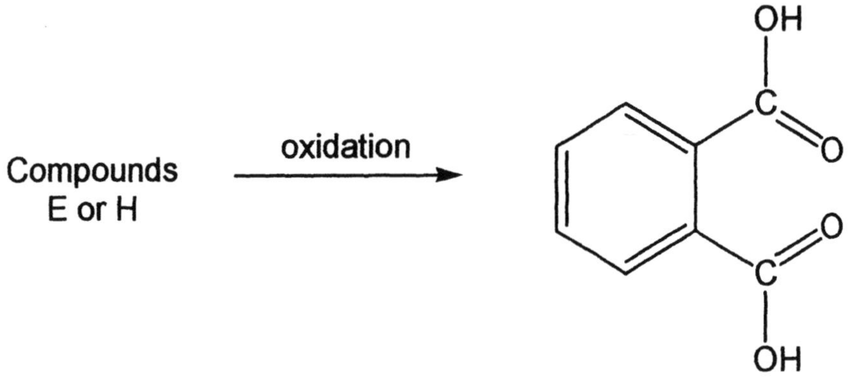

Compounds E or H →(oxidation)

1,2-benzenedicarboxylic acid,
phthalic acid
MW = 166, NE = 83

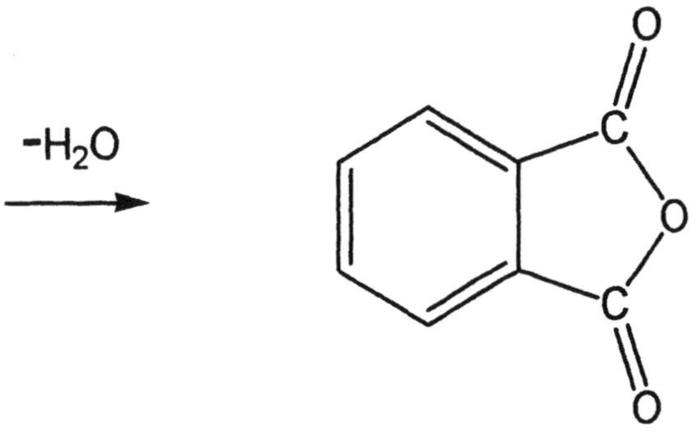

−H$_2$O →

1,2-benzenedicarboxylic anhydride,
phthalic anhydride

66351944R00228

Made in the USA
Lexington, KY
11 August 2017